Processing and Properties Analysis of Grain Foods

Processing and Properties Analysis of Grain Foods

Editors

Yonghui Li
Xiaorong (Shawn) Wu

MDPI • Basel • Beijing • Wuhan • Barcelona • Belgrade • Manchester • Tokyo • Cluj • Tianjin

Editors
Yonghui Li
Kansas State University
Manhattan
USA

Xiaorong (Shawn) Wu
USDA, ARS
Manhattan
USA

Editorial Office
MDPI
St. Alban-Anlage 66
4052 Basel, Switzerland

This is a reprint of articles from the Special Issue published online in the open access journal *Processes* (ISSN 2227-9717) (available at: https://www.mdpi.com/journal/processes/special_issues/Grain_Foods_Processing).

For citation purposes, cite each article independently as indicated on the article page online and as indicated below:

LastName, A.A.; LastName, B.B.; LastName, C.C. Article Title. *Journal Name* **Year**, *Volume Number*, Page Range.

ISBN 978-3-0365-7066-2 (Hbk)
ISBN 978-3-0365-7067-9 (PDF)

Contents

About the Editors

Yonghui Li

Dr. Yonghui Li is an Associate Professor of Grain Chemistry and the Director of the Wheat Quality Lab at Kansas State University. He also serves as a graduate faculty member of the Food Science Institute. He teaches undergraduate and graduate courses in Grain Analysis and Protein Chemistry and Technology. Dr. Li's research focuses on the structure, chemistry, modification, and functionality of grain proteins and bioactive peptides with the aim of developing high-quality, functional grain-based foods, ingredients, and nutraceuticals. Since joining the faculty in 2016, Dr. Li has received over USD 4.2 million in research funding, with USD 3.3 million awarded as the lead PI, from various federal, state, and industrial sources. Dr. Li has published 110 journal articles, 130 presentations, 3 book chapters, and 2 patents. He is an associate editor for both the Journal of Food Science and Cereal Chemistry and serves on the editorial boards of several other international journals.

Xiaorong (Shawn) Wu

Dr. Xiaorong (Shawn) Wu joined the Grain Quality and Structure Research Unit (GQSRU) at CGAHR, USDA-ARS, Manhattan, KS as a research chemist in 2020. His research focuses on carbohydrates in grain sorghum and hard winter wheat and their effects on the application quality of sorghum and wheat. Before joining the USDA-ARS, Shawn worked as a research scientist and later director of R&D unit for the Mennel Milling Co. for a few years. During his time with the Mennel Milling Co., he was involved with many R&D projects ranging from wheat quality and milling performance, flour functionality, baking products, process validation, and the production of RTE flours. Currently, Dr. Wu collaborates with research scientists, engineers, university professors, and crop breeders on developing quick methods to improve efficiency in screening early breeder lines, studying single-kernel NIR sorting and its effects on the quality and properties of sorted sorghum and wheat, and their applications in foods.

Preface to "Processing and Properties Analysis of Grain Foods"

Foods made from grains and grain-derived ingredients are essential for providing humans with vital nutrients and energy. Through grain and food processing, raw grains are transformed into functional ingredients that are further formulated and processed to create palatable and nutritious end-products. We collected 16 papers in the Special Issue of *Processes* "Processing and Property Analysis of Grain Foods". This book is a reprint of that Special Issue, making the valuable insights and research presented in those papers more widely accessible to interested readers.

Yonghui Li and Xiaorong (Shawn) Wu
Editors

Editorial

Processing and Properties Analysis of Grain Foods

Yonghui Li [1,*] **and Xiaorong Wu** [2]

[1]   Department of Grain Science and Industry, Kansas State University, Manhattan, KS 66506, USA
[2]   Grain Quality and Structure Research Unit, CGAHR, USDA-ARS, 1515 College Ave.,
      Manhattan, KS 66502, USA
*   Correspondence: yonghui@ksu.edu; Tel.: +1-785-532-4061; Fax: +1-785-532-7010

Citation: Li, Y.; Wu, X. Processing and Properties Analysis of Grain Foods. *Processes* **2023**, *11*, 95. https://doi.org/10.3390/pr11010095

Received: 25 December 2022
Accepted: 27 December 2022
Published: 29 December 2022

Foods from grains and grain-derived ingredients are among the most important energy and nutrient source for humans. Cereals (e.g., wheat, rice, corn) have been the conventional food materials, while interest is growing in utilizing pseudocereals (e.g., buckwheat, quinoa, amaranth), pulses (e.g., dry peas, chickpea, dry beans), oilseeds (soybean, peanut, hempseed), and other grains for the development of various foods and food ingredients. Grain and food processing converts raw grains to functional ingredients and produces palatable and nutritious end-products. Examples of grain-related processes include, but are not limited to, drying, milling, fractionation, hydration, fermentation, extrusion, cooking, baking, frying, steaming, freezing, etc. Various physical and chemical changes and interactions are expected during these processes, which further affect the nutritional, textural, sensory, and many other quality properties of the products. Methods, protocols, and equipment have been developed to process, monitor, and control the processing parameters to achieve the desired end-product quality or functionality. This Special Issue of *Processes* on "Processing and Property Analysis of Grain Foods", and the associated Special Issue reprint, published 15 research papers from lead scientists and researchers in the area and covered diverse topics related to grain-processing innovations and the effect of both conventional and innovative grain processes on the properties of grain-derived ingredients, intermediates, and end products. The Special Issue is available online at: https://www.mdpi.com/journal/processes/special_issues/Grain_Foods_Processing (accessed on 25 December 2022).

Sorghum (*Sorghum bicolor* (L.) Moench) ranks fifth in global cereal grain production. Starch is the primary constituent of sorghum grain, and its content and properties, especially the amount of amylose content in sorghum starch, influence the suitability of sorghum cultivars for specific end uses. Peiris et al. successfully developed partial least squares NIR (near infrared spectra) models to estimate starch and amylose contents in intact grain sorghum samples [1]. The newly developed calibrations can be used as a rapid screening tool for characterizing sorghum starch composition in segregating populations and to identify germplasms for developing new cultivars and hybrids for specific end uses.

Reducing the particle size of cereal grains is often the first step in the food and feed manufacturing process. Pulses are attracting an increasing interest due to their multiple agronomic and nutritional advantages. Siliveru and colleagues evaluated roller milling to produce chickpea, lentil, and yellow pea flours and characterized the particle and physicochemical properties of the produced flours [2,3]. The findings may assist millers to adapt pulse-milling technologies with minor modifications to their existing wheat milling facilities and provide guidance for more suitable uses of pulses.

Hammermills are among the most commonly used size-reduction equipment in the feed industry because of their high throughput and versatility in grinding different materials. Yellow dent #2 corn is a common grain type used for feed in the U.S. Paulk and colleagues investigated the effects of whole-corn moisture before grinding and hammermill screen size on subsequent ground corn moisture, particle size, and flowability [4]. They further evaluated the effects of hammermill tip speed, assistive airflow, and screen sizes on hammermill throughput and characteristics of ground corn [5]. The studies provide useful references for processes involving grinding operations.

Innovations in food extrusion technology are enabling its rapid expansion and applicability in diverse areas related to bioprocessing and value addition. McGuire et al. compared and characterized the flowability of some common powder ingredients (corn, wheat, and sucrose) in the extrusion process [6]. The study related raw material particulate rheology to the granular flow in a pilot-scale single=screw food extruder, and some findings were visualized using a transparent plexiglass window during extrusion.

Dough mixing is an important process that can affect the quality of end products. Elucidating the mechanisms underlying gluten formation and structure remains challenging. Iwaki et al. employed the front-face fluorescence method to assess changes in hydrophobic interactions among gluten proteins during dough formation by extracting proteins in different concentrations of 1-proponol solution [7].

Pulse flours are commonly added to food products to improve their functional properties, nutritional quality, and health benefits. Zhang et al. investigated the effects of the partial replacement (0–25%) of whole wheat flour with diversified whole pulse flours (yellow pea, green pea, red lentil, and chickpea) on dough properties and bread quality [8]. Among all the tested pulse flours, the composite flour containing yellow pea flour or chickpea flour showed overall better potential for bread-making with good dough-handling properties and product quality. Beans are also important pulse grains for food uses. Mariscal-Moreno et al. investigated the effects of the partial substitution of wheat flour by ayocote bean and black bean in bread production on the thermal properties of composite flours, the bread's proximal composition, in vitro protein digestibility (IPD), as well as color and sensorial parameters of the final products [9]. These studies on composite flours could lead to the development of more nutritious bread products by complementing cereal and pulse ingredients.

Noodles are a staple food in many Asian countries and are widely consumed throughout the world because of their convenience and palatability. Cheng et al. investigated the effects of gaseous chlorine dioxide treatment on the physicochemical properties of buckwheat-based composited flour (buckwheat–wheat–gluten) and shelf-life, textural qualities, and sensory properties of fresh buckwheat noodles [10]. The study reveals the effects of gaseous chlorine dioxide treatment on Tartary buckwheat flour properties and the shelf-life of the noodles. Park et al. explored the noodle-making performance of flour blends with different particle sizes and blending ratios of purple-colored wheat bran and their antioxidant properties [11]. Noodle quality and antioxidant activity were more significantly enhanced by small bran particles at higher blending ratios.

Last but not least, this Special Issue also published interesting papers addressing the physical characteristics of typical maize seeds in a cold area of North China based on principal component analysis [12], the spatiotemporal characteristics of heat resource effectiveness in the southern rice cropping area in China and relationships between heat resource effectiveness and rice potential yield, as well as grain yield reduction rate [13], the effect of different alternative sweeteners that contained sugar alcohols or bulking agents on the physiochemical properties of rolled "sugar" cookie [14], and the effects of different amylose contents of foxtail millet varieties on textural properties of Chinese steamed bread [15].

Author Contributions: Both authors contributed equally. All authors have read and agreed to the published version of the manuscript.

Funding: This research was supported in part by the USDA Pulse Crop Health Initiative projects (Grant Accession No. 0439205 and No. 0439200) and the USDA National Institute of Food and Agriculture Hatch project (Grant Accession No. 7003330).

Acknowledgments: This is contribution No. 23-156-J from the Kansas Agricultural Experimental Station. Mention of trade names or commercial products in this publication is solely for the purpose of providing specific information and does not imply recommendation or endorsement by the U.S. Department of Agriculture. The USDA is an equal opportunity provider and employer.

Conflicts of Interest: The authors declare that they have no known conflict of interest.

References

1. Peiris, K.H.S.; Wu, X.; Bean, S.R.; Perez-Fajardo, M.; Hayes, C.; Yerka, M.K.; Jagadish, S.V.K.; Ostmeyer, T.; Aramouni, F.M.; Tesso, T.; et al. Near Infrared Spectroscopic Evaluation of Starch Properties of Diverse Sorghum Populations. *Processes* **2021**, *9*, 1942. [CrossRef]
2. Martin, R.; Siliveru, K.; Watt, J.; Blodgett, P.; Alavi, S. Pilot Scale Roller Milling of Chickpeas into a De-Hulled Coarse Meal and Fine Flour. *Processes* **2022**, *10*, 2328. [CrossRef]
3. Pulivarthi, M.K.; Nkurikiye, E.; Watt, J.; Li, Y.; Siliveru, K. Comprehensive Understanding of Roller Milling on the Physicochemical Properties of Red Lentil and Yellow Pea Flours. *Processes* **2021**, *9*, 1836. [CrossRef]
4. Braun, M.; Dunmire, K.; Evans, C.; Stark, C.; Woodworth, J.; Paulk, C. Effects of Grinding Corn with Different Moisture Content on Subsequent Particle Size and Flowability. *Processes* **2021**, *9*, 1372. [CrossRef]
5. Braun, M.; Wecker, H.; Dunmire, K.; Evans, C.; Sodak, M.W.; Kapetanovich, M.; Shepherd, J.; Fisher, R.; Coble, K.; Stark, C.; et al. Evaluation of Hammermill Tip Speed, Air Assist, and Screen Hole Diameter on Ground Corn Characteristics. *Processes* **2021**, *9*, 1768. [CrossRef]
6. McGuire, C.; Siliveru, K.; Ambrose, K.; Alavi, S. Food Powder Flow in Extrusion: Role of Particle Size and Composition. *Processes* **2022**, *10*, 178. [CrossRef]
7. Iwaki, S.; Hayakawa, K.; Fu, B.-X.; Otobe, C. Changes in Hydrophobic Interactions among Gluten Proteins during Dough Formation. *Processes* **2021**, *9*, 1244. [CrossRef]
8. Zhang, Y.; Hu, R.; Tilley, M.; Siliveru, K.; Li, Y. Effect of Pulse Type and Substitution Level on Dough Rheology and Bread Quality of Whole Wheat-Based Composite Flours. *Processes* **2021**, *9*, 1687. [CrossRef]
9. Mariscal-Moreno, R.M.; Chuck-Hernández, C.; Figueroa-Cárdenas, J.D.; Serna-Saldivar, S.O. Physicochemical and Nutritional Evaluation of Bread Incorporated with Ayocote Bean (*Phaseolus coccineus*) and Black Bean (*Phaseolus vulgaris*). *Processes* **2021**, *9*, 1782. [CrossRef]
10. Cheng, Z.; Li, X.; Hu, J.; Fan, X.; Hu, X.; Wu, G.; Xing, Y. Effect of Gaseous Chlorine Dioxide Treatment on the Quality Characteristics of Buckwheat-Based Composite Flour and Storage Stability of Fresh Noodles. *Processes* **2021**, *9*, 1522. [CrossRef]
11. Park, G.; Cho, H.; Kim, K.; Kweon, M. Quality Characteristics and Antioxidant Activity of Fresh Noodles Formulated with Flour-Bran Blends Varied by Particle Size and Blend Ratio of Purple-Colored Wheat Bran. *Processes* **2022**, *10*, 584. [CrossRef]
12. Tang, H.; Xu, C.; Jiang, Y.; Wang, J.; Wang, Z.; Tian, L. Evaluation of Physical Characteristics of Typical Maize Seeds in a Cold Area of North China Based on Principal Component Analysis. *Processes* **2021**, *9*, 1167. [CrossRef]
13. Ye, Q.; Yang, X.; Xie, W.; Yao, J.; Cai, Z. Effect of Heat Resource Effectiveness Change on Rice Potential Yield in Southern China. *Processes* **2021**, *9*, 896. [CrossRef]
14. Heermann, M.L.; Brown, J.; Getty, K.J.K.; Yucel, U. Assessing Functionality of Alternative Sweeteners in Rolled "Sugar" Cookies. *Processes* **2022**, *10*, 868. [CrossRef]
15. Li, S.; Zhao, W.; Min, G.; Li, P.; Zhang, A.; Zhang, J.; Wang, Y.; Liu, Y.; Liu, J. Effects of Different Amylose Contents of Foxtail Millet Flour Varieties on Textural Properties of Chinese Steamed Bread. *Processes* **2021**, *9*, 1131. [CrossRef]

 *processes*

Article

Near Infrared Spectroscopic Evaluation of Starch Properties of Diverse Sorghum Populations

Kamaranga H. S. Peiris [1], Xiaorong Wu [1,*], Scott R. Bean [1], Mayra Perez-Fajardo [1], Chad Hayes [2], Melinda K. Yerka [3], S. V. Krishna Jagadish [4], Troy Ostmeyer [4], Fadi M. Aramouni [1], Tesfaye Tesso [4], Ramasamy Perumal [5], William L. Rooney [6], Mitchell A. Kent [6] and Brent Bean [7]

[1] Grain Quality and Structure Research Unit, Center for Grain and Animal Health Research, USDA-ARS, Manhattan, KS 66502, USA; Shantha.Peiris@usda.gov (K.H.S.P.); scott.bean@usda.gov (S.R.B.); mayra.perez-fajardo@usda.gov (M.P.-F.); fadi.aramouni@usda.gov (F.M.A.)

[2] Plant Stress and Germplasm Development Research Unit, Cropping Systems Research Lab, USDA-ARS, Lubbock, TX 79401, USA; chad.hayes@usda.gov

[3] Department of Agriculture, Veterinary & Rangeland Science, University of Nevada, Reno, NV 89557, USA; myerka@unr.edu

[4] Department of Agronomy, Kansas State University, Manhattan, KS 66506, USA; kjagadish@ksu.edu (S.V.K.J.); tjostmeyer@ksu.edu (T.O.); ttesso@ksu.edu (T.T.)

[5] Agricultural Research Center, Department of Agronomy, Kansas State University, Hays, KS 67601, USA; perumal@ksu.edu

[6] Department of Soil and Crop Sciences, Texas A&M University, College Station, TX 77843, USA; William.Rooney@ag.tamu.edu (W.L.R.); makent1995@tamu.edu (M.A.K.)

[7] United Sorghum Checkoff Program, Lubbock, TX 79403, USA; brentb@sorghumcheckoff.com

* Correspondence: shawn.wu@usda.gov

Citation: Peiris, K.H.S.; Wu, X.; Bean, S.R.; Perez-Fajardo, M.; Hayes, C.; Yerka, M.K.; Jagadish, S.V.K.; Ostmeyer, T.; Aramouni, F.M.; Tesso, T.; et al. Near Infrared Spectroscopic Evaluation of Starch Properties of Diverse Sorghum Populations. *Processes* **2021**, *9*, 1942. https://doi.org/10.3390/pr9111942

Academic Editor: Bernd Hitzmann

Received: 24 September 2021
Accepted: 27 October 2021
Published: 29 October 2021

Publisher's Note: MDPI stays neutral with regard to jurisdictional claims in published maps and institutional affiliations.

Abstract: Starch, mainly composed of amylose and amylopectin, is the major nutrient in grain sorghum. Amylose and amylopectin composition affects the starch properties of sorghum flour which in turn determine the suitability of sorghum grains for various end uses. Partial least squares regression models on near infrared (NIR) spectra were developed to estimate starch and amylose contents in intact grain sorghum samples. Sorghum starch calibration model with a coefficient of determination (R^2) = 0.87, root mean square error of cross validation (RMSECV) = 1.57% and slope = 0.89 predicted the starch content of validation set with R^2 = 0.76, root mean square error of prediction (RMSEP) = 2.13%, slope = 0.93 and bias = 0.20%. Amylose calibration model with R^2 = 0.84, RMSECV = 2.96% and slope = 0.86 predicted the amylose content in validation samples with R^2 = 0.76, RMSEP = 2.60%, slope = 0.98 and bias = −0.44%. Final starch and amylose cross validated calibration models were constructed combining respective calibration and validation sets and used to predict starch and amylose contents in 1337 grain samples from two diverse sorghum populations. Protein and moisture contents of the samples were determined using previously tested NIR spectroscopy models. The distribution of starch and protein contents in the samples of low amylose (<5%) and normal amylose (>15%) and the overall relationship between starch and protein contents of the sorghum populations were investigated. Percent starch and protein were negatively correlated, low amylose lines tended to have lower starch and higher protein contents than lines with high amylose. The results showed that NIR spectroscopy of whole grain can be used as a high throughput pre-screening method to identify sorghum germplasm with specific starch quality traits to develop hybrids for various end uses.

Keywords: near infrared spectroscopy; sorghum; starch; amylose; amylopectin; high throughput phenotyping; genetic diversity; plant breeding

1. Introduction

Sorghum (*Sorghum bicolor* (L.) Moench) ranks fifth in global cereal grain production after maize (*Zea mays* (L.)), wheat (*Triticum aestivum* (L.)), rice (*Oryza sativa* (L.)) and barley

(*Hordeum vulgare* (L.)) with 57.9 million megagrams (Mg) of grain sorghum harvested from 40.1 million ha in 2019 with an average yield of 1.44 Mg/ha. As the third most important cereal grain in the USA after wheat and maize, the USA produced 8.7 million Mg of grain sorghum harvested from 1.9 million ha with an average yield of 4.6 Mg/ha in 2019 (FAO STAT http://faostat.fao.org, accessed on 20 October 2021). Grain sorghum is used as food, feed, fodder and as a feedstock for bioethanol production [1–5].

In the USA, after exports, sorghum is mostly used as an ingredient in animal feed and as a bio-fuel feedstock. However, since sorghum has potential human health benefits in the prevention of chronic diseases [6–8] and as a gluten free food, it is also being increasingly used for preparation of various foods [9,10].

Starch is the primary constituent of sorghum grain. Starch comprises two types of macromolecules, the relatively small (up to 10^6 Da) and linear amylose with few long branches and the large (10^7–10^9 Da) and highly-branched amylopectin with many short branches. These macromolecules form starch granules with alternative crystalline and amorphous layers [11,12]. Amylose and amylopectin have different physiochemical properties. Starch content and starch properties, especially the amount of amylose content in sorghum starch, influence the suitability of sorghum cultivars for specific end uses [13–15] and the digestibility of sorghum starch [16]. For example, higher starch contents are important for grains used for ethanol fermentation. However, just the starch content itself is not sufficient to select the best varieties as the ethanol fermentation efficiency depends on the amylose levels in starch. Likewise, when the suitability of high starch sorghum as an animal feed ingredient is evaluated, the amylose levels should also be considered as it affects the digestibility of starch. Therefore, it is imperative to measure starch and amylose contents for developing cultivars for specific uses.

For plant breeding purposes, it is necessary to analyze starch and amylose contents of a large number of samples in breeding populations. Currently there are many methods for starch analysis [17]. However, laboratory starch analysis methods are laborious, vary in cost per test, and are time consuming. Near infrared (NIR) spectroscopy has been used as a rapid analytical method for the evaluation of numerous traits of cereal grains in plant breeding programs [18], including starch and amylose contents [19]. Most NIR spectroscopy methods developed for sorghum starch and/or amylose content have been for samples from ground grain [20,21]. In some studies where NIR has been used for intact grain, details of the NIR method used were not available [22]. De Alencar Figueiredo et al., 2006 used NIR spectroscopy for the analysis of amylose content in both intact and ground sorghum grain samples and found that prediction is poor when intact grains are used [23].

However, using intact grain for analysis avoids the need to grind samples, which is laborious and time consuming, and grinding has the potential to contaminate samples without proper cleaning of grinding equipment between samples. In addition, when using intact grains for non-destructive NIR analysis, grains can be saved and used as seed. Thus, using intact grain for NIR analysis allows for large sample sets to be scanned and analyzed within a short period of time with only minor sample preparation. The primary objective of this work was to develop NIR starch and amylose calibration models for use as a non-destructive, rapid, robust, and cost-effective method to estimate starch and amylose contents in intact grain sorghum for screening breeding and genetically diverse populations.

2. Materials and Methods

2.1. Grain Samples

Grains harvested from several sorghum breeding populations and agronomic trials were collected from the 2018 through 2020 growing seasons from different locations in California, Kansas, and Texas. Grain samples used for the starch calibration were selected from five populations and four different populations were used to select samples for the amylose calibration. For the starch calibration, Population 1 (Starch Population 1, SP1) samples were drawn from the sorghum association panel (SAP) described by Casa et al., (2008) [24] grown in Kansas. Samples from Population 2 (SP2) came from seven lines within the SAP grown in Kansas that were harvested at a higher moisture content of around 18% where samples were scanned as samples dried to introduce moisture variability to calibration. Population 3 (SP3) samples were from a single hybrid grown under 10 different nitrogen fertilization treatments grown in Kansas. Population 4 (SP4) was from hybrids and inbred lines grown in Kansas and Texas and Population 5 (SP5) was from a breeding population grown in California. Samples for the amylose calibrations were selected from four different populations distinct from SP1–5 consisting of hybrids, inbreds, and segregating early F2 generation plant selections grown in Kansas and Texas (designated as amylose populations 1 through 4, or AP1, AP2, AP3 and AP4). A summary of the sorghum sample populations used for starch and amylose calibrations is given in Table 1.

Table 1. Description of the sorghum grain sample population used in the study.

Sample Population	N	Year	Location(s)	Type of Sample Population
Amylose				
AP1	22	2018	Texas	Breeding Population
AP2	63	2019	Kansas/Texas	Breeding Population
AP3	31	2020	Texas	Breeding Population
AP4	37	2020	Texas	Breeding Population
Starch				
SP1	29	2018	Kansas	Diversity panel
SP2	61	2021	Kansas	Hybrid
SP3	39	2019	Kansas	Diversity panel
SP4	56	2019	Kansas/Texas	Breeding Population
SP5	26	2020	California	Breeding Population
Predictions				
Breeding 1	946	2020	Texas	Breeding Populations
Breeding 2	391	2020	California/Argentina/Mexico	Breeding Populations

Samples from two additional breeding populations harvested in California, Texas and in winter nurseries in Argentina and Mexico were scanned and used for the prediction of starch, amylose and protein contents and moisture to study the relationship between these traits in sorghum grain in genetically diverse materials. The sample populations used in generating the starch and amylose calibrations had a high degree of phenotypic diversity for pericarp color (red, white, yellow, etc.), tannin contents, grain sizes and kernel hardness, as these samples were from a diverse genetic and geographic background of several growing regions in North and South America, capturing a wide range of environmental variability in addition to different nitrogen fertilization treatments.

Preliminary starch and amylose calibrations constructed using the populations scanned in early years were used to predict starch and amylose contents in subsequent grain populations. Those predicted starch and amylose values were used to identify candidate lines across the constituent range for laboratory analysis of starch or amylose in order to use in calibration improvement. This approach enabled the efficient use of resources available for laboratory analysis to obtain samples with starch and amylose reference data more or less equally distributed along the available range of both constituents.

2.2. NIR Scanning

Grain samples were scanned as they were received at the laboratory. First, samples were screened to remove small broken pieces and dust, and then glumes and other debris were removed and cleaned seeds were used for scanning. A Perten DA7250 (Perten Instruments, Springfield, IL, USA) spectrometer was used to scan grain samples in reflectance mode. Samples were scanned using a Teflon cup (60 mm diameter and 10 mm deep) that can hold about 20 g of grains. A micromirror cup (Perten Instruments, Springfield, IL, USA) was used if the quantity of seeds available were less. The cup was filled with grains and excess grains were removed by levelling so that the distance from the surface of grains to the collecting optics of the instrument was uniform for all samples. The spectrometer recorded NIR absorbance data from 950 to 1650 nm in 5 nm intervals. Each sample was scanned in triplicate by mixing the grains and repacking the sample cup after each scan.

2.3. Starch and Amylose Content Determination

Grain samples were ground for total starch and amylose measurement using a cyclone mill equipped with a 0.5 mm screen (Udy Corp, Fort Collins, CO, USA). Total starch content was measured colorimetrically using a commercially available kit (Megazyme K-TSTA-100A kit, Bray, Ireland) and following the total starch assay procedure (amyloglucosidase/α-amylase method), procedure example (b), "Determination of total starch content of samples containing resistant starch (RTS-NaOH Procedure -Recommended)." [25]. Briefly, 100 mg grain meal in 16 $\times$ 120 mm glass tubes was wetted with 0.2 mL of 80% ethanol and dissolved in 2 mL 1.7 M sodium hydroxide for 15 min. Eight mL sodium acetate buffer (pH 3.8) was added into the glass tube to adjust pH to 5.0. The samples were hydrolyzed with thermostable α-amylase and amyloglucosidase (0.1 mL each) at 50 °C for 30 min. After centrifugation at 1300 rpm for 5 min, 0.1 mL of the hydrolysate was mixed with 3.0 mL GOPOD reagent and incubated at 50 °C for 20 min. The absorbance of the mixture was measured against the reagent blank and used to calculate the percent starch content in the grain meal sample.

Apparent amylose in the whole grain meal samples was quantitated colorimetrically taking advantage of amylose forming polyiodide-amylose complex with iodine, which has a maximum absorbance at around 620 nm [26,27]. Briefly, 25–30 mg of grain meal (alternatively 30–35 mg low amylose samples) were weighed (to $\pm$ 0.1 mg accuracy) in a 15 mL glass test tube and the samples were dispersed with 0.1 mL 80% ethanol to prevent them from forming clumps at the bottom. Next, 1 mL of 90% DMSO:0.6 M urea solution was added to the glass tubes while vortexing. The glass tubes were brought to 100 °C in a heat block until the starch was dissolved, another 5 mL of 90% DMSO was added, and samples were incubated at 100 °C for 30 min with vortexing every 5 min. The heated dissolved samples were allowed to cool to room temperature, and an aliquot (0.1 mL) was transferred into a test tube with 5.0 mL of 0.5% trichloroacetic acid and mixed with 0.1 mL 0.01 N I_2-KI solution (300 mg KI in 1–2 mL of deionized water with 127 mg iodine in 100 mL). Finally, the absorbance at 620 nm was read against a reagent blank after 30 min without disturbing the precipitates when transferring the solution into a cuvette. A standard curve was established using reference amylose (potato, Megazyme # P-AMYL, Bray, Ireland) and amylopectin (maize, Sigma #10120, St. Louis, MI, USA) to make mixtures with different amylose contents (0, 5, 15, 30, 50, 100% amylose) for calculating the apparent amylose content in the samples. Note, apparent amylose contents were reported as% amylose in the ground whole meal ("flour"), not as a percent of total starch (i.e., flour basis rather than starch basis). Both starch and amylose content data were converted to dry basis using moisture values obtained from NIR ground whole meal sorghum moisture calibration (R^2 = 0.98, RMSECV = 0.37%, Slope = 0.98).

2.4. Spectral Data Acquisition and Data Analysis

Spectral data from the Perten DA7250 spectrometer were retrieved in JCAMP-DX format [28] and the JCAMP-DX spectral data files were imported to the Unscrambler software Version 10.5.1 (CAMO Software AS, Oslo, Norway) for handling and subsequent pre-processing of spectra, calibration model development, validation, and prediction in new samples.

Spectral data in Unscrambler in the form of spectral identity and raw absorbance values from 950–1650 nm in 5 nm intervals were exported to Microsoft Excel. NIR spectra from three replicate sample scans were averaged. The spectra of the samples used for starch and amylose analysis by standard laboratory method for calibration and validation data sets were selected and the respective constituent values were appended. Lab-measured dry weight basis starch and amylose contents were converted to an 'as is' basis of the samples at the time of scanning, using the NIR predicted moisture content of the same samples.

Sample spectral data were then sorted by constituent value and samples were selected for use in the calibration and validation data sets. Samples from SP2 population for the starch calibration was divided such that the calibration included four lines scanned at different moisture contents while three lines were used in the validation set. Therefore, those sample spectra of lines scanned for multiple times at different moisture contents remained either in the calibration or the validation set, but not in both. Starch calibration spectra for SP3 came from one hybrid grown under five nitrogen fertilizer treatments, while the validation set included spectra from the same hybrid grown under five different treatments (10 treatments total). The rest of the spectra from the remaining populations were used in the ratio of 2:1 for calibration and validation sets, respectively. The spectral data and starch and amylose contents were imported to Unscrambler for analysis, calibration model development, and validations.

Raw spectral data of the starch and amylose datasets were subjected to principal component analysis to investigate similarity/diversity of spectra among sample populations. Spectra of calibration sample sets were pre-processed with extended multiplicative scatter correction (EMSC) [29] and mean centering. Resulting pre-processed and mean centered NIR spectral data were used to build partial least squares calibration models with leave-one-out cross validation. The number of PLS factors for the calibration models were selected considering the Root Mean Squared Error Cross Validation (RMSECV) and coefficient of determination (R^2) of calibration models and Root Mean Squared Error Prediction (RMSEP), R^2, slope and bias of the validation tests. After calibrations were validated, the spectra in the calibration and validation datasets were combined and a final cross validated model was developed using all spectra each for starch and amylose predictions.

2.5. Prediction of Moisture, Starch, Amylose and Protein Contents of New BREEDING Populations

The starch and amylose contents of samples from two diverse breeding populations grown in California, Texas, Argentina, and Mexico that had not contributed to the starch or amylose calibrations or validation sets were predicted using the above-mentioned combined starch and amylose calibrations. In addition to amylose and starch contents, moisture and protein contents of these two populations were also predicted using previously developed NIR calibrations for moisture (R^2 = 0.99, RMSECV = 0.23%, Slope = 0.99) and protein (R^2 = 0.92, RMSECV = 0.45%, Slope = 0.93) in intact grains [30]. Subsequently, dry weight basis starch, amylose and protein contents of the samples were calculated. Based on the predicted dry weight basis amylose contents, samples were grouped as low amylose (<5% amylose), intermediate amylose (5–15% amylose), and normal amylose (>15% amylose). The frequency distribution of the starch and protein contents of the low and normal amylose groups in the breeding populations were calculated. The relationship between starch and protein contents in this sample population was tested with Pearson correlation coefficient. Note that the breeding population used for these predictions contained early generation material which was still genetically segregating for various traits including starch, amylose, and protein contents. Thus, the wide range of intermediate amylose contents observed in this dataset may be due to the fact that each seed on a panicle could have a different starch, amylose, and/or protein content that would be averaged during NIR scans conducted on a per-panicle basis.

3. Results and Discussion

3.1. Diversity of Sample Populations

NIR spectra of intact sorghum grain samples from the populations used for starch and amylose calibrations are shown in the Figure 1. NIR spectra of the grain samples contributing to starch and amylose datasets were subjected to principal component analysis. The principal component (PC) score plot of PC1 against PC2 for raw NIR spectral data of different grain populations for starch and amylose spectral data sets are presented in

Figure 2. First and second principal components of both starch and amylose datasets explained 99% of the variance of spectra. PC scores of different populations showed that the individual populations were diverse. The observed diversity may be due to changes in spectra caused by different starch and amylose contents in the samples, as well as other factors such as variations in chemical and physical properties resulting from differences in genetics, growing seasons, locations, or other unknown causes. The least diversity was observed in the SP3 dataset, which came from a single hybrid grown under different nitrogen fertilizer treatments wherein the starch content varied from 63.93–69.55%. The use of samples from very diverse and heterozygous populations grown at different locations in different years and under various management regimes helped develop calibrations which can be more robust in predicting grain starch and amylose contents in new populations.

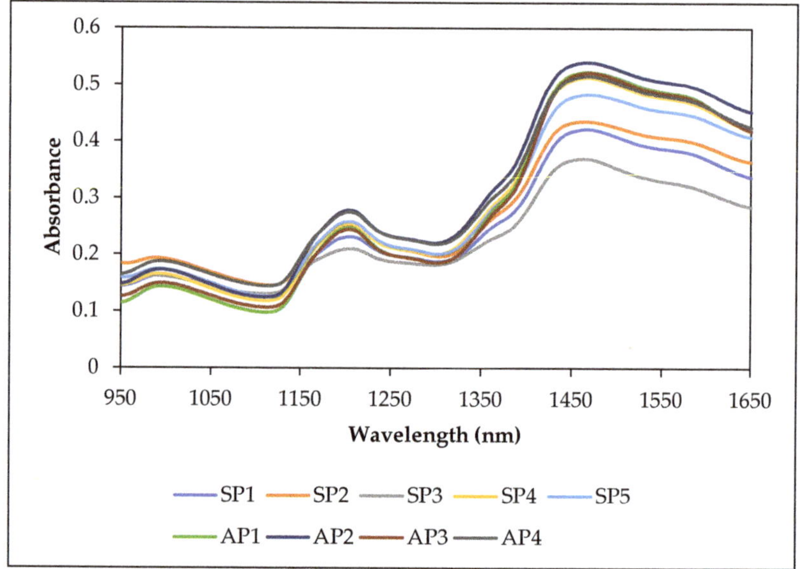

Figure 1. NIR spectra of some intact grain samples from populations used for starch and amylose calibrations.

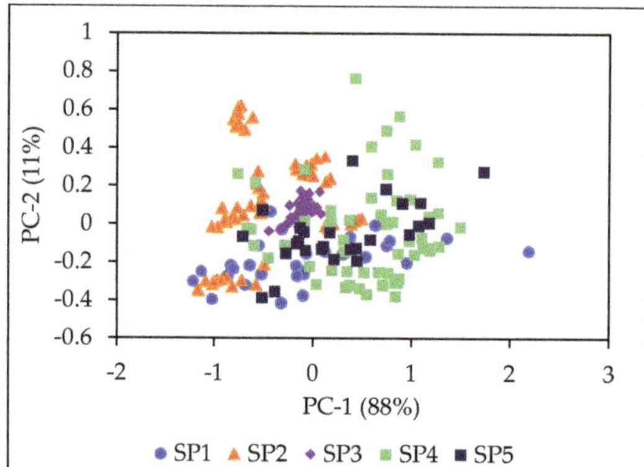

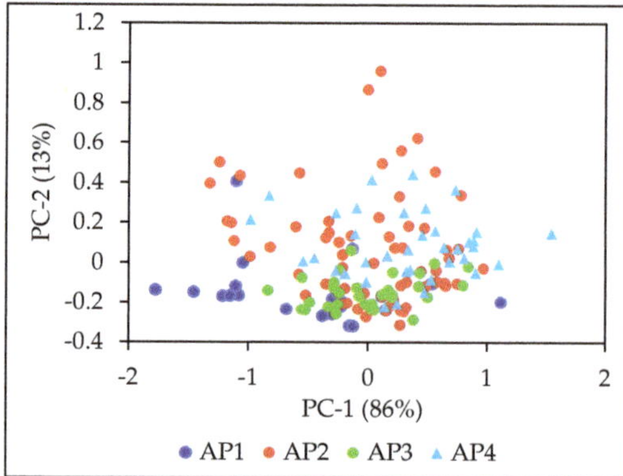

Figure 2. Principal component score plots of NIR spectra of different grain sorghum populations including starch (**left**) and amylose (**right**) data sets.

3.2. Starch and Amylose Contents in Grain Samples

Starch calibration was developed using five while amylose calibration was developed using four populations. The distribution of starch and amylose contents on fresh weight basis in grain populations and in selected calibration and validation data sets are presented in Table 2. The starch content of combined populations varied from 50.73–74.17% with an average of 62.99% and standard deviation of 4.31%. The range of starch contents of individual populations were narrow and use of multiple populations increased the range and variability of starch content. Selection of samples for calibration and validation datasets were conducted manually such that the calibration dataset had the total range of starch content while the validation set had a slightly lower starch content of 53.46–72.70%. Likewise, the amylose calibration set covered the widest range of amylose contents in the assayed populations, ranging from 0.25–27.90%, while the validation set included samples with amylose contents ranging from 0.28–27.25%.

Table 2. Starch and amylose content variability in grain sorghum populations and in calibration and validation sets.

Sample Set	Population	N *	Min	Max	Avg	SD
Starch samples						
All		211	50.73	74.17	62.99	4.31
	SP1	29	55.79	65.54	61.44	2.56
	SP2	61	50.73	64.56	58.82	2.96
	SP3	39	63.93	69.55	67.23	1.40
	SP4	56	58.69	68.10	63.14	2.28
	SP5	26	59.26	74.17	67.85	3.95
Calibration set		119	50.73	74.17	63.12	4.34
	SP1	16	58.12	65.54	62.46	2.16
	SP2	33	50.73	64.56	58.82	3.27
	SP3	19	63.93	69.29	67.31	1.60
	SP4	34	58.69	68.10	62.96	2.09
	SP5	17	59.26	74.17	67.75	4.38
Validation set		92	53.46	72.70	62.83	4.29
	SP1	13	55.79	63.42	60.18	2.52
	SP2	28	53.46	63.95	58.82	2.59
	SP3	20	64.84	69.55	67.17	1.21
	SP4	22	58.74	67.00	63.42	2.58
	SP5	9	62.81	72.70	68.03	3.21
Amylose samples						
All		153	0.25	27.90	9.17	7.14
	AP1	22	4.00	27.90	14.68	10.91
	AP2	63	0.29	18.21	8.27	5.71
	AP3	31	0.25	12.27	3.70	3.56
	AP4	37	1.40	18.48	12.01	4.98
Calibration set		102	0.25	27.90	9.20	7.19
	AP1	16	4.00	27.90	14.36	10.89
	AP2	41	0.29	18.21	8.74	5.92
	AP3	24	0.25	12.27	4.07	3.77
	AP4	21	1.40	18.00	12.03	4.89
Validation set		51	0.28	27.25	9.12	7.09
	AP1	6	4.02	27.25	15.52	11.95
	AP2	22	0.86	16.75	7.41	5.33
	AP3	7	0.28	6.29	2.44	2.52
	AP4	16	1.99	18.48	11.98	5.25

* N: Number of samples; Min: Minimum; Max: Maximum; Avg: Average; SD: Standard deviation.

3.3. Starch Calibration Development and Model Validation

Starch calibration model constructed with 119 samples were validated with 92 samples that were not used for the construction of the calibration model. Starch calibration model with 11 PLS factors had a R^2 = 0.87, RMSECV = 1.57% and a slope of 0.89. The number of PLS factors for the calibration was selected by taking into consideration the cross-validation statistics including R^2, RMSECV, the slope of the curve and regression coefficient plots. This calibration model predicted the starch content in the validation sample set with R^2 = 0.76, RMSEP = 2.13%, slope = 0.93 and bias = 0.20% (Figure 3).

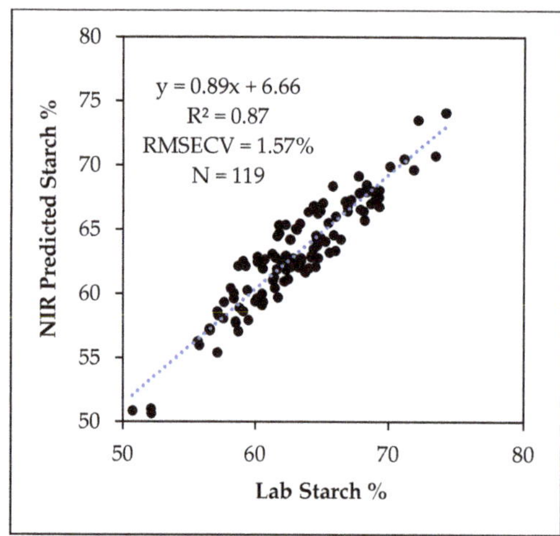

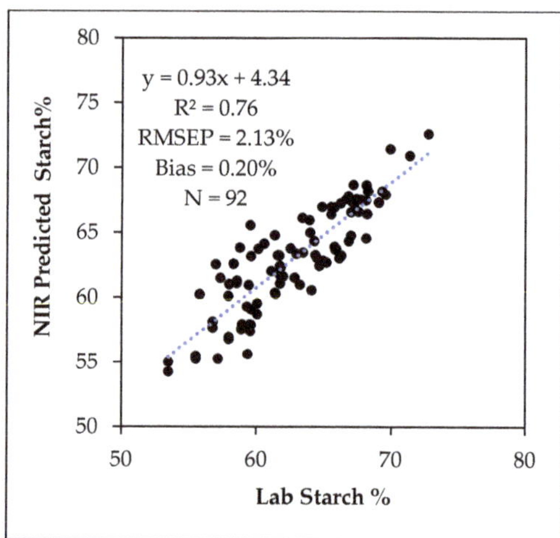

Figure 3. The relationship between laboratory determined and NIR predicted starch content for NIR starch calibration (**left**) and validation (**right**).

Analysis of the regression coefficient plots of the PLS models is important to make sure that the key wavelengths of the model are related to the spectroscopic signal of the interested constituent molecule to ensure the validity of the NIR spectroscopy model [31,32]. The regression coefficient plot for the starch calibration model with 11 PLS factors is shown in Figure 4. Some of the key regression peaks, both positive and negative, in the regression coefficient plot that may have direct or indirect relation with the sorghum grain starch content may be due to second overtone of C-H stretch (peaks around 1160, 1205, 1240 nm), C-H stretch + C-H deformation (1365 and 1390 nm), first overtone of O-H stretch of starch (1580 nm) and first overtone of C-H stretch (1645 nm) vibrations of different C-H and O-H groups of starch [33,34].Therefore, it is possible that the starch model is capable of predicting the starch content of whole grain samples by using the interactions between some key NIR wavelengths and starch molecules in the grain. Hence, these results suggest that NIR spectroscopy can be used to predict starch contents of intact grain samples.

3.4. Amylose Calibration Development and Model Validation

The amylose calibration curve from 102 grain samples had 11 PLS factors with R^2 = 0.84, RMSECV = 2.96% and a slope of 0.86. This amylose calibration model predicted the amylose content in an independent set of 51 samples with R^2 = 0.76, RMSEP = 2.60%, slope = 0.98 and bias = −0.44% (Figure 5). The regression coefficient plot of the amylose calibration with 11 PLS factors is shown in Figure 6. The dominant regression peak in this plot is at 1235 nm and this may be due to C-H stretch second overtone of CH_2 vibration [33]. Starch is a glucose polymer composed of straight chain amylose, a linear α (1–4) linked glucan, and branched amylopectin, an α (1–4) linked glucan that contains around 5% α (1–6) linkages resulting in a branched molecule [12]. Therefore, amylopectin is chemically different from amylose in that the sixth C atom of the α (1–6) linkage contain a CH_2 group attached to O in one end and to the 5th C atom of a glucose unit at the branching point. The vibrational frequency of this CH_2 group may differ from the vibrational frequency of other CH_2 groups of the sixth C atom of glucose units in a linear chain. The second overtone C-H stretch vibration of this particular CH_2 group in amylopectin around 1235 nm may be the primary wavelength that the calibration model uses to distinguish and quantify amylose from amylopectin in sorghum starch or flour samples. Fertig et al., (2004) found the best correlation of amylose content in amylose/amylopectin binary mixtures was around 1730–1750 nm which corresponds to the C-H stretch first overtone vibration of CH_2 group [35]. Since the spectral range of 950–1650 nm we used mostly covered the second overtone region of C-H vibrations, our model apparently works using the difference of second overtone C-H vibrations of amylose and amylopectin in sorghum starch.

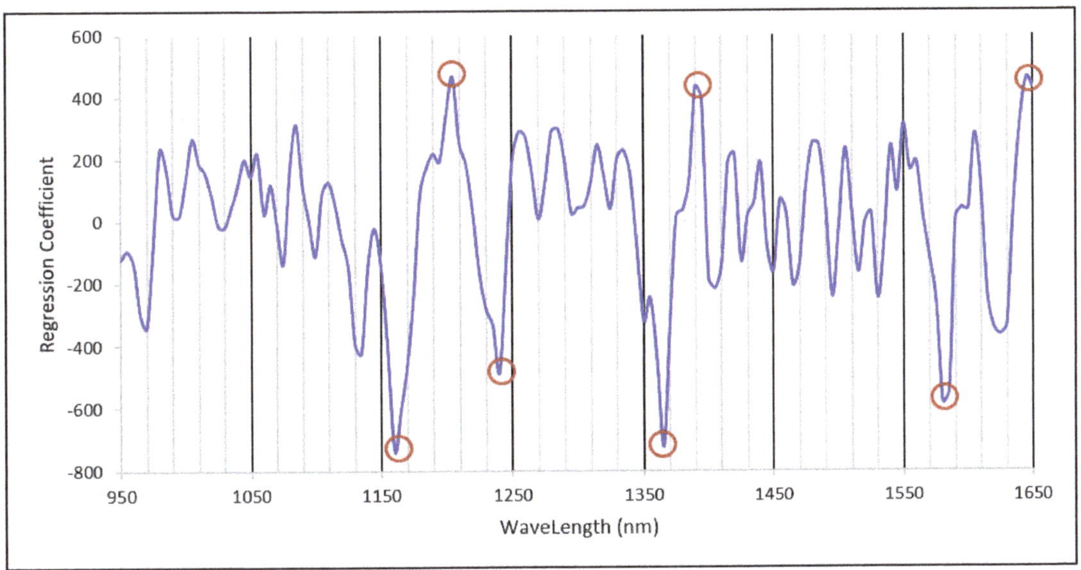

Figure 4. Regression coefficient plot of the 11 PLS factor starch calibration with important regression peaks marked.

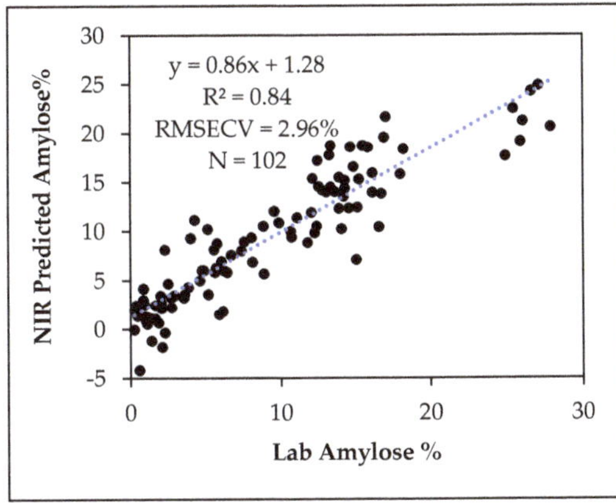
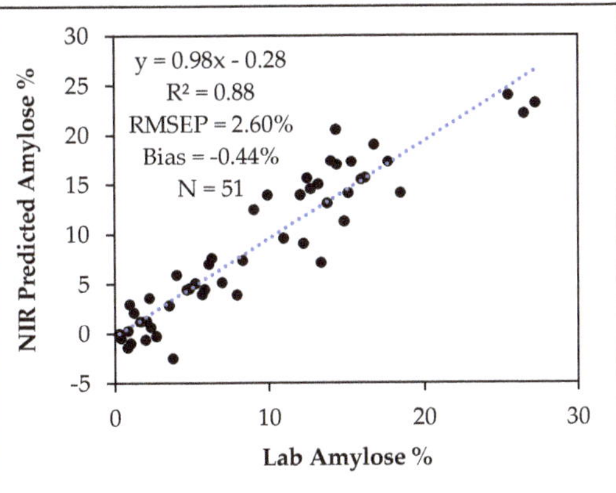

Figure 5. The relationship between laboratory determined and NIR predicted amylose contents for the amylose NIR calibration (**left**) and validation (**right**).

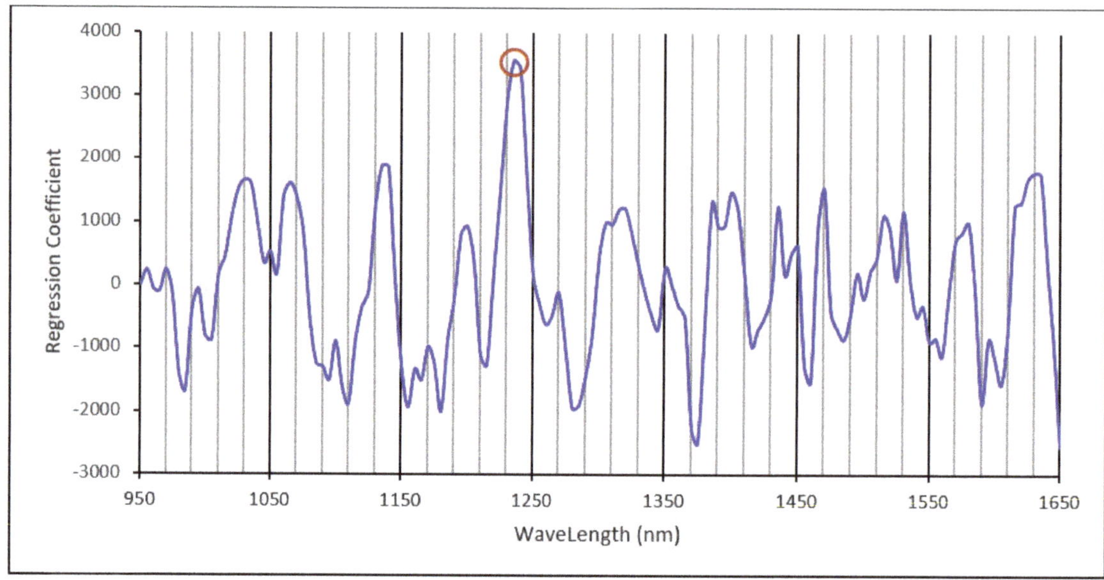

Figure 6. Regression coefficient plot of the 11 PLS factor amylose calibration with important regression peaks marked.

3.5. NIR Prediction of Starch and Amylose Contents in Breeding Populations

Spectra of the calibration and validation sets were combined to construct cross validated starch and amylose calibration curves for predicting starch and amylose contents in two breeding populations not used in calibration development. The starch calibration with 11 PLS factors from 211 sample spectra used for the calibration and validation of the starch curve had R^2 = 0.85, RMSECV = 1.67% and a slope of 0.86. Likewise, the combined 11 PLS factor amylose calibration from 153 grain samples had R^2 = 0.86, RMSECV = 2.66% and a slope of 0.87 (Figure 7). After a calibration is developed and tested with an external validation set, calibration and validation sets may be combined to make a model incorporating maximum information into the final calibration model, thereby improving the robustness of the model so that the accuracy of the predictions of future samples will generally be better [36]. Prediction of the starch and amylose contents in the breeding populations by the respective calibrations before and after validation set was combined showed that the average Mahalanobis distance (MD) [37] was reduced. Average MD of starch predictions was reduced from 2.78 to 2.48 while the average MD of amylose predictions was reduced from 4.19 to 3.83, suggesting that combining the calibration and validation datasets was helpful to improve the robustness of the calibrations.

3.6. Relationship between, Starch, Amylose and Protein Contents in Grain Sorghum Populations

The relationship between the dry weight basis starch and protein contents in grain sorghum based on NIR predictions of 1337 grain samples from the two breeding populations is shown in Figure 8. There was a negative relationship between starch and protein percent in grain sorghum (r = −0.755, $p < 10^{-181}$). Previous studies have shown mixed results regarding the association between starch and protein contents of sorghum grain. Subramanian and Jambunathan (1981) [38] found a strong negative correlation between starch and protein, while Buffo et al., (1998) [39] found no relationship. However, Buffo et al., (1998) [39] evaluated only 45–46 commercial hybrids and the range of starch and protein was narrow compared to the current study. Rhodes et al., (2017) evaluated 265 accessions and also reported a strong negative relationship between starch and protein contents in sorghum grain [40]. We evaluated a large number of samples covering a very wide range of starch and protein contents, and our results further confirm that there is a strong negative relationship between starch and protein content (on a percentage basis) in grain sorghum.

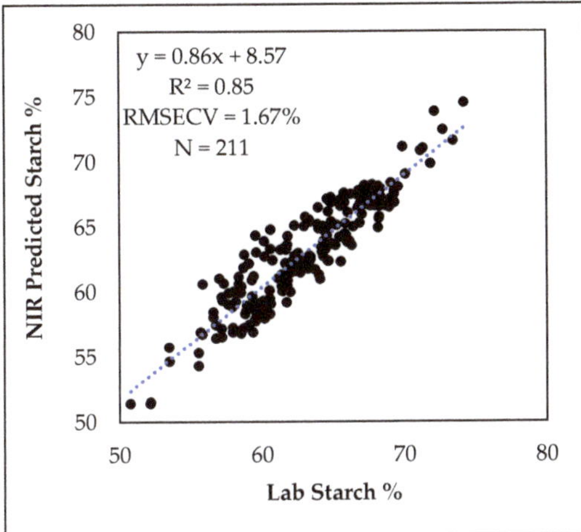

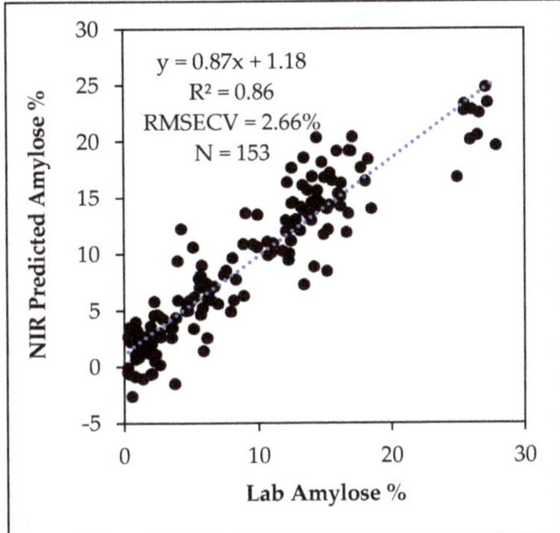

Figure 7. Lab-determined versus NIR predicted final starch (**left**) and amylose (**right**) cross validated calibrations constructed by combining all spectra in the calibration and validation datasets.

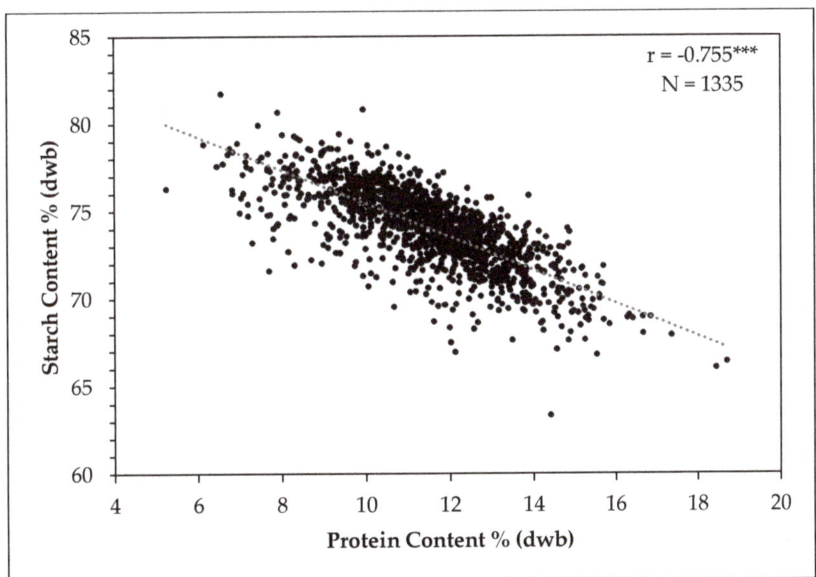

Figure 8. Scatter plot between dry weight basis starch and protein contents of sorghum grain. *** = $p < 0.001$.

The amylose calibration was used to estimate amylose contents in grain samples and based upon this, samples were divided into low amylose (<5% amylose) and normal amylose groups (>15% amylose). The frequency distribution of starch and protein contents in the selected groups are given in Figure 9. In these specific sample populations, low amylose samples tended to have less starch compared to that of normal samples. Accordingly, the reverse was observed with the protein contents of low amylose samples tending to be higher than in samples with normal amylose contents, partly due to the negative relationship between grain starch and protein. Multi-location trials with pedigreed populations segregating for variability in starch, amylose, and protein contents would be beneficial to further investigate the relationship between these constituents of sorghum grain. Such effort may be necessary to identify potential germplasm that minimizes negative interactions among starch, amylose, and protein contents by breaking up deleterious genetic linkages, similar to how historically low yield in "waxy" (low amylose) sorghum was overcome [41–44]. Since starch chemical composition is important for different end uses of sorghum grain, these new NIR calibrations can be used to pre-screen and select parent lines for specific uses; for example, to develop waxy hybrids having higher starch contents for ethanol fermentation and gluten-free frozen foods, or hybrids with optimum starch and protein contents for use in animal feed.

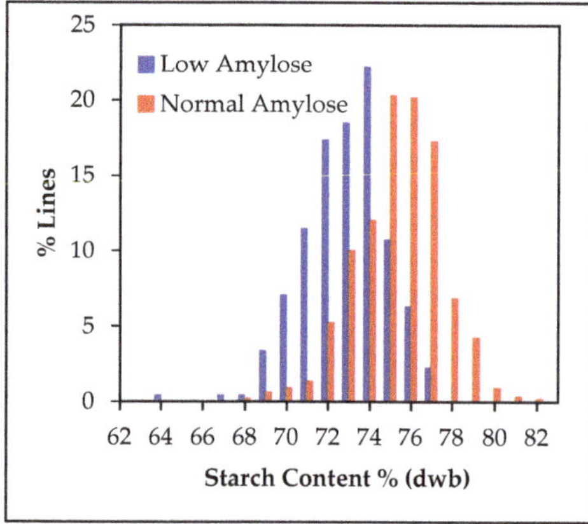
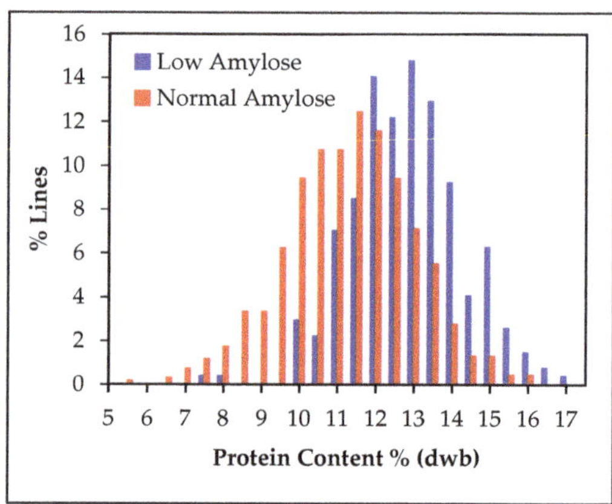

Figure 9. Variability of NIR predicted dry weight basis starch and protein contents in the low amylose (Amylose < 5%) and normal amylose (Amylose > 15%) grain samples of the two breeding populations.

3.7. NIR Spectroscopy for High Throughput Phenotyping of Segregating Sorghum Populations

Osborne (2006) has reviewed the application of NIR spectroscopy for quality evaluation of early generation materials in cereal breeding programs [18]. New high throughput techniques such as near-infrared spectroscopy are greatly lowering the cost per data point of phenotypic analysis. High throughput phenotyping of grain composition by NIR spectroscopy can be valuable for screening breeding populations, but also for use in genetic studies of grain composition. Genetic locus detection was improved more by increasing phenotyping throughput over accuracy [45] and NIR spectroscopic analysis of intact sorghum grain is one avenue to increase phenotypic analysis of grain composition. Amylose content and starch properties of sorghum are significantly affected by both genetic and environmental factors [46,47]. Therefore, in breeding programs selection for starch properties at a single location may be misleading [14] and the throughput of analyzing intact sorghum grain can assist in screening sorghum from multi-location trials.

A single scan of a grain sample takes about 2–3 min including sample handling and scanning, depending on the purity of sample. Thus, analysis of starch, amylose, protein, and moisture contents in large segregating breeding populations could be conducted with a much shorter time and at a fraction of the cost, compared to wet chemical analysis of similar number of samples. As long as NIR calibration for other traits are available, other interested traits can also be predicted simultaneously. Therefore, the use of NIR spectroscopy to screen germplasm may enable plant breeders to have more replicates or locations to better estimate the genetic potential of segregating plants in each generation.

Goncalves et al., 2021 used a genomics model combining single nucleotide polymorphisms (SNPs) and NIR spectroscopy data to predict fiber and sucrose content in a sugarcane population [48]. They found that the NIR model showed the highest prediction accuracy of phenotypes compared to using genomics alone or genomics + NIR models suggesting that NIR-based predictions are an effective strategy for predicting the genetic merits of sugarcane clones. This shows that NIR spectroscopy data is worth testing for its ability to enhance genomic prediction for sorghum grain end-use quality traits in the future.

4. Conclusions

NIR spectroscopy calibrations presented in this report can be used as a rapid analysis pre-screening tool for characterizing sorghum starch composition in segregating populations and to identify germplasm for developing new cultivars and hybrids for specific end uses. This whole grain NIR methodology can therefore be used to support the improvement of sorghum at the genetic level by identifying potentially useful sorghum lines based on starch and/or amylose levels. The information about the starch and protein contents in waxy (low amylose) and nonwaxy (normal amylose) groups of breeding populations can be used to identify lines with a desired combination of starch, amylose, and protein contents to match specific end use quality requirements.

Author Contributions: Conceptualization, S.R.B., K.H.S.P. and X.W.; methodology, X.W., S.R.B. and K.H.S.P.; software, K.H.S.P. and S.R.B.; formal analysis, S.R.B., X.W. and K.H.S.P.; resources, M.P.-F., C.H., M.K.Y., S.V.K.J., T.O., F.M.A., T.T., R.P., W.L.R., M.A.K. and B.B.; writing—original draft preparation, K.H.S.P. and S.R.B.; writing—review and editing, K.H.S.P., S.R.B., X.W., M.P.-F., C.H., M.K.Y., S.V.K.J., T.O., F.M.A., T.T., R.P., W.L.R., M.A.K. and B.B.; supervision, S.R.B., X.W. and F.M.A.; project administration, S.R.B. and X.W.; All authors have read and agreed to the published version of the manuscript.

Funding: MKY was funded by USDA-AFRI Plant Breeding Program Award# 2019-67014-29174.

Institutional Review Board Statement: Not applicable.

Informed Consent Statement: Not applicable.

Data Availability Statement: Data is contained within the article.

Acknowledgments: Mention of trade names or commercial products in this publication is solely for the purpose of providing specific information and does not imply recommendation or endorsement by the U.S. Department of Agriculture. The USDA is an equal opportunity provider and employer. This work was supported in part by the U.S. Department of Agriculture, Agricultural Research Service. Contribution number 22-082-J from Kansas Agricultural Research Station. For M.K.Y., this research was funded by USDA-AFRI Plant Breeding Program Award# 2019-67014-29174.

Conflicts of Interest: The authors declare no conflict of interest.

References

1. Anglani, C. Sorghum for human food–A review. *Plant Foods Hum. Nutr.* **1998**, *52*, 85–95. [CrossRef] [PubMed]
2. Aruna, C.; Visarada, K.B. Other Industrial Uses of Sorghum. In *Breeding Sorghum for Diverse End Uses*; Aruna, C., Visarada, K.B.R.S., Venkatesh Bhat, B., Tonapi, V.A., Eds.; Woodhead Publishing: Cambridge, MA, USA, 2019; pp. 271–292, ISBN 9780081018804.
3. Dahlberg, J.; Janoš, B.; Sikora, V.; Latković, D. Assessing sorghum [*Sorghum bicolor* (L.) Moench] germplasm for new traits: Food, fuels & unique uses. *Maydica* **2012**, *56*, 85–92.
4. Taylor, J.R.; Schober, T.J.; Bean, S.R. Novel food and non-food uses for sorghum and millets. *J. Cereal Sci.* **2006**, *44*, 252–271. [CrossRef]
5. Wang, D.; Bean, S.; McLaren, J.; Seib, P.; Madl, R.; Tuinstra, M.; Shi, Y.; Lenz, M.; Wu, X.; Zhao, R. Grain sorghum is a viable feedstock for ethanol production. *J. Ind. Microbiol. Biotechnol.* **2008**, *35*, 313–320. [CrossRef] [PubMed]
6. de Morais Cardoso, L.; Pinheiro, S.S.; Martino, H.S.D.; Pinheiro-Sant'Ana, H.M. Sorghum (*Sorghum bicolor* L.): Nutrients, bioactive compounds, and potential impact on human health. *Crit. Rev. Food Sci. Nutr.* **2017**, *57*, 372–390. [CrossRef] [PubMed]
7. Xiong, Y.; Zhang, P.; Warner, R.D.; Fang, Z. Sorghum grain: From genotype, nutrition, and phenolic profile to its health benefits and food applications. *Compr. Rev. Food Sci. Food Saf.* **2019**, *18*, 2025–2046. [CrossRef] [PubMed]

8. McGinnis, M.J.; Painter, J.E. Sorghum: History, use, and health benefits. *Nutr. Today* **2020**, *55*, 38–44. [CrossRef]
9. Stefoska-Needham, A.; Tapsell, L. Considerations for progressing a mainstream position for sorghum, a potentially sustainable cereal crop, for food product innovation pipelines. *Trends Food Sci. Technol.* **2020**, *97*, 249–253. [CrossRef]
10. Khoddami, A.; Messina, V.; Vadabalija Venkata, K.; Farahnaky, A.; Blanchard, C.L.; Roberts, T.H. Sorghum in foods: Functionality and potential in innovative products. *Crit. Rev. Food Sci. Nutr.* **2021**, 1–17. [CrossRef] [PubMed]
11. Taylor, J.R.; Duodu, K.G. Sorghum and millets: Grain-quality characteristics and management of quality requirements. In *Cereal Grains*, 2nd ed.; Wrigley, C., Batey, I., Miskelly, D., Eds.; Woodhead Publishing: Cambridge, MA, USA, 2017; pp. 317–351, ISBN 978-0-08-100719-8.
12. Tester, R.F.; Karkalas, J.; Qi, X. Starch—composition, fine structure and architecture. *J. Cereal Sci.* **2004**, *39*, 151–165. [CrossRef]
13. Ai, Y.; Medic, J.; Jiang, H.; Wang, D.; Jane, J.L. Starch characterization and ethanol production of sorghum. *J. Agric. Food Chem.* **2011**, *59*, 7385–7392. [CrossRef] [PubMed]
14. Beta, T.; Corke, H. Noodle quality as related to sorghum starch properties. *Cereal Chem.* **2001**, *78*, 417–420. [CrossRef]
15. Miller, O.H.; Burns, E.E. Starch characteristics of selected grain sorghums as related to human foods. *J. Food Sci.* **1970**, *35*, 666–668. [CrossRef]
16. Lichtenwalner, R.E.; Ellis, E.B.; Rooney, L.W. Effect of incremental dosages of the waxy gene of sorghum on digestibility. *J. Anim. Sci.* **1978**, *46*, 1113–1119. [CrossRef]
17. McCleary, B.V.; Charmier, L.M.; McKie, V.A. Measurement of starch: Critical evaluation of current methodology. *Starch-Stärke* **2019**, *71*, 1800146. [CrossRef]
18. Osborne, B.G. Applications of near infrared spectroscopy in quality screening of early-generation material in cereal breeding programmes. *J. Near Infrared Spec.* **2006**, *14*, 93–101. [CrossRef]
19. Cozzolino, D.; Degner, S.; Eglinton, J. A review on the role of vibrational spectroscopy as an analytical method to measure starch biochemical and biophysical properties in cereals and starchy foods. *Foods* **2014**, *3*, 605–621. [CrossRef]
20. Boyles, R.E.; Pfeiffer, B.K.; Cooper, E.A.; Rauh, B.L.; Zielinski, K.J.; Myers, M.T.; Brenton, Z.; Rooney, W.L.; Kresovich, S. Genetic dissection of sorghum grain quality traits using diverse and segregating populations. *Theor. Appl. Genet.* **2017**, *130*, 697–716. [CrossRef]
21. Li, J.; Danao, M.G.C.; Chen, S.F.; Li, S.; Singh, V.; Brown, P.J. Prediction of starch content and ethanol yields of sorghum grain using near infrared spectroscopy. *J. Near Infrared Spec.* **2015**, *23*, 85–92. [CrossRef]
22. Griebel, S.; Adedayo, A.; Tuinstra, M.R. Genetic diversity for starch quality and alkali spreading value in sorghum. *Plant Genome* **2021**, *14*, e20067. [CrossRef] [PubMed]
23. De Alencar Figueiredo, L.F.; Davrieux, F.; Fliedel, G.; Rami, J.F.; Chantereau, J.; Deu, M.; Courtois, B.; Mestres, C. Development of NIRS equations for food grain quality traits through exploitation of a core collection of cultivated sorghum. *J. Agric. Food Chem.* **2006**, *54*, 8501–8509. [CrossRef] [PubMed]
24. Casa, A.M.; Pressoir, G.; Brown, P.J.; Casa, A.M.; Pressoir, G.; Brown, P.J. Community resources and strategies for association mapping in sorghum. *Crop Sci.* **2008**, *48*, 30–40. [CrossRef]
25. Megazyme. Total Starch Assay Procedure (Amyloglucosidase/α-Amylase Method). K-TSTA-100A, Procedure (b). 2020. Available online: https://www.megazyme.com/documents/Assay_Protocol/K-TSTA-100A_DATA.pdf (accessed on 9 January 2021).
26. Morrison, W.R.; Laignelet, B. An improved colorimetric procedure for determining apparent and total amylose in cereal and other starches. *J. Cereal Sci.* **1983**, *1*, 9–20. [CrossRef]
27. Chrastil, J. Improved colorimetric determination of amylose in starches or flours. *Carbohydr. Res.* **1987**, *159*, 154–158. [CrossRef]
28. McDonald, R.S.; Wilks, P.A., Jr. JCAMP-DX: A standard form for exchange of infrared spectra in computer readable form. *Appl. Spectrosc.* **1988**, *42*, 151–162. [CrossRef]
29. Martens, H.; Stark, E. Extended multiplicative signal correction and spectral interference subtraction: New preprocessing methods for near infrared spectroscopy. *J. Pharmaceut. Biomed. Anal.* **1991**, *9*, 625–635. [CrossRef]
30. Peiris, K.H.; Bean, S.R.; Chiluwal, A.; Perumal, R.; Jagadish, S.K. Moisture effects on robustness of sorghum grain protein near-infrared spectroscopy calibration. *Cereal Chem.* **2019**, *96*, 678–688. [CrossRef]
31. Dowell, F.E.; Maghirang, E.B.; Graybosch, R.A.; Berzonsky, W.A.; Delwiche, S.R. Selecting and sorting waxy wheat kernels using near-infrared spectroscopy. *Cereal Chem.* **2009**, *86*, 251–255. [CrossRef]
32. Wu, D.; Yong, H.; Shuijuan, F. Short-wave near-infrared spectroscopy analysis of major compounds in milk powder and wavelength assignment. *Anal. Chim. Acta* **2008**, *610*, 232–242. [CrossRef] [PubMed]
33. Osborne, B.G.; Fearn, T.; Hindle, P.H. *Practical NIR Spectroscopy with Applications in Food and Beverage Analysis*; Longman Scientific and Technical: Singapore, 1993; 227p.
34. Williams, P.C. Implementation of near-infrared technology. In *Near-Infrared Technology in the Agricultural and Food Industries*; Williams, P., Norris, K.H., Eds.; American Association of Cereal Chemist: St. Paul, MN, USA, 2001; pp. 145–169.
35. Fertig, C.C.; Podczeck, F.; Jee, R.D.; Smith, M.R. Feasibility study for the rapid determination of the amylose content in starch by near-infrared spectroscopy. *Eur. J. Pharm. Sci.* **2004**, *21*, 155–159. [CrossRef] [PubMed]
36. Dardenne, P. Some considerations about NIR spectroscopy: Closing speech at NIR-2009. *NIR News* **2010**, *21*, 8–14. [CrossRef]
37. De Maesschalck, R.; Jouan-Rimbaud, D.; Massart, D.L. The mahalanobis distance. *Chemometr. Intell. Lab. Syst.* **2000**, *50*, 1–18. [CrossRef]

38. Subramanian, V.; Jambunathan, R. Properties of sorghum grain and their relationship to roti quality. Sorghum in the eighties. In Proceedings of the International Symposium on Sorghum Grain Quality, ICRISAT Center, Patanchuru, India, 28–31 October 1981; pp. 281–288.
39. Buffo, R.A.; Weller, C.L.; Parkhurst, A.M. Relationships among grain sorghum quality factors. *Cereal Chem.* **1998**, *75*, 100–104. [CrossRef]
40. Rhodes, D.H.; Hoffmann, L.; Rooney, W.L.; Herald, T.J.; Bean, S.; Boyles, R.; Brenton, Z.W.; Kresovich, S. Genetic architecture of kernel composition in global sorghum germplasm. *BMC Genom.* **2017**, *18*, 15. [CrossRef]
41. Jampala, B.; Rooney, W.L.; Peterson, G.C.; Bean, S.; Hays, D.B. Estimating the relative effects of the endosperm traits of waxy and high protein digestibility on yield in grain sorghum. *Field Crops Res.* **2012**, *139*, 57–62. [CrossRef]
42. Yerka, M.K.; Toy, J.J.; Funnell-Harris, D.L.; Sattler, S.E.; Pedersen, J.F. Registration of A/BN641 and RN642 waxy grain sorghum genetic stocks. *J. Plant Regist.* **2015**, *9*, 258–261. [CrossRef]
43. Yerka, M.K.; Toy, J.J.; Funnell-Harris, D.L.; Sattler, S.E.; Pedersen, J.F. Registration of N619 to N640 grain sorghum lines with waxy or wild-type endosperm. *J. Plant Regist.* **2015**, *9*, 244–248. [CrossRef]
44. Yerka, M.K.; Toy, J.J.; Funnell-Harris, D.L.; Sattler, S.E.; Pedersen, J.F. Evaluation of interallelic waxy, heterowaxy, and wild-type grain sorghum hybrids. *Crop Sci.* **2016**, *56*, 1–9. [CrossRef]
45. Lane, H.M.; Murray, S.C. High Throughput can produce better decisions than high accuracy when phenotyping plant populations. *Crop Sci.* **2021**, *61*, 3301–3313. [CrossRef]
46. Beta, T.; Corke, H. Genetic and environmental variation in sorghum starch properties. *J. Cereal Sci.* **2001**, *34*, 261–268. [CrossRef]
47. Tester, R.F.; Karkalas, J. The effects of environmental conditions on the structural features and physico-chemical properties of starches. *Starch-Stärke* **2001**, *53*, 513–519. [CrossRef]
48. Gonçalves, M.T.V.; Morota, G.; Costa, P.M.D.A.; Vidigal, P.M.P.; Barbosa, M.H.P.; Peternelli, L.A. Near-infrared spectroscopy outperforms genomics for predicting sugarcane feedstock quality traits. *PLoS ONE* **2021**, *16*, e0236853. [CrossRef] [PubMed]

Article

Pilot Scale Roller Milling of Chickpeas into a De-Hulled Coarse Meal and Fine Flour

Randall Martin, Kaliramesh Siliveru *, Jason Watt, Paul Blodgett and Sajid Alavi *

Department of Grain Science & Industry, Kansas State University, Manhattan, KS 66506, USA
* Correspondence: kaliramesh@ksu.edu (K.S.); salavi@ksu.edu (S.A.)

Abstract: Chickpeas and other high protein plants are becoming increasingly popular. Traditionally, attrition or hammer mills are used for milling chickpeas. However, the use of roller mills on chickpeas has not been extensively researched. This study compared pilot-scale milling trials involving whole Kabuli compared to split and de-hulled Desi chickpeas. A flow sheet was designed and optimized for meal production with minimal co-product flour produced. Milling yields, particle size, and proximate analysis data were recorded. The optimum flow sheet consisted of 4 break passages, 2 smooth roll passages, and 4 purifiers. Results showed whole Kabuli chickpeas had a higher meal yield, at 63.8%, than split Desi seeds, at 54.1%; with both percentages proportional to the weight of milled seed. The remaining 36.2% or 45.9% consisted of co-product flour, feed streams and process losses. Both meals had an average particle size between 600 and 850 microns and both flours had a bimodal particle size distribution with peaks at 53 and 90–150 microns. The use of purifiers facilitated better separation of hull and resulted in lower crude fiber levels in the Kabuli meal. Proximate analysis trends were similar for both chickpea meals with higher protein (~2% more), crude fiber (~1% more) and ash (0.1–0.3% more) in the meal compared to the co-product flour. The co-product flour had substantially higher total starch (~15% more) than the meal. The results of this research can be used to modify wheat mills to process chickpeas.

Keywords: chickpea; kabuli; milling; de-hull; roller mill

Citation: Martin, R.; Siliveru, K.; Watt, J.; Blodgett, P.; Alavi, S. Pilot Scale Roller Milling of Chickpeas into a De-Hulled Coarse Meal and Fine Flour. *Processes* **2022**, *10*, 2328. https://doi.org/10.3390/pr10112328

Academic Editor: Dariusz Dziki

Received: 30 June 2022
Accepted: 19 October 2022
Published: 9 November 2022

Publisher's Note: MDPI stays neutral with regard to jurisdictional claims in published maps and institutional affiliations.

1. Introduction

Pulses have recently been gaining more consumer attention for being a good source of plant-based protein, without being an allergen such as soybeans. Sustainability is another reason for the growing popularity of pulses. Chickpeas are a pulse crop that contain high levels of proteins (23–26%), while also containing substantial levels of carbohydrates (37–39%) [1]. Many varieties are commercially available and can thrive in diverse regions and climates. The wide growing region has led chickpeas to become an important food crop worldwide [1]. Chickpeas require fewer resources, such as land and water, to grow than animal protein [2]. As global demand for protein increases the demand for more sustainable plant-based ingredients will increase, and chickpeas fit this criterion well. New processing methods and novel products are needed to fulfill the consumer demand for plant-based proteins. Proteins isolated from chickpeas and other pulses such as peas are important ingredients for such products, for example, plant-based meat alternatives [3].

Chickpeas are beneficial as they exhibit antioxidant properties [4]. Unfortunately, polyphenols in chickpeas also reduce digestibility due to their anti-nutritional properties, the mitigation of which is a bigger priority. Previous research found that 75% of these polyphenolic compounds are found in the hull (seed coat or testa) [5]. Some of the anti-nutritional properties can be reduced by cooking [6]. Some traditional processing methods include roasting, boiling or slurry cooking. Milling and removing the hull can also be very effective in lowering the anti-nutritional factors, reducing the need for intense cooking.

Chickpeas are often sold as whole seeds, but milling is also used to break down chickpeas into smaller particles, such as split chickpeas or flour, that require lower cooking times. This allows for a wide variety of food applications. Chickpea milling has primarily been carried out using dahl mills, which use attrition forces to de-hull the seeds and split the cotyledons apart [7]. Hammer milling is another method used, which grinds the chickpeas into flour [7]. These milling and dehulling methods typically require soaking, dehulling, drying, and then milling. The overall process can be very energy-, water- and time-intensive and limits capacity by using batch hydration methods.

The use of roller milling for chickpeas can lead to substantial process efficiency, although research and industry usage has not focused much on this technique. Previous milling research with chickpeas has been focused on attrition mills for de-branning and splitting [8]. Roller milling has a long history and is used extensively for grains such as wheat to produce many products, such as de-hulled or whole flours, coarser meals, etc. Improvements have been made but the same principle remains of passing the grain, between two rotating rollers to break it open, remove the hull and reduce the particle size. The removal of bran from wheat using roller mills requires minimal water for tempering, and the technique could be used with adaptations to de-hull chickpeas. Roller mills also provide a more controlled particle size range than hammer mills. In a recent study, Pulivarthi et al. [9] adopted roller milling for processing lentils and yellow pea flours. The usage of roller mills for processing chickpeas will not only facilitate efficiencies in energy and water consumption but also gives products with a uniform particle size distribution compared to the hammer and attrition mills. Energy efficiency in processing operations including milling is increasingly coming into focus for a sustainable food supply in both emerging and developed regions of the world [10–13]. Milling is a very energy-intensive process, depending on several factors, including the technology used, type of grain, throughput and fineness of desired particle size. Limited research exists on the milling of pulses, but energy usage has been studied for cereal grain milling in various food and feed applications and different size reduction technologies, with roller milling reported to be more energy-efficient than hammer milling and attrition milling [12,14–16].

The primary hypotheses of this study were that roller milling can be used for dehulling of chickpeas and will allow for the production of coarse chickpea meal and fine flour from the same equipment. A pilot-scale facility designed to mill wheat was used after necessary modification in unit operations and flows for the evaluation of efficiency of dehulling and particle size reduction in chickpeas. Two varieties of chickpeas with different pre-processing levels were evaluated: whole Kabuli and split and de-hulled Desi. These varieties are common and commercially available. Whole Kabuli seeds are typically larger and have twice the 100-seed weight of whole Desi seeds [7]. The use of split and de-hulled versus whole chickpeas provided insight into whether starting with whole or pre-processed seeds was more efficient. The milled chickpea samples were evaluated for yields, proximate analysis, particle size and flowability. The results of this research can be used and adapted to process chickpeas on equipment typically used in commercial wheat mills and allow for the diversification of their milling portfolio.

2. Materials and Methods
2.1. Materials

Two varieties of chickpeas were studied: Kabuli and Desi. Kabuli chickpeas were received as whole seeds and Desi chickpeas received spilt and de-hulled seeds. These were chosen to compare differences in chickpea varieties and preprocessing levels. The main hardware parameters or unit operations used in the pilot-scale milling are mentioned in Table 1.

Table 1. Summary of pilot-scale roller milling parameters or unit operations.

Parameter	Number
Grain type	2
Number of Break rolls	4
Number of Grinding rolls	1
Number of Middling rolls	2
Number of Purifiers	4

2.2. Pilot Scale Milling

Chickpeas were milled on a pilot-scale research wheat flour mill (Hal Ross Flour Mill, Kansas State University, Manhattan, KS, USA). The mill was modified to produce the maximum amount of chickpea meal and minimal co-product flour. This decision was made to create a novel ingredient, chickpea meal, and because designing the flow for meal would be a more difficult task than for flour. The milling flow shown in Figure 1 could easily be modified to produce only flour by regrinding the meal to flour, if desired.

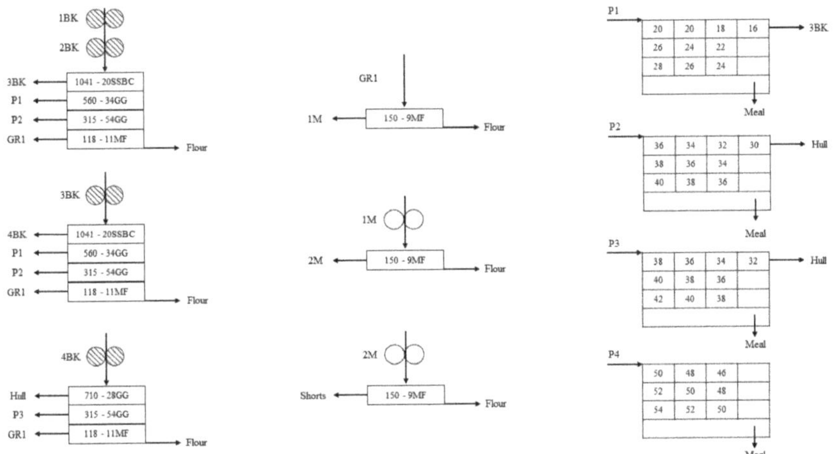

Figure 1. Roller milling flow sheet to obtain Chickpea Meal.

2.2.1. Cleaning House

Chickpeas were processed through the "cleaning house" system of the flour mill, which consisted of a combi-cleaner, scourer and infestation destroyer. The combi-cleaner used aspiration and gyratory vibration to separate based on density. The combi-cleaner removed light materials, such as loose hull or discolored seeds, and denser materials, such as stones. The scourer used abrasion to remove dust and debris from the outside of the seeds. The scourer was originally tested to remove the hull of the chickpea, by increasing intensity, but was found to be not effective. The infestation destroyer used impact force to remove hollowed-out grain from insect pests. The chickpea seeds did not show any evidence of insects, but this equipment was used to split some seeds and break off loose hull. All cleaning equipment had a capacity of 6000 lbs/h (2751.6 kg/h).

2.2.2. Storage and Tempering

After cleaning, the chickpeas were stored in tempering bins. Benchtop laboratory milling trials (unpublished data), showed that tempering to 11% moisture content (MC) on wet basis (w.b.) for 12 h was ideal for hull separation. The chickpeas arrived at the flour mill at 10.9% MC, so tempering was not needed.

2.2.3. Roller Milling

Chickpeas were milled using roller mills to create a coarse meal between 300 and 600 microns. Particles that were produced finer than 300 microns were milled to flour less than 150 microns. The roller mills and sifters had a capacity of 2.5 MT/h but were operated at 1 MT/h Milling equipment included a double high-break roll stand (MDDO, Bühler Group, Uzwil, Switzerland) and a single high-break roll (MDDM, Bühler Group, Uzwil, Switzerland). Samples were sifted with sifter boxes (MPAJ, Bühler Group, Uzwil, Switzerland) to remove larger particles of hull (>1041 microns), finer particles too small for meal (<315 microns) and flour (<150 microns). Air separation was used to separate light hull from the meal through the use of four purifiers (MQRF, Bühler Group, Uzwil, Switzerland). The final flow sheet used for both Kabuli and Desi chickpeas is shown in Figure 1.

The roll gaps were adjusted, but not recorded between samples to ensure similar milling intensity between chickpea samples. The break system was similar to the benchtop trial flow sheet for process 2. The major difference between benchtop and pilot scale was that there was no sieve separation between the 1st and 2nd break because a double high-roll stand was used. The pilot-scale reduction rolls (2M and 3M) were gear-driven and could grind with more force than the belt driven benchtop mills, so fewer roll stands were needed.

Chickpeas were first pneumatically transported from the tempering tanks to the 1st and 2nd break (BK) double high-roll stand. At this double high-roll stand, the chickpeas were ground twice without sifting in-between to break open the seeds and remove the hull in large pieces. After grinding, particles were pneumatically transferred sifters where particles were separated according to size. Particles larger than 1041 microns were sent to the third break (3BK), particles larger than 560 microns were sent to the 1st Purifier (P1), particles larger than 315 microns were sent to the 2nd Purifier (P2), particles larger than 118 microns were sent to the grader sifter (GR-1), and particles smaller than 118 microns was collected as flour (F1-2).

The next break roll stand, 3BK, the material from the 1/2 BK roll stands and P1was processed. After grinding on 3BK particles were sent to be sifted. Particles were sifted with the same separation as before; the only difference was that particles larger than 1041 were sent to the fourth break roll (4BK) stand.

The 4BK roll stand processed material from 3BK and P1. After grinding on 4BK separation differed from the other break roll stands. Particles larger than 710 microns were separated as hull or feed. Particles larger than 315 microns were sent to the third purifier (P3). Particles larger than 118 microns were sent to GR-1 and those smaller than 118 microns were separated as flour.

The grader sifter separated material from 1/2BK, 3BK, 4BK and the 4th purifier (P4). Particles less than 150 and 118 microns were separated as flour. Particles larger than 150 microns were sent to the 4th purifier (P4).

Smooth roll stands were used to grind particles too small for meal (<315 microns) but larger than flour (>150 microns). These stands were labeled as they would be in a commercial flour mill as 1M and 2M for processing "middlings". After grinding on the first smooth roll, 1M, particles less than 150 microns were separated as flour and particles larger than flour were sent to 2M. The same separation was used after 2M but with larger particles removed as feed and labeled as overs of 2M or shorts.

The purifiers had 3 decks and used screens with the openings recorded in grit gauze (GG). The particles pass along each screen, aided with air agitation, from left to right, with larger GG numbers corresponding to smaller openings. As particles fall through one deck, they pass onto the lower deck. Empty squares in the purifier figure show where screens were removed. These screens were removed to reduce the amount of purified chickpea meal being sent to be reground. Screen size could be manipulated in future research or milling facilities to achieve the desired particle ranges.

For P1, particles that did not fall though the first deck were sent to 3BK to be reground. For P2 and P3, particles that did not fall through to the second deck were collected as hull. Particles that fell through the first deck but remained on the second deck and fell through the open screen at the end were collected as meal. Particles that fell through the first two decks and remained on the 3rd deck until falling through the open screen were also collected as meal. P4 differed from the other purifiers by having all right end sieves removed, as no hull contamination was found in any of the outlets. The last purifier, P4, was not needed for sifting but was necessary to ensure the product flowed through the correct pneumatic lines for separation.

2.2.4. Packaging and Sealing

Final products, meal and flour, were bagged in plastic lined kraft paper bags and sealed with thread. Each bag weighed approximately 50 lbs (22.7 kg) and was palleted with 5 bags per row and was no more than 10 rows high.

2.3. Proximate Analysis

Proximate analysis of samples was carried out by SDK laboratories (Hutchinson, KS, USA) in duplicate. The crude protein (AOAC 976.06) [17] was measured by combusting samples and analyzing the amount of nitrogen in the sample. Crude protein for chickpeas was calculated using N × 6.25. Crude fiber (AOAC 962.09) [17] was determined after digestion of sample with sulfuric acid and sodium hydroxide and then combusted. Crude fat by acid hydrolysis (AOCS Ba 3-38) [17] was measured by extracting oil with petroleum ether and given as a percentage of original sample weight. Total starch (AOAC 979.10) was measured by digesting starch with enzymes and the glucose level was measured through spectrophotometry. Whole-chickpea moisture contents were determined using a laboratory oven at 105 °C for 72 h (ASABE S352.2) [18]. Moisture contents of ground samples were determined using a laboratory oven at 135 °C for 2 h (AACC44-19.01) [17].

2.4. Particle Size

Particle size analysis on final flour and meal streams was carried out using an Alpine jet sieve analyzer (e200LS, Hosokawa Alpine, Germany) or a Ro-Tap sieve (RX-29, W.S. Tyler, Mentor, OH, USA). Samples were measured in triplicate.

2.5. Flow Properties

Bulk density of the meal and flour was determined using a Winchester cup arrangement that ensured consistent filling into a cup of known volume, and the mass was recoded. Bulk density was calculated using the equation below.

$$\frac{Sm}{Sv} \tag{1}$$

where Sm is sample mass (g) and Sv is sample volume (4.732 × 10^{-4} m^3).

Tapped density was calculated using an Autotap density analyzer (Quanta Chrome Instruments, Boynton Beach, FL, USA). A graduated cylinder was filled with a known mass of sample and then the cylinder was then tapped 720 times (260 taps/min) and the volume was recorded after tapping. The tap density was calculated using the sample mass and tapped volume and recoded in g/mL.

Angle of repose was measured using a Winchester cup arrangement. The funnel was filled with 100 g of sample and allowed to pour from a height of 10 cm to a flat aluminum tray. The diameter of the pile was recorded three times from varying angles and averaged. The height was measured once. Angle of repose was calculated using the following equation

$$\theta = tan^{-1}\left(\frac{2H}{D}\right) \tag{2}$$

where θ is the angle of repose, H is the height of the pile formed, and D is the average diameter of the pile.

The flowability of the flour and meal was measured by comparing the Hausner ratio (HR) shown in Equation (3) and the Compressibility Index (CI), as shown in Equation (4). These values, along with the angle of repose, will be able to categorize the flow properties of the product as excellent to very, very poor, as shown in Table 2.

$$\text{Hausner ratio} = \frac{TD}{BD} \tag{3}$$

where TD is the tapped density and BD is the bulk density.

$$\text{Compressibility Index} = 100 \times \left(\frac{TD - BD}{DD} \right) \tag{4}$$

where TD is the tapped density and BD is the bulk density.

Table 2. Physicochemical properties of chickpea.

Flow Characteristic	HR	CI (%)	Angle of Repose
Excellent	1.00–1.11	<10	25–30
Good	1.12–1.18	11–15	31–35
Fair	1.19–1.25	16–20	36–40
Passable	1.26–1.34	21–25	41–45
Poor	1.35–1.45	26–31	46–55
Very Poor	1.46–1.59	32–37	56–65
Very, Very Poor	>1.6	>38	>66

2.6. Statistical Analyses

One-way ANOVA was performed on data to ascertain statistical significance of treatment effects (chickpea variety, milling process, flour stream, etc.). ANOVA was followed by Tukey's test for comparison of means and determine statistical significance of differences. Analyses was conducted using the SAS analysis software (Version 9.4, SAS, Cary, NC, USA) and statistical significance of $p < 0.05$.

3. Results and Discussion

3.1. Milling Yields

An overview of the milling yields can be found in Table 3. Almost twice the amount of Kabuli chickpeas was milled compared to Desi chickpeas. This was important to note because process loss was higher for the Desi chickpeas and this loss was most likely attributed to the shorter runtime. The process loss for both chickpeas would be expected to be lower if the mill was running for a longer production time.

Table 3. Milling yields (% as-is) of kabuli and desi chickpeas from roller mill.

	Starting Material (kg)	Meal (%)	Flour (%)	Overs of 4th BK Hull (%)	Overs of 2M Shorts (%)	Process Loss (%)
Kabuli	3641.44	63.8	23.9	6.0	1.9	4.37
Desi	2147.31	54.07	16.03	17.37	0.7	11.8

Kabuli chickpeas milled more efficiently into meal than Desi chickpeas, with almost 10% more meal. Desi chickpeas produced less meal and less flour but produced more "overs of 4th BK"; this was coarser material that would be hull if it were Kabuli chickpeas. Upon visual inspection, this stream was coarse cotyledon. This larger section showed that the break system should have had smaller roll gaps to increase grinding action to produce smaller particles. Another solution could have been to send this section back into the break system. However, due to the capabilities of the Hal Ross Flour mill, it was not possible to re-route this stream back into the process.

Comparing only the products of each process, (meal/(meal + flour) or flour/meal + flour), Desi chickpeas actually produced more meal than flour. Kabuli chickpeas, using the previous formulas produced 72.7% meal and 27.3% flour compared to Desi Chickpeas yields of 77.1% meal and 22.9% flour. If the "Overs of 4th BK" from process 2 (Desi) could have been re-entered into the break system; this could have increased the meal and flour production. This could have made split Desi chickpeas more efficient to mill. Both meal yields were still low (<80%) and, therefore, could not be considered an efficient process when designed for meal. Selling the co-product flour would increase the cost-effectiveness of the process.

The hull stream, "Overs of 4th BK", yielded 6% hull removal from the Kabuli chickpeas. This is similar to reported previously levels, where chickpeas contained 5% hull [7]. The 1% difference could have been from cotyledon still in the hull stream or due to varietal differences. The remaining 1% could be separated by density or air separation, but since the amount is so small, it would likely not be cost-effective. The "overs of 2M" represent any material that was not ground to flour through the reduction system. This material was harder than the rest of the flour and, therefore, stayed intact longer. This increased strength was a result of increased protein level. It had previously been reported that the outer layer of the chickpea cotyledon had higher protein levels, while the center of the cotyledon consisted of more starch [19].

3.2. Proximate Analysis

The proximate analysis data for both processes were recorded in Table 4. The raw material section showed the similarities and differences between the two chickpea types. They both had similar protein levels of around 24%. Crude fiber was lower in the Desi seeds because they had previously been dehulled. The dehulling partially explained the increased fat and starch in the Desi seeds as chickpea hull is known to contain low levels of fat and protein while having high levels of fiber [20]. Direct correlations cannot be made solely between the raw materials as they can have varietal differences. When all proximate contents were combined, it was clear that a portion of the composition was missing. The large difference (~25%) was believed to be oligosaccharides that would have not shown up in any of the other analysis [21].

Table 4. Proximate Analysis of kabuli and desi chickpeas.

% Dry Basis	Sample	Crude Protein	Crude Fiber	Fat	Ash	Total Starch
Raw Materials	Whole Kabuli Seeds	23.99 cd	3.84 b	5.54 c	2.53 f	39.90 d
	Split Desi Seeds	24.36 cd	1.44 f	6.47 b	2.45 fg	42.25 c
Process 1: Whole Kabuli Seeds	Meal	24.63 b	2.72 c	4.71 d	2.74 d	33.25 g
	Flour	22.41 e	1.44 g	6.45 b	2.41 gh	49.70 b
	Hull	14.65 f	20.90 a	3.42 f	4.52 a	14.9 i
	Shorts	29.09 a	2.66 c	6.77 b	3.29 b	31.75 h
Process 2: Split Desi Seeds	Meal	24.64 c	2.09 e	4.53 de	2.59 fg	37.15 f
	Flour	22.86 e	0.94 g	5.91 c	2.34 h	51.95 a
	Overs of 4th Break	23.87 c	2.43 d	4.27 e	2.64 e	38.95 e
	Shorts	27.50 b	2.01 e	7.40 a	2.89 c	37.55 f

Letters denote significant difference for every column $p < 0.05$.

The proximate analysis of process 1 showed the effect of de-hulling process as well as differences in meal and flour. The hull stream contained the largest amount of crude fiber and ash of all the process 1 streams. The meal stream had higher crude fiber and ash than the flour stream showing signs of higher hull contamination in the meal. The ash content of the hull contradicted the benchtop work (unpublished data) with Kabuli chickpeas that found that ash content was not significantly lowered by hull removal. One possibility was that the use of the purifiers increased hull separation and reduced cotyledon in the hull stream that allowed for proximate analysis differences to be observed. Further studies would need to be carried out to correlate the ash content and hull contamination in chickpeas before ash could be used as an indicator, similar to wheat milling.

The significantly higher protein and lower starch in the meal stream from both processes further enforced the theory that protein content increases strength. The outside of the cotyledon that contained higher protein remained in larger granules than the inner cotyledon that contained more starch [19]. Similar trends are seen with corn, horny vs. floury cotyledon, and wheat, hard vs. soft wheat [22]. The "Overs of 2M"/shorts had significantly higher protein than the flour stream, again reinforcing the strength theory.

Process 2 results followed similar patterns as process 1. The meal stream again contained significantly more protein and lower levels of starch. The flour stream from process 2 had the highest starch content of any tested sample. The "Overs of 4th BK" proximate results were practically similar to the meal stream. This similarity further emphasizes that the "Overs of 4th BK" stream should be re-milled to create meal and not discarded as feed.

The differences in starch and protein showed additional benefits to focusing on meal. The separation based on particle size resulted in different proximate results. These differences, such as in the flour, could lead to different functionalities in the raw flour, making it more beneficial in products or processes where increased protein is not desired. The decreased protein was also found without the use of water or chemicals, further increasing the sustainability of the ingredient.

3.3. Particle Size

Particle size data are recorded in Figures 2–4. The particles retained on each sieve were represented as a percent of each sample and the magnitude of the distributions was not related to yield.

The particle sizes of the final products from both processes were recorded in Figure 2. The flour for both chickpeas follows similar particle size ranges, with a bimodal distribution of around 53 and 90–150 microns. These bimodal peaks were also seen in benchtop milling trials [23]. The previously mentioned strength theory that compared meal and flour could explain the bimodal peaks. The larger micron peak could contain more protein than the smaller micron peak. Further testing could be carried out to confirm this.

The wider range of Kabuli meal was previously observed in benchtop trials and was due to the grinding intensity of the break system. However, at pilot scale, the larger range was more likely attributed to the removed purifier screens. The range of meal could be further adjusted by a change of purifier screens. The majority (~92%) of Kabuli meal was less than 850 microns, which ensured it was still an acceptable range of meal.

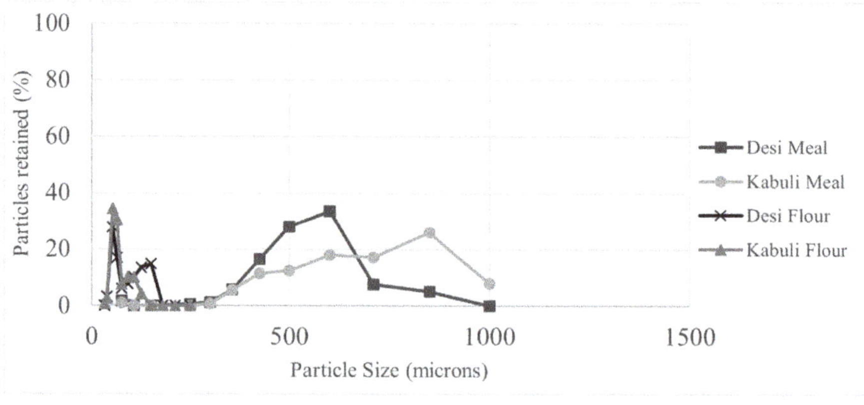

Figure 2. Particle Size of Chickpea Meal and Co-Product Flour Streams.

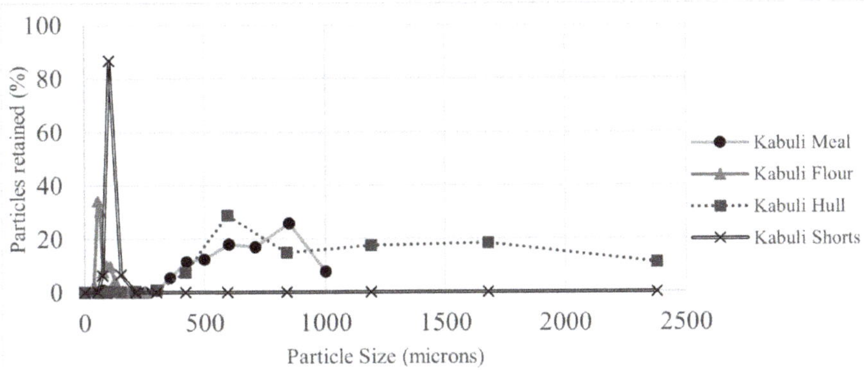

Figure 3. Particle Size of Process 1, (Kabuli) Milling Streams.

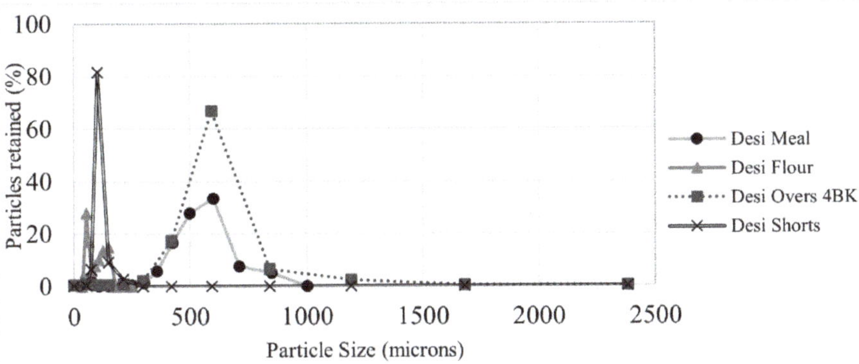

Figure 4. Particle Size of Process 2, (Desi) Milling Streams.

All products from process 1 are shown in Figure 3. The extremely large range of the hull showed that the break system was efficient at removing the hull in as large pieces as possible. The parts of the meal and hull distribution that overlap showed the importance of the purifiers in the milling flow. The purifier separated the bran with air separation that could not be separated by only particle size differences. The smaller particle range of the shorts stream showed that this could have been collected as flour. It was not clear if blinding of the screens led to these smaller particles not being included in the flour stream. However, because this stream was so small, at 1.9%, it was not a major concern that warranted further research.

Figure 4 shows that the "Overs of 4th BK" for Desi chickpeas needs only minimal grinding to convert it into meal. As mentioned previously, there was not a way to transport this stream back into the break system. In a commercial mill, this would not be wasted and would increase meal and flour yields. Again, the shorts stream should have been collected as flour, but this stream only contributed to 0.7% of the total products and was not considered practically significant.

3.4. Flow Properties

The flow properties of the chickpea products were recorded in Table 5. As expected, both flour streams flowed less easily compared to the meal streams. Kabuli flour from process 1 had better flow properties than Desi Flour. Kabuli flour was labeled as "excellent" by the Hausner Ratio and Compressibility Index, but only "passable" by the Angle of

Repose test. Desi Flour was labeled "good" by the Hausner Ratio and Compressibility Index, but on the upper end of "passable" by the Angle of Repose" test. These tests show the importance of using more than one flowability measurement when testing a sample, as they can often times result in different classifications. Fortunately, by this scale, neither of the flour streams showed any major flowability or transportation issues.

Table 5. Flow properties' data of kabuli and desi chickpea meal and flour.

	Sample	Bulk Density (g/mL)	Tap Density (g/mL)	Hausner Ratio	Compressibility Index (%)	Angle of Repose (Degrees)
Process 1	Kabuli Meal	0.719 [b]	0.689 [b]	0.958 [c]	−4.337 [c]	28.9 [b]
	Kabuli Flour	0.508 [c]	0.556 [c]	1.095 [b]	8.703 [b]	42.1 [a]
Process 2	Desi Meal	0.775 [a]	0.719 [a]	0.927 [c]	−7.826 [d]	29.0 [b]
	Desi Flour	0.448 [d]	0.520 [d]	1.161 [a]	13.896 [a]	45.2 [a]

Letters denote significant difference for every column $p < 0.05$.

The meal streams flowed better than the flour samples. This was expected due the larger particle size [24]. Kabuli and Desi meal were labeled as "excellent" or better. The Hausner ratio and Compressibility Index values were better than shown in the table due to the tapped density levels. Both meals flowed so well that it caused negative values for the Compressibility index.

4. Conclusions

The results of this research showed that chickpeas can be de-hulled and can be ground into meal and flour using roller (wheat) mills. Whole Kabuli chickpeas milled more efficiently (63.8% meal yield and only 4.4% losses) than split and de-hulled Desi chickpeas (only 54.1% meal yield and 11.8% losses). Differences in the protein and starch levels reinforced that protein is not distributed evenly across then cotyledon and led to higher protein (24.6% versus 22.4–22.8%) and lower starch (33.2–37.2% versus 49.7–52.0%) levels in the meal streams as compared to flour streams. Flowability tests confirmed that the produced chickpea products would not lead to any major flow concerns (good to excellent flow properties based on Hausner ratio of 0.927–1.161). Usage of the wheat (roller) mills for processing chickpeas will not only facilitate efficiencies in energy and water consumption but also gives products (meal and flour) with a uniform particle size distribution compared to the reported hammer and attrition mills. More research will be needed to increase the meal yields of both processes; however, selling the flour as a desirable product would increase profitability.

Author Contributions: Conceptualization, S.A.; methodology, R.M., K.S., J.W., P.B. and S.A.; formal analysis, R.M. and K.S.; investigation, R.M., S.A. and K.S.; resources, J.W., P.B. and S.A.; data curation, R.M., S.A. and K.S.; writing—original draft preparation, R.M.; writing—review and editing, K.S. and S.A.; supervision, S.A.; project administration, S.A.; funding acquisition, S.A. All authors have read and agreed to the published version of the manuscript.

Funding: This research received no external funding.

Institutional Review Board Statement: Not applicable.

Informed Consent Statement: Not applicable.

Data Availability Statement: Data is contained within the article.

Conflicts of Interest: The authors declare no conflict of interest.

References

1. Gil, J.; Nadal, S.; Luna, D.; Moreno, M.T.; Haro, A.D. Variability of Some Physicochemical Characters in Desi and Kabuli Chickpea Types. *J. Sci. Food Agric.* **1996**, *71*, 179–184. [CrossRef]
2. Mekonnen, M.M.; Hoekstra, A.Y. Value of Water Research Report. Series No. 48. In *Value of Water*; UNESCO-IHE Institute for Water Education: Delft, The Netherlands, 2010.
3. Van der Weele, C.; Feindt, P.; van der Goot, A.J.; van Mierlo, B.; van Boekel, M. Meat alternatives: An integrative comparison. *Trends Food Sci. Technol.* **2019**, *88*, 505–512. [CrossRef]
4. Heiras-Palazuelos, M.J.; Ochoa-Lugo, M.I.; Gutiérrez-Dorado, R.; López-Valenzuela, J.A.; Mora-Rochín, S.; Milán-Carrillo, J.; Garzón-Tiznado, J.A.; Reyes-Moreno, C. Technological properties, antioxidant activity and total phenolic and flavonoid content of pigmented chickpea (*Cicer arietinum* L.) cultivars. *Int. J. Food Sci. Nutr.* **2013**, *64*, 69–76. [CrossRef] [PubMed]
5. Singh, U. Antinutritional factors of chickpea and pigeonpea and their removal by processing. *Plant Foods Hum. Nutr.* **1988**, *38*, 251–261. [CrossRef] [PubMed]
6. Adamidou, S.; Nengas, I.; Grigorakis, K.; Nikolopoulou, D.; Jauncey, K. Chemical Composition and Antinutritional Factors of Field Peas (*Pisum sativum*), Chickpeas (*Cicer arietinum*), and Faba Beans (*Vicia faba*) as Affected by Extrusion Preconditioning and Drying Temperatures. *Cereal Chem. J.* **2011**, *88*, 80–86. [CrossRef]
7. Ravi, R.; Harte, J.B. Milling and physicochemical properties of chickpea (*Cicer arietinum* L.) varieties. *J. Sci. Food Agric.* **2009**, *89*, 258–266. [CrossRef]
8. Wood, J.A.; Knights, E.J.; Harden, S. Milling performance in desi-type chickpea (*Cicer arietinum* L.): Effects of genotype, environment and seed size. *J. Sci. Food Agric.* **2008**, *88*, 108–115. [CrossRef]
9. Pulivarthi, M.K.; Nkurikiye, E.; Watt, J.; Li, Y.; Siliveru, K. Comprehensive understanding of roller milling on the physico chemical properties of red lentil and yellow pea flours. *Processes* **2021**, *9*, 1836. [CrossRef]
10. Akinoso, R.; Lawal, I.A.; Aremu, A.K. Energy requirements of size reduction of some selected cereals using attrition mill. *Int. Food Res. J.* **2013**, *20*, 1205–1209.
11. Sanchez, G.C.; Monteagudo Yanes, J.P.; Perez, M.M.; Cabrera Sanchez, J.L.; Padron, A.P.; Haeseldonckx, D. Efficiency in electromechanical drive motors and energy performance indicators for implementing a management system in balanced animal feed manufacturing. *Energy* **2020**, *194*, 116818. [CrossRef]
12. Eras, J.J.C.; Gutierrez, A.S.; Ulloa, M.J.C. The temperature gradient of cereals as an optimization parameter of the milling process in hammermills. *J. Clean. Prod.* **2021**, *297*, 126685. [CrossRef]
13. Sannik, U.; Pappel, T.K. Complete milling technologies of different valuable materials. In Proceedings of the 4th International DAAAM Conference "Industrial Engineering–Innovation as Competitive Edge for SME", Tallinn, Estonia, 29–30 April 2004; pp. 146–148.
14. Koch, K.B. Feed Mill Efficiency. Engormix. 2012. Available online: https://en.engormix.com/feed-machinery/articles/feed-mill-efficiency-t35290.htm (accessed on 11 September 2022).
15. Smejtkova, A.; Vaculik, P. Comparison of power consumption of a two-roll mill and a disc mill. *Agron. Res.* **2018**, *16* (Suppl. S2), 1486–1492.
16. Ziggers, B.D. Hammering or rolling the grain. *Feed. Technol.* **2001**, *5*, 9–17.
17. *Official Methods of Analysis*, 17th ed.; AOAC International: Gaithersburg, MD, USA, 2006.
18. *S3194.4*; Method of Determining and Expressing Fineness of Feed Materials by Sieving. ASABE Standards: St. Joseph, MI, USA, 2008.
19. Wood, J.A.; Knights, E.J.; Chocty, M. Morphology of chickpea seeds (*Cicer arietinum* L.): Comparison of desi and kabuli types. *Int. J. Plant Sci.* **2011**, *172*, 632–643. [CrossRef]
20. Soni, B.; Sarita, S. Comparative evaluation of milling bi-products: Finger millet seed coat (FMSC), chick pea husk (CPH) and wheat bran (WB) for their nutritional, nutraceutical potential. *Int. J. Basic Appl. Agric. Res.* **2014**, *12*, 104–110.
21. Xu, Y.; Thomas, M.; Bhardwaj, H.L. Chemical composition, functional properties and microstructural characteristics of three kabuli chickpea (*Cicer arietinum* L.) as affected by different cooking methods. *Int. J. Food Sci. Technol.* **2014**, *49*, 1215–1223. [CrossRef]
22. Hoseney, R.C. Principles of cereal science and technology. A general reference on cereal foods. In *Principles of Cereal Science and Technology. A General Reference on Cereal Foods*; American Association of Cereal Chemists, Inc.: St. Paul, MN, USA, 1986.
23. Martin, R.C. Physico-Chemical Properties of Chickpea Flour Obtained Using Roller Milling and Extrusion Pre-Cooking. Master's Thesis, Kansas State University, Manhattan, KS, USA, 2000.
24. Riley, R.E.; Hausner, H.H. Effect of particle size distribution on the friction in a powder mass. *Int. J. Powder Met.* **1970**, *6*, 17–22.

Article

Comprehensive Understanding of Roller Milling on the Physicochemical Properties of Red Lentil and Yellow Pea Flours

Manoj Kumar Pulivarthi, Eric Nkurikiye, Jason Watt, Yonghui Li and Kaliramesh Siliveru *

Department of Grain Science and Industry, Kansas State University, Manhattan, KS 66506, USA; pmanoj@ksu.edu (M.K.P.); enkurikiye@ksu.edu (E.N.); jwatt1@ksu.edu (J.W.); yonghui@ksu.edu (Y.L.)
* Correspondence: kaliramesh@ksu.edu; Tel.: +1-785-532-4071

Abstract: The development of convenience foods by incorporating nutrient-rich pulses such as peas and lentils will tremendously alter the future of pulse and cereal industries. However, these pulses should be size-reduced before being incorporated into many food products. Therefore, an attempt was made to adapt roller mill settings to produce de-husked yellow pea and red lentil flours. The milling flowsheets unique to yellow peas and red lentils were developed in producing small, medium, and large flours with maximum yield and flour quality. This study also investigated the differences in chemical composition, physical characteristics, and particle size distributions of the resultant six flour fractions. The kernel dimensions and physicochemical properties of the whole yellow pea and red lentils were also studied to develop customized mill settings. Overall, the mill settings had a significant effect on the physical properties of different particle-sized flours. The geometric mean diameters of different particle-sized red lentil flours were 56.05 µm (small), 67.01 µm (medium), and 97.17 µm (large), while for yellow pea flours they were 41.38 µm (small), 60.81 µm (medium), and 98.31 µm (large). The particle size distribution of all the flour types showed a bimodal distribution, except for the small-sized yellow pea flour. For both the pulse types, slightly more than 50% flour was approximately sizing 50 µm, 75 µm, and 100 µm for small, medium, and large settings, respectively. The chemical composition of the flour types remained practically the same for different-sized flours, fulfilling the objective of this current study. The damaged starch values for red lentil and yellow pea flour types increased with a decrease in flour particle size. Based on the Hausner's ratios, the flowability of large-sized flour of red lentils could be described as passable; however, all the remaining five flour types were indicated as either poor or very poor. The findings of this study assist the millers to adapt yellow pea and red lentil milling technologies with minor modifications to the existing facilities. The study also helps in boosting the production of various baking products using pulse and wheat flour blends to enhance their nutritional quality.

Keywords: pulses; yellow pea; red lentils; protein; roller milling; particle size

Citation: Pulivarthi, M.K.; Nkurikiye, E.; Watt, J.; Li, Y.; Siliveru, K. Comprehensive Understanding of Roller Milling on the Physicochemical Properties of Red Lentil and Yellow Pea Flours. *Processes* **2021**, *9*, 1836. https://doi.org/10.3390/pr9101836

Academic Editor: Bernd Hitzmann

Received: 29 September 2021
Accepted: 11 October 2021
Published: 15 October 2021

Publisher's Note: MDPI stays neutral with regard to jurisdictional claims in published maps and institutional affiliations.

1. Introduction

The demand for healthier food choices has been growing with an increase in the consumer's interest for adapting nutrient-rich functional foods in their diets. Even though whole-grain cereal-based convenience products offer a good amount of fiber, vitamins, minerals, and phytochemicals owing to the presence of bran and germ portions, these foods still lack protein to some extent and contain few essential amino acids such as lysine and threonine [1]. Therefore, it is important to fortify the baked products with functional ingredients such as pulses, to further improve the nutritional and health benefits of the end products.

Yellow peas (*Pisum sativum* L.) and red lentils (*Lens culinaris* L.), like other pulses, are rich in protein and are highly nutritious, offering all the essential amino acids except for methionine, cysteine, and tryptophan [2]. The limiting amino acid in wheat is lysine, while all pulses have an adequate amount of lysine and lack sulfur-containing amino acids, which are essential for competing with the protein quality from animal sources. Therefore, fortification of wheat-based bakery products with pulses is a perfect supplement to offer a balanced proportion of essential amino acids [3].

In recent times, numerous studies have been driven towards the addition of composite flours containing pulses in baked foods [4,5]. The addition of pulse flour has also shown promising results, such as increased protein and antioxidant activity, compared to regular wheat products [6]. Research has shown that the amount/level of pulse flour substituted with wheat affects the quality of the baked product [7]. Depending on the end goal or the objective, the resultant pulse flours can be incorporated into wheat-based bakery products as whole/de-hulled pulse flour, or in the form of refined fractions of protein, fiber, or starch.

The development of convenience foods by incorporating nutrient-rich pulses such as peas and lentils will tremendously alter the future of pulse and cereal industries. However, a standard milling method is still unavailable for generating high-quality pulse flours from raw peas and lentils. These pulses are being milled using various milling techniques such as roller, hammer, ball, and pin mill. Among all these milling methods, roller milling is unique and gives the miller great control over the final quality of the flours as the size reduction takes place gradually over a series of corrugated and smooth set of rolls with varied rotational speeds [8]. A standard and widely used method for producing wheat flour is roller milling. Hence, altering the roller mill settings to produce pea and lentil flour is the best and most economic opportunity for wheat millers, as they could use the wheat-milling equipment to produce pulse flours as well. Additionally, a standard protocol for optimal pretreatments and mill settings for these pulses is needed for upscaling the milling process industrially.

The size reduction of pulses depends on various factors such as pulse kernel characteristics and milling operational parameters. The particle size of the resultant flour can be adjusted by optimizing the mill settings such as roll gaps and sieve sizes. It is an important parameter to predict the performance of milling [9].

Several researchers have already studied the effects of incorporating pulse flours of different particle sizes into wheat-based products [10–12]. It is evident that the final quality of the product is being affected by differences in particle size. However, the majority of these studies reported the use of a single milling process to obtain a single batch of flour with varied particle sizes. The obtained single flour batch was then sieved/segregated to produce multiple flour batches (small, medium, and large). However, a major shortcoming in separating a single batch of flour into different particle sizes led to a change in their chemical composition. Hence, it is important to obtain flours of different particle sizes with the same chemical composition to comprehend the original effect of particle size on the final quality of the convenience foods.

Hence, an attempt has been made to develop different mill settings in a roller mill to obtain small, medium, and large flour fractions from yellow peas and red lentils. The specific objectives were to (a) develop milling flowsheets unique to yellow peas and red lentils in producing small, medium, and large flour fractions with maximum yield and quality; and (b) understand the differences in chemical composition, physical characteristics, and particle size distributions of the resultant flour fractions.

2. Materials and Methods

2.1. Materials

Two types of pulses, namely, yellow peas from "D'allesandro gourmet online store" and red lentils from "Food to live online store", were received as whole grains with an initial moisture content of 11.88 ± 0.11% and 10.6 ± 0.58% in wet basis (w.b.), respectively. The moisture content of these grains was determined by drying 10 gram (g) of grain samples, in triplicate, at 105 °C for 72 hours (h) in a hot-air convection oven.

2.2. Methods

2.2.1. Kernel Characteristics

The kernel characteristics of red lentils and yellow peas were studied by randomly collecting 100 grains each from the sample lots. The linear dimensions such as length (*L*), width (*W*), and thickness (*T*) for each kernel were determined using a digital caliper (General Tools and Instruments LLC, Secaucus, NJ, USA) of 0.01 mm accuracy.

The equations given by Mohsenin [13] were used to further deduce the average diameter of the pulses with the help of linear dimensions (*L*, *W*, and *T*). The arithmetic mean diameter (D_a) and geometric mean diameter (D_g) of the pulse grains were calculated using the below Equations (1) and (2), respectively. Similarly, the sphericity values (Φ) were obtained from the Equation (3) [13].

$$D_a = (L + W + T)/3 \tag{1}$$

$$D_g = (LWT)^{\frac{1}{3}} \tag{2}$$

$$\Phi = \frac{(LWT)^{\frac{1}{3}}}{L} \tag{3}$$

2.2.2. Tempering

The preliminary milling trials (not shown) conducted at various tempering moisture and time combinations indicated 12% moisture content for 18 h to be ideal for both yellow peas and red lentils for maximum yield. The amount of water (*Q*, mL) to be added to achieve the above-mentioned moisture level was calculated from the following Equation (4).

$$Q = w\frac{(mf - mi)}{100 - mf} \tag{4}$$

where *w* = weight of grains to be tempered in grams, *mf* = desired grain moisture (%), and *mi* is initial grain moisture (%).

The initial conditioning or uniform mixing of water to the yellow peas and red lentils was obtained by mixing in power-driven rotary bins for 30 min and further storing these pulses in airtight resealable bags for 18 h at room temperature.

2.2.3. Milling

The yellow peas and red lentils were milled using a bench top stainless steel roller mill (Model 915, Ross Machine and Mill Supply, Oklahoma City, OK, USA). After conducting several preliminary trials, the milling flowsheets were developed to obtain three different-sized flours (small, medium, and large) for each of the pulse types. The particle sizes of these small, medium, and large flours were aimed to be ≤ 75 µm, ≤ 150 µm, and ≤ 200 µm, respectively. The number of break rolls and smooth rolls for milling varied with type of pulse used and desired flour size. The mill settings were designed to separate the hull portions completely from the flour fractions. The flour fractions were being sieved after each pass of the milling using a gyratory sifting box (Great Western Manufacturing, Leavenworth, KS, USA) with the sieves mentioned in the flowsheets below in results section (Figures 1 and 2). As we progress through each pass, the corrugations on the reduction rolls varied from 15 teeth/inch to 22 teeth/inch, whereas there were no corrugations on the reduction rolls. The differential speed between the pair of rolls also varied between reduction rolls and break rolls. The roll size, differential speed, number of corrugations per inch, and roller gap between the rolls were varied in the milling process as cited below in Figures 1 and 2. The milling of these pulses was conducted in triplicates, and the fractions such as hull, shorts, and flour were individually weighed to determine the milling performance. The milling yield (%) for all the six milling conditions was calculated using Equation (5) mentioned below.

$$\text{Milling Yield (\%)} = \frac{\text{Weight of flour fraction}}{\text{Weight of pulses}} \times 100 \tag{5}$$

2.2.4. Proximate Analysis

The moisture content of the flour fractions post milling was determined by heating 2–3 g of flour sample in hot air oven at 135 °C for 2 h according to AACC 44-19.01 method [14]. The moisture content is expressed in percent wet basis throughout the manuscript.

The proximate composition of the whole-grain and flour fractions of yellow peas and red lentils was determined through an external lab. The methods used include AACC Method 46–30 for crude protein [15], AACC method 30–10 for crude fat [16], AACC Method 32–10 for crude fiber [17], and AACC Method 08–01 for ash content [18]. The percent of total carbohydrate contents excluding fiber in the grain and flour samples was determined by subtracting the protein, fiber, fat, ash, and moisture content from 100.

The AACC 76–33.01 method [19] was followed to determine the damaged starch content of the yellow pea and red lentil flour fractions using an SDmaticTM (Chopin Technologies, Villeneuve la Garenne, France).

2.2.5. Particle Size Distribution

An alpine jet sieve (Hosokawa Micron, E200 LS) was used to measure the particle size distributions (PSD) of the small, medium, and large flours obtained from yellow peas and red lentils.

The ASTM standard sieve sizes used for understanding the granulation ranges from 25 μm to 250 μm. The analysis was carried out in triplicates by taking 100 g of flour for each run. The PSD was measured based on the amount of flour (g) passing through the sieves as described by Rivera et al. [20].

Additionally, by measuring the weights of flour retained over each sieve, the geometric mean diameter (d_{gw}) was calculated using the following expression for all the six types of flour fractions [21].

$$d_{gw} = log^{-1}\left[\frac{\sum_{i=1}^{n}\left(W_i \, log\overline{d}_i\right)}{\sum_{i=1}^{n} W_i}\right] \tag{6}$$

where W_i is amount of flour retained on i'th sieve (g), n is the total number of sieves, and d_i is aperture size of the i'th sieve.

2.2.6. Color Analysis

The color values (L*, a*, and b*) for the grain and flour fractions of yellow pea and red lentils were measured using a Miniscan EZ 4500 colorimeter (Hunter lab, Reston, VA, USA) [5]. The results were reported based on triplicates.

2.2.7. Bulk Density

The bulk density of the whole yellow peas, red lentils, and their respective flour fractions was measured using a standard Winchester cup arrangement (Seedburo Equipment Co., Des Plaines, IL, USA). The sample was allowed to fall from a height of 10 cm into an empty cup until it overflows. A scrapper was used to remove the excess sample over the cup, and the bulk density was measured by taking the final sample weight (g) and cup volume in triplicates using expression 7 below.

$$\text{Bulk Density} = \frac{\text{Weight of Sample (g)}}{\text{Volume (cm}^3)} \tag{7}$$

2.2.8. Tapped Density

The tapped density, also known as the compacted density of the whole yellow peas, red lentils and their respective flour fractions, was obtained using an Autotap density analyzer (Quantachrome Instruments, Boynton Beach, FL, USA). The setup consisted of a measuring cylinder of known volume connected to the tapping mechanism. The measuring cylinder was filled with a known amount of sample (g), and then the volume occupied by the sample after 750 taps was measured in triplicates using expression 8 below.

$$\text{Tap Density} = \frac{\text{Weight of Sample (g)}}{\text{Tapped Volume (cm}^3)} \tag{8}$$

2.2.9. True Density

The true density of the whole yellow peas, red lentils and their respective flour fractions was measured using helium gas pycnometer (AccuPync II 1340, Micromeritics, Norcross, Georgia). The helium gas was passed through the chamber to determine the volume occupied by solid particles by filling the pore spaces between and within the grain/flour fractions. The weight and volume occupied by the sample were used to obtain true density. The mean value of true densities was obtained from three replications for each sample.

The Hausner's ratio and porosity of the flour samples were determined from the following Equations (9) and (10). Hausner's ratio is a flow factor and can be used to approximate the flowability of flour particles.

$$\text{Hausner's ratio} = \frac{\text{Tap density}}{\text{Bulk density}} \tag{9}$$

$$\text{Porosity} = \left(1 - \frac{\text{Bulk density}}{\text{True density}}\right) \times 100 \tag{10}$$

2.3. Statistical Analysis

The analysis of variance (ANOVA) and Tukey's test at a significance level of 0.05 were performed using SAS studio (SAS Institute, Cary, NC, USA) to analyze the statistical significance and differences in means.

3. Results and Discussion

3.1. Grain Characteristics

The understanding of raw material composition and its properties is crucial in designing handling, processing, and milling processes. The relation between a grain's physical properties and the milling properties has been well established from the beginning of grain processing industries. The density of a grain is one of the widely used quality indexes for predicting the soundness of a kernel [22].

The physicochemical properties evaluated for whole-grain red lentils and yellow peas are summarized in Table 1. The differences among the kernel dimensions are obvious among the red lentils and yellow peas. The geometric and arithmetic diameters of the red lentils (3.77 mm and 3.42 mm, respectively) were much smaller than the yellow peas (6.61 mm and 6.48 mm, respectively). The arithmetic and geometric mean diameters, and sphericity values of red lentil variety: Seyran-96 from the study conducted by Gürsoy and Güzel [23] matches with the results of red lentils in present study. The higher sphericity value for yellow peas (92.99%) compared to red lentils (76.54%) suggests that the shape of yellow peas tends to be more spherical in nature. The bulk, tap, and true densities of red lentils were observed to be significantly ($p \leq 0.05$) higher than yellow peas. Additionally, density values of red lentils were found to be much higher than the values reported by Kumar et al. [24]. The moisture content of the red lentils and yellow peas before tempering was found to be 10.6% and 11.88% (w.b), respectively, with no significant

($p \leq 0.05$) difference. The protein content of red lentils (24.48%) was much higher compared to yellow peas (20.07%) and slightly lower than the values reported by Roy et al. [25]. The crude fiber contents for red lentils and yellow peas were 3.33% and 5.29%, respectively. The toughening effect of pulses during the size reduction process was due to the presence of crude fiber [26]. Finally, the physicochemical properties (except for moisture and carbohydrates) of both the pulses were significantly different and required distinct mill settings. Additionally, the overall proximate composition of the pulses reaffirms the importance of utilizing these pulses in baking sector to enhance the nutritional quality of final product.

Table 1. Physicochemical properties of pulses.

Test Variable	Red Lentils	Yellow Peas
Arithmetic mean diameter (mm)	3.77 ± 0.26 [b]	6.61 ± 0.37 [a]
Geometric mean diameter (mm)	3.42 ± 0.22 [b]	6.48 ± 0.37 [a]
Sphericity (%)	76.54 ± 1.82 [b]	92.99 ± 4.2 [a]
Bulk density (kg/m^3)	815.43 ± 3.76 [a]	759.73 ± 3.39 [b]
Tap density (kg/m^3)	866.41 ± 1.15 [a]	820.67 ± 2.08 [b]
True Density (kg/m^3)	1419.87 ± 1.33 [a]	1407 ± 0.36 [b]
L* (0- black to 100-white)	62.39 ± 0.92 [a]	42.40 ± 0.05 [b]
a* (-green to +red)	9.07 ± 0.14 [a]	6.94 ± 0.90 [b]
b* (-blue to +yellow)	13.84 ± 0.20 [b]	20.03 ± 1.26 [a]
1000 Kernel weight (g)	19.2 ± 0.38 [b]	224.8 ± 0.79 [a]
Moisture (% wet basis)	10.6 ± 0.58 [a]	11.88 ± 0.11 [a]
Crude protein (%)	24.48 ± 0.01 [a]	20.07 ± 0.06 [b]
Crude fat (%)	0.38 ± 0.01 [b]	0.9 ± 0.09 [a]
Crude fiber (%)	3.33 ± 0.12 [b]	5.29 ± 0.05 [a]
Ash (%)	2.32 ± 0.00 [b]	2.47 ± 0.01 [a]
Total carbohydrates excluding crude fiber (%)	59.9 ± 0.45 [a]	60.4 ± 0.07 [a]

The different super scripts for mean values in each row are significantly different ($p \leq 0.05$).

3.2. Development of Milling Flowsheets

The optimized roller milling flowsheets for obtaining de-husked yellow pea and red lentil flours in three different particle sizes, namely, small—S, medium—M, and large—L, are shown in Figures 1 and 2. The flour sizes for both red lentil flour and yellow pea flour are referred as S, M, and L throughout the manuscript. All the flowsheets were optimized after conducting numerous trials on roller mills to obtain maximum husk separation for a designated particle size. The major differences between the mill settings for obtaining red lentil and yellow pea flours of different sizes are highlighted (Figures 1 and 2) for easy identification. The changes w.r.t roller gaps are shown in red font, and sieves are marked in blue background.

Initially, tempering trials were conducted on both the pulses for optimizing the tempering moisture and time for achieving maximum flour yield with the highest bran separation. The tempering trials (data not shown) suggested that a 12% moisture content on wet basis for 18 hours (h) time period resulted in the highest flour recovery with efficient bran separation.

3.2.1. Red Lentil Flour Mill Settings

To obtain a small particle sized flour, the red lentils were milled step wise as indicated in the Figure 1a, i.e., the tempered grains were initially passed through the first break roll (1BK) adjusted to a roller gap (RG) of 0.025 inches. The roller gap was adjusted based on the size of the lentil kernels. It is important to make sure that the gap is not too large for the grain to pass through the break rolls without grinding. Similarly, the gap should not be too small to let the grains bounce off from the rolls. After the first pass, the milled output was passed through a set of three sieves: 1041 µm, 630 µm, and 75 µm, as mentioned in Figure 1a. The particles retained over 1041 µm were further passed through second break roll, adjusted to a roller gap of 0.012 inches. The particles above 630 μm were allowed to pass through a third break roll set at 0.006 inches roller gap, and the particles above 75 μm were passed into the first reduction roll or 1M (Middling) passage through a roller gap of 0.003 inches. The product collected in pan under the 75 μm sieve was separated as flour after every single pass. In similar fashion, different sieve sets were used as mentioned below in Figure 1a for all the subsequent passes. The coarse hull was obtained over the sieve 1041 μm after second pass (2BK), and fine hull was obtained after the third pass (3BK) and first reduction pass or 1M pass over the 630 μm sieve and 425 μm sieve, respectively. Shorts, the inseparable mixture of bran and endosperm were collected over 425 μm sieve from 2nd, 3rd, and 4th reduction passes (or 2M, 3M, and 4M passes). However, in case of medium and large flours, shorts were obtained right from 1M to 4M passes, as shown in Figure 1b,c.

Even though flour was being collected after each pass in the milling process, most of it was produced after the first reduction pass (1M). The reduction roll passes (2M, 3M and 4M) mentioned in the flowsheet below were individual passes through the smooth rolls to obtain maximum yield and reduce the size of particles to below 75 μm, 150 μm and 200 μm for small, medium, and large mill settings, respectively.

Similarly, the medium sized lentil flour was produced as per the mill settings shown below in Figure 1b. The number and type of passes used for milling remain constant for small, medium, and large mill settings. However, the key difference between the small and medium size flour settings includes altering the sieve size used at the bottom for flour collection i.e., 150 μm. Additionally, the roller gap used for reduction rolls remains 0.003 inches for all the four passes to get a medium sized flour.

The same series of rolls used in small and medium mill settings was followed in producing the large lentil flour with minor modifications in roller gaps and sieves as shown in Figure 1c. The sieve size used at the bottom of the sifter for separating flour was 200 μm for all the seven passes. Since the objective of this mill setting was to produce large flour, the roller gap for reduction roll passes was slightly bigger than the small and medium mill settings. However, identical roller gaps to the small and medium flour settings were used for all the break rolls since they play a major role in break opening the kernels and separating the husk from endosperm. As the average particle size of the flour has to be higher than small and medium flours, the roller gap for the first set of reduction rolls was adjusted to 0.004 inches and subsequent passes were run at gap of 0.003 inches.

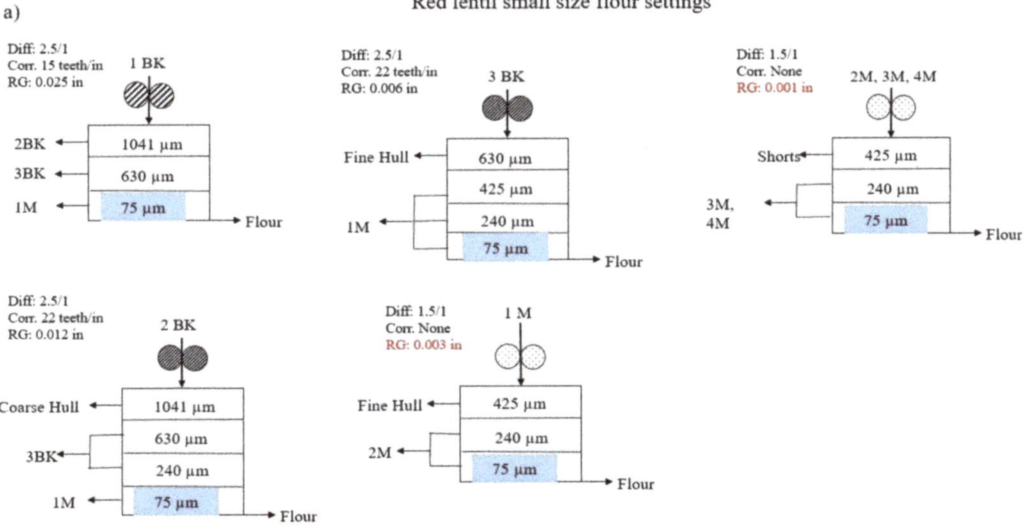

Figure 1. *Cont.*

b)   Red lentil medium size flour settings

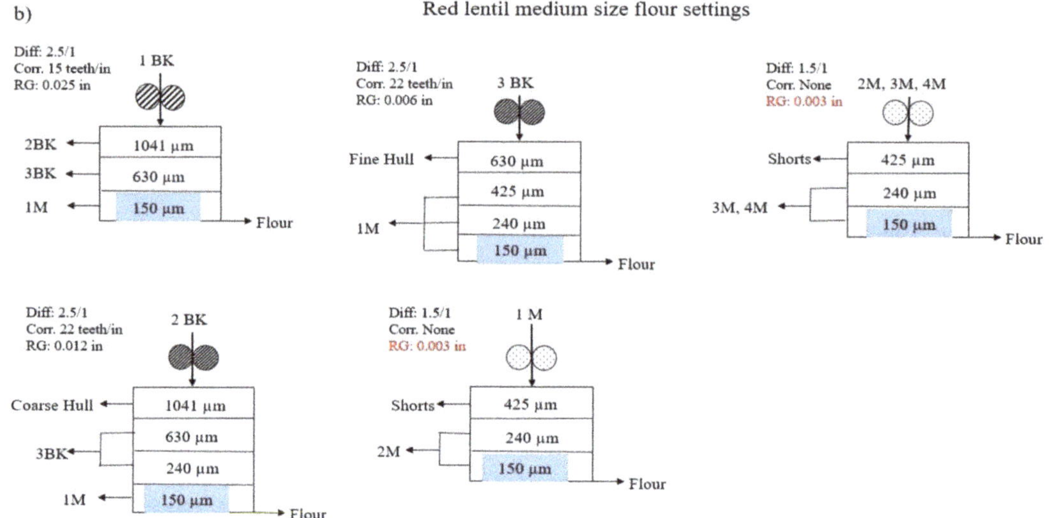

Figure 1. Roller milling flowsheet to obtain (**a**) small red lentil flour (≤75 µm), (**b**) medium red lentil flour (≤150 µm), and (**c**) large red lentil flour (≤200 µm); (Diff—Differential speed; Corr—Corrugations; RG—Roller Gap; BK—Break, M—Middling).

3.2.2. Yellow Pea Flour Mill Settings

The mill settings developed for obtaining de-husked yellow pea flours greatly differ from that of red lentil flours. The differences in the kernel and shape characteristics observed in Table 1 for both the pulses necessitates customized flow sheets based on individual grain requirements. Since the yellow peas are in different shape and much larger in size and diameter compared to red lentils (Table 1), the mill settings applied were more aggressive. For small, medium, and large yellow pea flours, the series of rolls and number of passes remain constant, as shown in Figure 2. Same set of sieves (75 µm, 150 µm, and 200 µm) similar to red lentil flour settings was used at the very bottom in each pass for separating flour fractions to achieve small, medium, and large flours. A total of four break roll passes were needed to bring down the particle size of milled fractions for subsequent reduction roll milling. Among these four break roll passes, the corrugations for the first two sets were 15 teeth/inch and the remaining two had 22 teeth/inch.

In all the three mill flowsheets of yellow peas, a larger roller gap of 0.11 inches was set for the break rolls in first pass (1BK) to break open the pea kernels. The resultant milled fraction obtained was further passed through a set of sieves, as indicated in Figure 2. Unlike red lentils, the first husk separation took place in third break roll pass (3BK) over a 1041 µm screen. The fine hull was separated from top of 630 µm sieve at the fourth break roll pass (4BK), whereas shorts were separated over a 240 µm size screen in all the reduction roll passes.

Roller gaps of 0.055 inches and 0.012 inches were set for second and third set of break rolls, respectively, in all the three milling flowsheets. However, the roller gap settings were altered from 4th to final step to obtain desired flour sizes. In cases of small and medium yellow pea flour types, roller gaps of 0.006 inches and 0.003 inches were used during the fourth (4BK) and fifth (1M) passes, respectively. The smallest roller gap, 0.001 inches, was used exclusively for producing small flour during the 2nd, 3rd, and 4th reduction passes. In case of medium flour settings, the roller gap (0.003 inches) remained constant throughout the reduction roll passes to avoid particles from excessive grinding. This is a crucial distinguishing step for achieving the medium flour particles set to pass through a 150 µm sieve.

In case of large flour settings, slightly bigger roller gaps (0.006 inches-1M pass and 0.004 inches-2M to 4M passes) were introduced, which is uncommon for a reduction roll milling step. Unlike red lentils, the yellow peas tend to grind more readily to a smaller particle size, even at a larger roller gap. Hence, the large yellow pea flour settings used in this study for a reduction roll pass are unique.

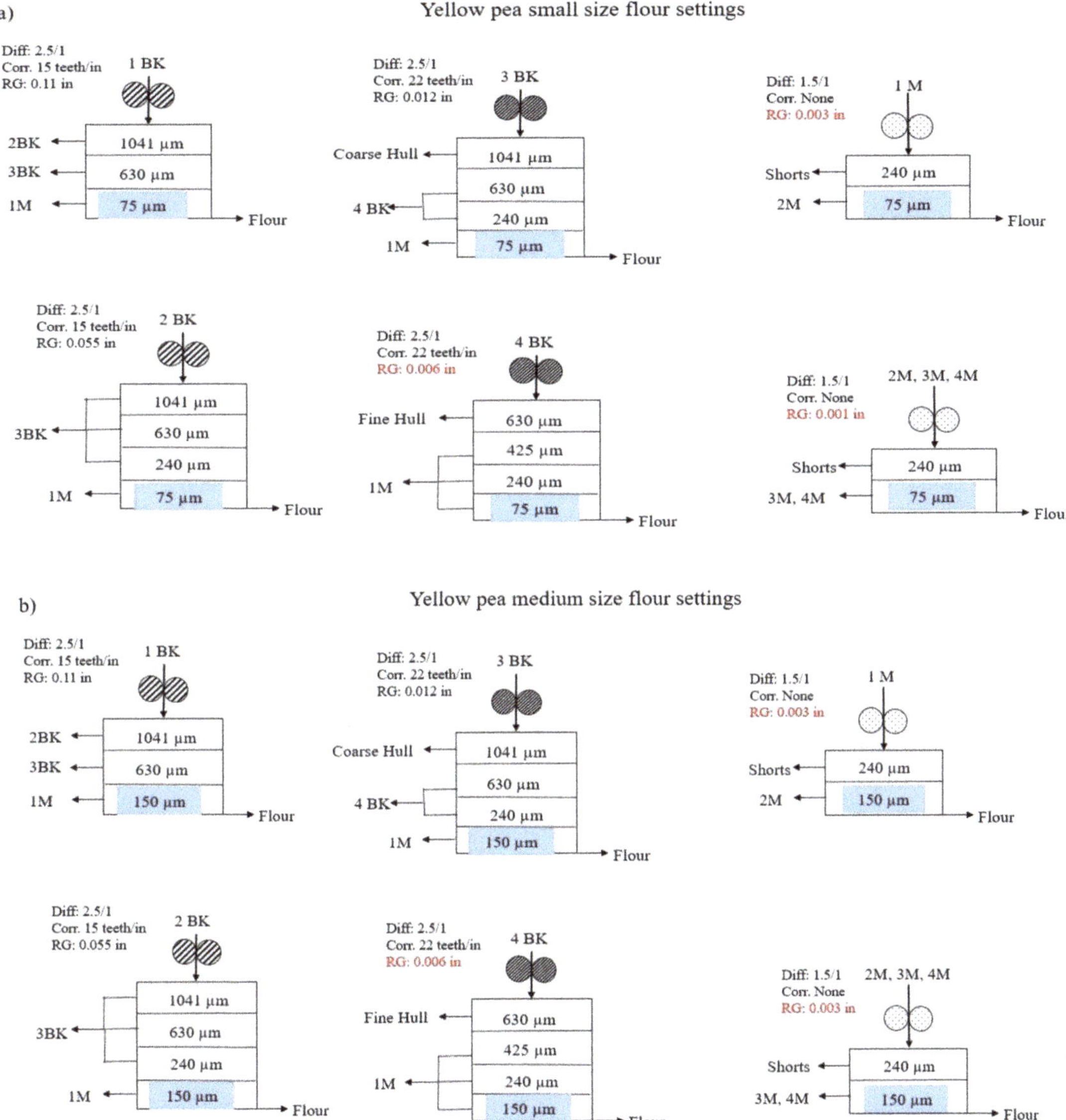

Figure 2. *Cont.*

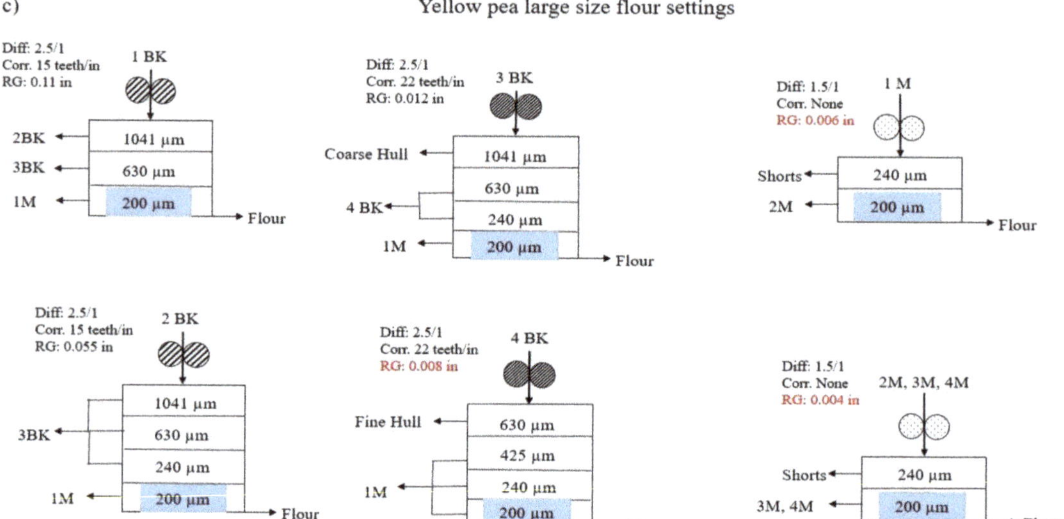

Figure 2. Roller milling flowsheet to obtain (**a**) small yellow pea flour (≤75 µm), (**b**) medium yellow pea flour (≤150 µm), and (**c**) large yellow pea flour (≤200 µm); (Diff—Differential speed; Corr—Corrugations; RG—Roller Gap; BK—Break, M—Middling).

3.3. Milling Yield

The milling yields in terms of bran, shorts, and flour are presented in Table 2. The highest flour recovery for red lentil flour was obtained for small flour settings. The target particle size had a significant ($p \leq 0.05$) effect on all three red lentil flour yields. Additionally, the flour yield (84.86–82.27 %) followed a decreasing trend by moving from small to large particle size settings. The bran portions of red lentils ranged from 9.71 to 10.15%, with no significant differences amongst the flour types. However, the amount of dietary fiber present in the lentil kernels was found to be small: 3.3%. This could be due to the presence of some portion of endosperm in bran yields. Additionally, the crude fiber value only signifies the amount of insoluble fiber. The shorts resulted from milling process increased with increase in particle sizes of flours.

Table 2. Milling yields (% as -is) of red lentils and yellow peas from roller mill.

Sample	Type	Flour	Bran	Shorts
Red Lentils	S	84.86 ± 0.19 [a]	10.15 ± 0.48 [a]	3.31 ± 0.2 [a]
	M	83.74 ± 0.42 [b]	9.92 ± 0.23 [a]	3.84 ± 0.59 [ab]
	L	82.27 ± 0.28 [c]	9.71 ± 0.22 [a]	4.26 ± 0.08 [b]
Yellow peas	S	82.55 ± 0.37 [C]	9.84 ± 0.12 [A]	3.84 ± 0.10 [B]
	M	84.07 ± 0.17 [B]	9.45 ± 0.07 [B]	4.85 ± 0.10 [A]
	L	86.86 ± 0.36 [A]	8.61 ± 0.10 [C]	3.45 ± 0.08 [C]

Mean values with different lower-case letters in each column are significantly different for red lentils ($p \leq 0.05$); mean values with different upper-case letters in each column are significantly different for yellow peas ($p \leq 0.05$).

In case of yellow pea flour settings, the highest flour recovery (86.86%) was achieved for large flour settings followed by medium flour (84.07%). The yellow pea flour yield significantly increased (82.55–86.86%) with increase in particle size, whereas bran content declined from 9.84 to 8.61%. Highest bran separation was achieved for small yellow pea flour settings. The change in particle size flour settings produced a significant difference among yields of all the three components.

3.4. Particle Diameter and Particle Size Analysis

The geometric mean diameter for all the three flour sizes of red lentil and yellow pea flours is given in Table 3. The performance and the resultant average size of the six flours milled were determined to understand the degree of difference between the flours with respect to particle sizes. This calculation was obtained from the amount of flour retained on each sieve during the granulation analysis in alpine jet sieve [21]. It was necessary to develop the flowsheets to produce ≥98% of the flour particles below 212 µm as per the definition given by Code of Federal Regulations for wheat flour [27]. However, the commercial pulse flours available in the market did not comply with the definition of flour as per the above criteria [26].

Table 3. Geometric mean diameter of red lentil and yellow pea flours.

Flour Type	Geometric Mean Diameter (µm)	
	Red Lentil Flour	**Yellow Pea Flour**
Small	56.05 ± 3.59	41.38 ± 0.56
Medium	67.01 ± 3.29	60.81 ± 2.02
Large	97.17 ± 2.6	98.31 ± 0.87

It is evident from Figures 3 and 4 that 100% of all the flour types from red lentils and yellow peas were below the 212 µm range. The two peaks observed in Figure 3 for all three flour types (S, M, and L) of red lentil flours indicates a bimodal distribution at 53 µm and 125 µm. The larger peaks at 53 µm might be attributed to the release of small starch particles (below 53 µm) from the kernel matrix, and the smaller peaks at 125 µm could be pulse protein or starch and protein aggregates. A similar pattern was observed in case of wheat, where other researchers have reported that starch granules ranging from 2 to 38 µm in diameter were found in the first peak of particle size distribution [28–30].

In case of yellow pea flours, the bimodal distribution was observed for medium and large flour types. Only the first peaks of all three flour types matched at 53 µm, whereas the second peaks for medium and large flours were obtained near 125 µm and 175 µm, respectively. The small flour showed a unique pattern of unimodal distribution with a single peak at 53 µm, as shown in Figure 4. The amount of flour that passed through the 53 µm sieve for small flour was more than 80%, leading to a unimodal distribution. Overall, for both the pulse types, slightly more than 50% flour was sized around 50 µm, 75 µm, and 100 µm for small, medium, and large settings, respectively, fulfilling the objective of the study.

3.5. Pulse Flour Quality Evaluation

3.5.1. Red Lentil Flour Quality

A physicochemical analysis of small, medium, and large red lentil flours as a function of particle size is presented in Table 4. From Tables 1 and 4, the effect of milling on the physicochemical properties of lentils can be clearly observed. The moisture of the grains decreased from 12% to 9.9% (small), 10.04 (medium) and 9.39 (large) post milling. A higher reduction of moisture content was observed in case of large mill settings. This loss of moisture could be due to the generation of heat during the milling process. There was an increase in fat content in resultant flours (0.68 to 0.98%) compared to the grain (0.38%). The removal of husk and potential contamination with germ portions might have caused this slight increase in fat values.

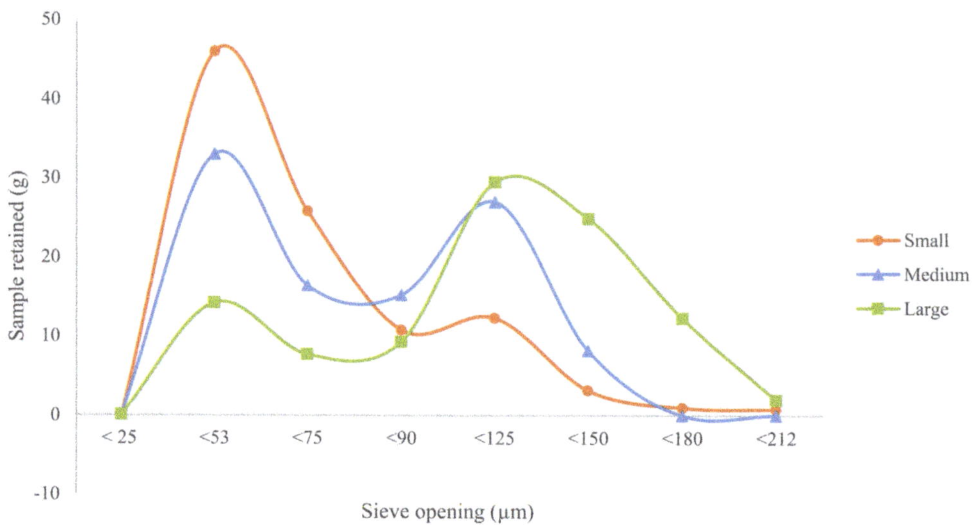

Figure 3. Particle size distribution of small, medium, and large red lentil flour.

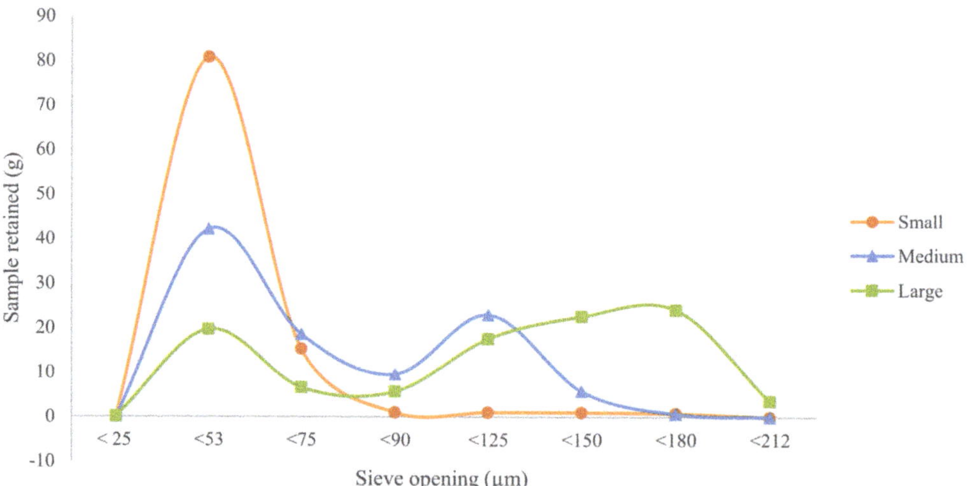

Figure 4. Particle size distribution of small, medium, and large yellow pea flour.

The proximate composition (except for carbohydrates) of the flours did not alter significantly ($p \leq 0.05$) with varying size of flours. The protein values of all the lentil flours ranged from 25.47 to 25.71%. The study conducted by Marchini et al. [10] showed that the protein content of the red lentil flours increased (21.21% to 24.06%) with decrease in particle sizes of the flour from $\geq$200 µm fraction to below $\leq$100 µm fraction. However, in the present study, the designed mill settings have helped in achieving significantly similar values supporting the objective of this study. The crude fiber content varied from 0.85 to 0.97%, a significant reduction compared to the whole lentils (Table 1) due to the removal of husk portions from the flour. The ash content of the flours altered between 2.04 and 2.19%, close to the range indicated by Ahmed et al. [11]. However, the fat contents (0.68–0.98%) of the lentil flours obtained were lower compared to the values reported in literature [10,11]. This could be due to the greater separation of germ portions in the form of shorts during the sifting process. As expected, the starch damage significantly increased from 3.41 to 5.47% with decrease in the particle size of flour. This trend agrees with the study reported by Bourré et al. [12] on yellow pea and red lentil flours of different particle sizes. The proximate composition of all lentil flours agreed with the results reported by Bourré et al. [12] and Hall et al. [31], except for crude fiber and fat. The lightness values of the red lentil flour varied significantly ($p \leq 0.05$) for small and large mill settings. Conversely, a* and b* values of the medium and large remained unchanged, showing no statistical significance ($p \leq 0.05$).

Table 4. Physicochemical properties of red lentil flours.

Test Variable	Red Lentil Flour		
	S	**M**	**L**
Moisture (% wet basis)	9.90 ± 0.1 [a]	10.04 ± 0.04 [a]	9.39 ± 0.22 [b]
Crude protein (%)	25.50 ± 0.11 [a]	25.47 ± 0.11 [a]	25.71 ± 0.12 [a]
Crude fat (%)	0.98 ± 0.14 [a]	0.98 ± 0.04 [a]	0.68 ± 0.22 [a]
Crude fiber (%)	0.85 ± 0.1 [a]	0.90 ± 0.02 [a]	0.97 ± 0.08 [a]
Total carbohydrates excluding crude fiber (%)	60.73 ± 0.2 [ab]	60.41 ± 0.13 [b]	61.08 ± 0.35 [a]
Ash (%)	2.04 ± 0.33 [a]	2.19 ± 0.05 [a]	2.16 ± 0.01 [a]
Damaged starch (AACC)	5.47 ± 0.13 [a]	4.19 ± 0.30 [b]	3.41 ± 0.13 [b]
L* (0- black to 100-white)	80.54 ± 0.51 [a]	79.14 ± 0.61 [ab]	77.96 ± 0.79 [b]
a* (-green to +red)	11.09 ± 0.05 [b]	13.88 ± 0.12 [a]	14.33 ± 0.52 [a]
b* (-blue to +yellow)	26.57 ± 0.04 [b]	29.88 ± 0.29 [a]	30.31 ± 0.90 [a]
Bulk density (kg/m^3)	487 ± 1.41 [c]	566 ± 1.41 [b]	576 ± 1.41 [a]
Tap density (kg/m^3)	735 ± 0.00 [b]	773.5 ± 3.53 [a]	764 ± 2.83 [a]
True density (kg/m^3)	1456.6 ± 0.37 [a]	1453 ± 0.15 [c]	1454.3 ± 0.2 [b]
Hausner ratio	1.51 ± 0.00 [a]	1.37 ± 0.00 [b]	1.33 ± 0.00 [c]
Porosity	66.56 ± 0.11 [a]	61.05 ± 0.1 [b]	60.39 ± 0.1 [c]

The different super scripts for mean values in each row are significantly different ($p \leq 0.05$).

The bulk density values of the flours significantly ($p \leq 0.05$) increased with increase in particle size. The smaller the particle size of the flour, higher the inter-particle spaces, resulting in lower test weight or bulk densities. The true density values of the powders significantly changed ($p \leq 0.05$) with varying particle size. The specifications based on Hausner's ratio for flowability from Lebrun et al.'s [32] study suggest that the small (1.51) and medium flours (1.37) fall under the category of "very poor" and "poor", respectively. However, large (1.33) flours fall into "passable" category.

3.5.2. Yellow Pea Flour Quality

The proximate composition and physical properties of the small, medium, and large yellow pea flours as a function of particle size are shown in Table 5. The differences in proximate compositions of the grain and flour samples are evident from Tables 1 and 5. The moisture contents of the yellow peas (12%) were reduced to 11.1–11.22% for the yellow pea flours. The highest moisture loss was observed for small mill settings. The reason for this loss of moisture is same as discussed above for lentils. The crude fiber content reduced from 5.29% (grains) to a range of 1.11–1.19% (flours) post milling. The removal of husk during milling process led to these decreased fiber values in the final flour fractions.

Table 5. Physicochemical properties of yellow pea flours.

Test Variable	Yellow Pea Flour		
	S	**M**	**L**
Moisture (% wet basis)	11.10 ± 0.08 [a]	11.14 ± 0.05 [a]	11.22 ± 0.09 [a]
Crude protein (%)	21.58 ± 0.12 [b]	22.03 ± 0.01 [a]	21.25 ± 0.08 [c]
Crude fat (%)	0.91 ± 0.09 [a]	0.81 ± 0.15 [a]	0.50 ± 0.11 [b]
Crude fiber (%)	1.19 ± 0.03 [a]	1.18 ± 0.06 [a]	1.11 ± 0.04 [a]
Total carbohydrates excluding crude fiber (%)	62.78 ± 0.13 [b]	62.35 ± 0.13 [c]	63.42 ± 0.12 [a]
Ash (%)	2.43 ± 0.02 [a]	2.51 ± 0.06 [a]	2.50 ± 0.01 [a]
Damaged starch (AACC)	9.05 ± 0.28 [a]	7.10 ± 0.02 [b]	5.90 ± 0.66 [c]
L* (0- black to 100-white)	83.61 ± 0.53 [a]	82.6 ± 0.62 [a]	83.11 ± 1.23 [a]
a* (-green to +red)	1.56 ± 0.04 [b]	1.65 ± 0.09 [b]	2.19 ± 0.06 [a]
b* (-blue to +yellow)	22.22 ± 0.28 [b]	22.70 ± 0.32 [b]	25.27 ± 0.58 [a]
Bulk density (kg/m^3)	440.5 ± 4.95 [b]	451.5 ± 4.96 [b]	468.5 ± 0.71 [a]
Tap density (kg/m^3)	687.5 ± 3.54 [a]	681.5 ± 4.95 [a]	678 ± 5.66 [a]
True density (kg/m^3)	1469.63 ± 1.76 [a]	1458.07 ± 0.51 [b]	1459.6 ± 0.92 [b]
Hausner ratio	1.56 ± 0.01 [a]	1.51 ± 0.01 [b]	1.45 ± 0.01 [c]
Porosity	70.02 ± 0.29 [a]	69.03 ± 0.35 [a]	67.89 ± 0.07 [b]

The different super scripts for mean values in each row are significantly different ($p \leq 0.05$).

The proximate composition of the flours altered significantly ($p \leq 0.05$), with varying flour sizes, except for crude fiber, ash, and moisture values. Even though there is statistically significant difference ($p \leq 0.05$) between protein, fat, and carbohydrates, the variation can be considered to be negligible in a practical industry setting. The proximate composition of yellow pea flour was in concordance with the values of de-husked yellow pea flour obtained through roller milling by Motte et al. [33]. The protein values of yellow pea flours ranged from 21.25 to 22.03%. Similar to the red lentil flours, the crude fiber content was reduced considerably compared to the whole yellow pea kernels (Table 1) and can be attributed to the de-husking of pulses. The ash content of the yellow pea flours lay between 2.43 and 2.51%, higher than the red lentil flours reported in Table 4. These results closely agreed with ash values reported for untreated yellow pea flours by Xu et al. [34].

The fat contents of the yellow pea flour ranged from 0.50–0.91%, with no statistical difference between small and medium flours. The reason for significantly lower fat content in large yellow pea flour (0.50%) compared to small (0.91%) and medium flours (0.81%) could be due to the efficient separation of germ portions for large mill settings. The starch damage significantly increased from 5.9 to 9.05% as particle size of the flours decreased from large to smaller size. The red lentil flours also followed a similar pattern in damaged starch values with increase in fineness of the flour. A similar trend of higher damaged starch was attributed to smaller particle size of yellow pea and lentil flours by Motte et al. [33]. This same effect was most commonly reported by various researchers for wheat milling [35–37]. The increase in damaged starch values for finer flours can be explained by the greater impact of mechanical damage due to decreased roller gaps. The damaged starch values in milling play an important role in understanding the effectiveness of milling. Lower starch damage is more preferable in the final flour fraction because the dough mixing and pasting properties are greatly affected, resulting in the poor color and firmness of final food products [38,39]. The starch damage values of large flour match with the study conducted by Gu et al. [38] on yellow peas.

The lightness values of the yellow pea flour did not show any statistical significance ($p \leq 0.05$) with respect to particle size as a function. Additionally, the a* and b* values of small and medium flours remained constant with no statistical significance ($p \leq 0.05$).

The bulk density values of the small and medium flours remained unchanged ($p \leq 0.05$) with increase in particle size. However, the large flour exhibited a significantly ($p \leq 0.05$) higher bulk density value. The tap density values of all the three types of flours observed no statistical significance ($p \leq 0.05$), whereas the true density value of small flour was found to be significantly ($p \leq 0.05$) higher than the other two flours. Based on Hausner's ratio, the small (1.56) and medium (1.51) flours fall into "very poor" category and the large (1.45) flour can be considered as "poor" in terms of flowability [32].

4. Conclusions

The current study explored the possibility of adapting roller milling to produce dehusked red lentil and yellow pea flours. The particle size of all the flour sizes were restricted to below 212 µm as per the wheat flour definition given by the Code of Federal Regulations. The individual flowsheets were developed for each type of pulse with three different desired particle sizes. The milling flowsheets helped in achieving similar proximate compositions for the three different flours. The flour quality evaluation was also conducted to understand the effect of milling on some important physicochemical properties. The particle size analysis and geometric mean diameter showed a clear understanding of the particle sizes of all flour types for further end applications. These findings assist the wheat millers to adapt yellow pea and red lentil milling technologies with minor modifications to the existing facilities. The study also helps in boosting the production of various baking products using pulse and wheat flour blends to enhance the nutritional quality.

Author Contributions: Conceptualization, M.K.P., Y.L. and K.S.; methodology, M.K.P., J.W., Y.L. and K.S.; software, M.K.P., Y.L. and K.S.; validation, M.K.P., E.N., Y.L. and K.S.; formal analysis, M.K.P. and K.S.; investigation, M.K.P., E.N., Y.L. and K.S.; resources, J.W., Y.L. and K.S.; data curation, M.K.P.; writing—original draft preparation, M.K.P.; writing—review and editing, M.K.P., E.N., Y.L. and K.S.; visualization, M.K.P.; lead supervision, K.S.; supervision, Y.L.; project administration, K.S. and Y.L.; funding acquisition, K.S. and Y.L. All authors have read and agreed to the published version of the manuscript.

Funding: This research was in part funded by the USDA Pulse Crop Health Initiative (PCHI), grant number 58-3060-0-051 and also KS00-0056-NC213 USDA HATCH project" and "The APC was funded by the USDA Pulse Crop Health Initiative (PCHI), grant number 58-3060-0-051.

Institutional Review Board Statement: Not applicable.

Informed Consent Statement: Not applicable.

Data Availability Statement: The data presented in this study are available on request from the corresponding author.

Acknowledgments: This is contribution No. 22-094-J from the Kansas Agricultural Experimental Station.

Conflicts of Interest: The authors declare no conflict of interest.

References

1. Rutherfurd, S.M.; Bains, K.; Moughan, P.J. Available lysine and digestible amino acid contents of proteinaceous foods of India. *Br. J. Nutr.* **2012**, *108*, S59–S68. [CrossRef] [PubMed]
2. Tiwari, B.K.; Singh, N. *Pulse Chemistry and Technology*; Royal Society of Chemistry: Cambridge, UK, 2012.
3. Temba, M.C.; Njobeh, P.B.; Adebo, O.A.; Olugbile, A.O.; Kayitesi, E. The role of compositing cereals with legumes to alleviate protein energy malnutrition in Africa. *Int. J. Food Sci. Technol.* **2016**, *51*, 543–554. [CrossRef]
4. Herranz, B.; Canet, W.; Jiménez, M.J.; Fuentes, R.; Alvarez, M.D. Characterisation of chickpea flour-based gluten-free batters and muffins with added biopolymers: Rheological, physical and sensory properties. *Int. J. Food Sci. Technol.* **2016**, *51*, 1087–1098. [CrossRef]
5. Giménez, M.A.; Drago, S.R.; De Greef, D.; Gonzalez, R.J.; Lobo, M.O.; Samman, N.C. Rheological, functional and nutritional properties of wheat/broad bean (Vicia faba) flour blends for pasta formulation. *Food Chem.* **2012**, *134*, 200–206. [CrossRef]
6. Zucco, F.; Borsuk, Y.; Arntfield, S.D. Physical and nutritional evaluation of wheat cookies supplemented with pulse flours of different particle sizes. *LWT-Food Sci. Technol.* **2011**, *44*, 2070–2076. [CrossRef]
7. Zhang, Y.; Hu, R.; Tilley, M.; Siliveru, K.; Li, Y. Effect of Pulse Type and Substitution Level on Dough Rheology and Bread Quality of Whole Wheat-Based Composite Flours. *Processes* **2021**, *9*, 1687. [CrossRef]
8. Maskus, H.; Bourre, L.; Fraser, S.; Ashok, S.; Malcolmson, L. Effects of grinding method on the compositional, physical, and functional properties of whole and split yellow pea flours. *Cereal Foods World* **2016**, *61*, 59–64. [CrossRef]

9. Campbell, G.M.; Bunn, P.J.; Webb, C.; Hook, S.C.W. On predicting roller milling performance: Part II. The breakage function. *Powder Technol.* **2001**, *115*, 243–255. [CrossRef]

10. Marchini, M.; Carini, E.; Cataldi, N.; Boukid, F.; Blandino, M.; Ganino, T.; Vittadini, E.; Pellegrini, N. The use of red lentil flour in bakery products: How do particle size and substitution level affect rheological properties of wheat bread dough? *LWT-Food Sci. Technol.* **2021**, *136*, 110299. [CrossRef]

11. Ahmed, J.; Taher, A.; Mulla, M.Z.; Al-Hazza, A.; Luciano, G. Effect of sieve particle size on functional, thermal, rheological and pasting properties of Indian and Turkish lentil flour. *J. Food Eng.* **2016**, *186*, 34–41. [CrossRef]

12. Bourré, L.; Frohlich, P.; Young, G.; Borsuk, Y.; Sopiwnyk, E.; Sarkar, A.; Nickerson, M.T.; Ai, Y.; Dyck, A.; Malcolmson, L. Influence of particle size on flour and baking properties of yellow pea, navy bean, and red lentil flours. *Cereal Chem.* **2019**, *96*, 655–667. [CrossRef]

13. Mohsenin, N.N. *Physical Properties of Plant and Animal Materials*; Gordon and Breach Science Publishers: New York, NY, USA, 1970.

14. Cereals & Grains Association. *AACC Approved Methods of Analysis*, 11th ed.; Method 44-19.01: Moisture—Air-Oven Method, Drying at 135; Cereals & Grains Association: St. Paul, MN, USA, 1999.

15. Cereals & Grains Association. *AACC Approved Methods of Analysis*, 11th ed.; Method 46-30.01: Crude Protein—Combustion Method; Cereals & Grains Association: St. Paul, MN, USA, 1999.

16. Cereals & Grains Association. *AACC Approved Methods of Analysis*, 11th ed.; Method 30-10.01: Crude Fat in Flour, Bread, and Baked Cereal Products Not Containing Fruit; Cereals & Grains Association: St. Paul, MN, USA, 1999.

17. Cereals & Grains Association. *AACC Approved Methods of Analysis*, 11th ed.; Method: 32-10.01: Crude Fiber in Flour, Bread, and Baked Cereal Products Not Containing Fruit; Cereals & Grains Association: St. Paul, MN, USA, 1999.

18. Cereals & Grains Association. *AACC Approved Methods of Analysis*, 11th ed.; Method 08-01.01: Ash—Basic Method; Cereals & Grains Association: St. Paul, MN, USA, 2009.

19. Cereals & Grains Association. *AACC Approved Methods of Analysis*, 11th ed.; Method 76-33.01: Damaged Starch—Amperometric Method by SDmatic; Cereals & Grains Association: St. Paul, MN, USA, 2011.

20. Rivera, J.; Deliephan, A.; Dhakal, J.; Aldrich, C.G.; Siliveru, K. Significance of Sodium Bisulfate (SBS) Tempering in Reducing the Escherichia coli O121 and O26 Load of Wheat and Its Effects on Wheat Flour Quality. *Foods* **2021**, *10*, 1479. [CrossRef] [PubMed]

21. Patwa, A.; Malcolm, B.; Wilson, J.; Ambrose, K.R. Particle size analysis of two distinct classes of wheat flour by sieving. *Trans. ASABE* **2014**, *57*, 151–159.

22. Dziki, D.; Laskowski, J. Wheat kernel physical properties and milling process. *Acta Agrophysica* **2005**, *6*, 59–71.

23. Gürsoy, S.; Güzel, E. Determination of physical properties of some agricultural grains. *Res. J. Appl. Sci. Eng. Technol.* **2010**, *2*, 492–498.

24. Kumar, M.M.; Prasad, K.; Chandra, T.S.; Debnath, S. Evaluation of physical properties and hydration kinetics of red lentil (Lens culinaris) at different processed levels and soaking temperatures. *J. Saudi Soc. Agric. Sci.* **2018**, *17*, 330–338.

25. Roy, F.; Boye, J.I.; Simpson, B.K. Bioactive proteins and peptides in pulse crops: Pea, chickpea and lentil. *Food Res. Int.* **2010**, *43*, 432–442. [CrossRef]

26. Thakur, S.; Scanlon, M.G.; Tyler, R.T.; Milani, A.; Paliwal, J. Pulse flour characteristics from a wheat flour miller's perspective: A comprehensive review. *Compr. Rev. Food Sci. Food Saf.* **2019**, *18*, 775–797. [CrossRef]

27. Food and Drug Administration (FDA). Code of Federal Regulations Title 21; Sec 137.105 Flour. Available online: https://www.ecfr.gov/current/title-21/chapter-I/subchapter-B/part-137 (accessed on 10 October 2021).

28. Kim, W.; Choi, S.G.; Kerr, W.L.; Johnson, J.W.; Gaines, C.S. Effect of heating temperature on particle size distribution in hard and soft wheat flour. *J. Cereal Sci.* **2004**, *40*, 9–16. [CrossRef]

29. Peng, M.; Gao, M.; Abdel-Aal, E.S.; Hucl, P.; Chibbar, R.N. Separation and characterization of A-and B-type starch granules in wheat endosperm. *Cereal Chem.* **1999**, *76*, 375–379. [CrossRef]

30. Raeker, M.Ö.; Gaines, C.S.; Finney, P.L.; Donelson, T. Granule size distribution and chemical composition of starches from 12 soft wheat cultivars. *Cereal Chem.* **1998**, *75*, 721–728. [CrossRef]

31. Hall, C.; Hillen, C.; Garden Robinson, J. Composition, nutritional value, and health benefits of pulses. *Cereal Chem.* **2017**, *94*, 11–31. [CrossRef]

32. Lebrun, P.; Krier, F.; Mantanus, J.; Grohganz, H.; Yang, M.; Rozet, E.; Boulanger, B.; Evrard, B.; Rantanen, J.; Hubert, P. Design space approach in the optimization of the spray-drying process. *Eur. J. Pharm. Biopharm.* **2012**, *80*, 226–234. [CrossRef] [PubMed]

33. Motte, J.C.; Tyler, R.; Milani, A.; Courcelles, J.; Der, T. Pea and lentil flour quality as affected by roller milling configuration. *Legume Sci.* **2021**, 1–15. [CrossRef]

34. Xu, M.; Jin, Z.; Simsek, S.; Hall, C.; Rao, J.; Chen, B. Effect of germination on the chemical composition, thermal, pasting, and moisture sorption properties of flours from chickpea, lentil, and yellow pea. *Food Chem.* **2019**, *295*, 579–587. [CrossRef] [PubMed]

35. Siliveru, K.; Ambrose, R.K.; Vadlani, P.V. Significance of composition and particle size on the shear flow properties of wheat flour. *J. Sci. Food Agric.* **2017**, *97*, 2300–2306. [CrossRef]

36. Kuakpetoon, D.; Flores, R.A.; Milliken, G.A. Dry mixing of wheat flours: Effect of particle properties and blending ratio. *LWT-Food Sci. Technol.* **2001**, *34*, 183–193. [CrossRef]

37. Wang, L.; Flores, R.A. Effects of flour particle size on the textural properties of flour tortillas. *J. Cereal Sci.* **2000**, *31*, 263–272. [CrossRef]

38. Gu, Z.; Jiang, H.; Zha, F.; Manthey, F.; Rao, J.; Chen, B. Toward a comprehensive understanding of ultracentrifugal milling on the physicochemical properties and aromatic profile of yellow pea flour. *Food Chem.* **2021**, *345*, 128760. [CrossRef]
39. Deng, L.; Manthey, F.A. Laboratory-scale milling of whole-durum flour quality: Effect of mill configuration and seed conditioning. *J. Sci. Food Agric.* **2017**, *97*, 3141–3150. [CrossRef]

Article

Effects of Grinding Corn with Different Moisture Content on Subsequent Particle Size and Flowability

Michaela Braun [1], Kara Dunmire [1], Caitlin Evans [1], Charles Stark [1], Jason Woodworth [2] and Chad Paulk [1,*]

1   Department of Grain Science & Industry, Kansas State University, Manhattan, KS 66503, USA;
    mbraun1@ksu.edu (M.B.); karadunmire@ksu.edu (K.D.); caitlinevans@ksu.edu (C.E.); crstark@ksu.edu (C.S.)
2   Department of Animal Science & Industry, Kansas State University, Manhattan, KS 66503, USA;
    jwoodworth@ksu.edu
*   Correspondence: cpaulk@ksu.edu

Abstract: The objective of this study was to determine the effects of whole-corn moisture and hammermill screen size on subsequent ground corn moisture, particle size and flowability. Treatments were arranged as a 2 × 2 factorial design with two moisture concentrations (14.5 and 16.7%), each ground using 2 hammermill screen sizes (3 mm and 6 mm). Corn was ground using a lab-scale 1.5 HP Bliss Hammermill at three separate timepoints to create three replications per treatment. Ground corn flowability was calculated using angle of repose (AOR), percent compressibility, and critical orifice diameter (COD) measurements to determine the composite flow index (CFI). There was no evidence for a screen size × corn moisture interaction for ground corn moisture content (MC), particle size, standard deviation, or flowability metrics. Grinding corn using a 3 mm screen resulted in decreased ($p < 0.041$) moisture content compared to corn ground using the 6 mm screen. There was a decrease ($p < 0.031$) in particle size from the 6 mm screen to the 3 mm, but no evidence of difference was observed for the standard deviation. There was a decrease ($p < 0.030$) in percent compressibility as screen size increased from 3 mm to 6 mm. Angle of repose tended to decrease ($p < 0.056$) when corn was ground using a 6 mm screen compared to a 3 mm screen. For the main effects of MC, 16.7% moisture corn had increased ($p < 0.001$) ground corn MC compared to 14.5%. The 14.5% moisture corn resulted in decreased ($p < 0.050$) particle size and an increased standard deviation compared to the 16.7% moisture corn. The increased MC of corn increased ($p < 0.038$) CFI and tended to decrease ($p < 0.050$) AOR and COD. In conclusion, decreasing hammermill screen size increased moisture loss by 0.55%, decreased corn particle size by 126 µm and resulted in poorer flowability as measured by percent compressibility and AOR. The higher moisture corn increased subsequent particle size by 89 µm and had improved flowability as measured by CFI.

Keywords: corn; hammermill; moisture content; particle size

Citation: Braun, M.; Dunmire, K.; Evans, C.; Stark, C.; Woodworth, J.; Paulk, C. Effects of Grinding Corn with Different Moisture Content on Subsequent Particle Size and Flowability. *Processes* **2021**, *9*, 1372. https://doi.org/10.3390/pr9081372

Academic Editors: Yonghui Li and Shawn/Xiaorong Wu

Received: 2 June 2021
Accepted: 30 July 2021
Published: 5 August 2021

1. Introduction

Reducing the particle size of cereal grains is often the first step in the feed manufacturing process. This process specifically ruptures the hard-outer shell, or hull, of the grain and exposes the interior nutrient-dense endosperm and germ. The grinding process alone typically consumes 70% of the total energy used during the feed production process [1]. Hammermills are the most commonly used size-reduction equipment because of their high throughput rates and versatility in the grinding of different materials [2].

Hammermills achieve particle size reduction by utilizing impact forces to shatter larger particles into smaller particles [3]. Overall mill performance is dependent on many different factors, such as initial particle size, material, feed rate, machine configuration, and moisture content (MC) [4]. Yellow dent #2 corn at approximately 15% moisture content is the most common grain used for feed in the US. However, the MC of corn can vary greatly based on the region of origin, weather patterns, harvest conditions, and corn genetics. Changes in the MC can affect the performance of the grinding equipment, energy usage, and the characteristics of the ground material [5]. Furthermore, MC influences the cohesion and adhesion of particles which, in turn, can alter the flowability of materials [6]. Therefore, the overall objective of this study was to determine the effects of whole-corn moisture prior to grinding and hammermill screen size on subsequent ground corn moisture, particle size and flowability.

2. Materials and Methods

Whole yellow dent #2 corn with an initial MC of 14.5% was used in this experiment. Treatments were arranged as a 2 × 2 factorial design with 2 moisture levels (14.5 and 16.7%) and ground using 2 hammermill screen sizes measuring 3 mm and 6 mm in diameter. Increasing initial whole corn moisture was accomplished by adding 5% water and heating at 55 °C for 3 h in sealed glass jars using a Fisherbrand Isotemp Oven (Model 15-103-051). Whole-corn moisture was analyzed using a Dickey-John GAC 2500-UGMA. Corn was then ground using a lab-scale 1.5 HP Bliss Hammermill (Model 6K630B, Bliss Industries, LLC, Ponca City, OK, USA) 3 separate times to create 3 replications per treatment. Samples of each treatment were collected and stored in vacuum-sealed bags to minimize moisture loss and later analyzed for moisture, particle size, and flowability characteristics. Ground corn samples were analyzed for moisture according to AOAC 930.15 [7].

Particle size analysis was conducted according to the ANSI/ASAE S319.2 standard particle size analysis method with dispersing agent [8]. A 100 ± 5 g sample was sieved with a 13-sieve stainless steel sieve stack containing sieve agitators with bristle sieve cleaners and rubber balls measuring 16 mm in diameter. Each sieve was individually weighed with the sieve agitators to obtain a tare weight. An additional 0.5 g of silicon dioxide dispersing agent was mixed into the sample and then placed on the top sieve. The sieve stack was placed in a Ro-Tap machine (Model RX-29, W. S. Tyler Industrial Group, Mentor, OH, USA) and shaken for 10 min. Once completed, each sieve was individually weighed with the sieve agitator(s) to obtain the weight of sample on each sieve. The amount of material on each sieve was used to calculate the geometric mean diameter (dgw) and geometric standard deviation (Sgw) according to the equations described in ANSI/ASAE standard S319.2 [9]. The weight of the dispersing agent was not subtracted from the weight of the pan, as specified in the ANSI/ASAE S319.2. Sieves were cleaned after each analysis with compressed air and a stiff bristle sieve cleaning brush.

The flowability characteristics of ground corn samples were evaluated using the results of percent compressibility, angle of repose, and critical orifice diameter, which were then compiled into a composite flow index (CFI) using equations [10].

$$CFI = y_1 + y_2 + y_3$$

where:

Y_1 to y_3 are the transformed scores for test 1 to 3.
Critical orifice diameter value (COD; y_1) = $-1.111 \times$ COD + 37.778.
Compressibility (y_2) = $-0.667 \times$ Compresibility + 36.667/
Angle of repose (AoR; y_3) = $-0.667 \times$ AoR + 50.

Angle of repose was determined by allowing a sample to flow from a vibratory conveyor above a free-standing platform until it reached its maximum piling height. The angle between the free-standing platform of the sample pile and the height of the pile was calculated by taking the inverse tangent of the height of the pile divided by the platform radius [11].

The critical orifice diameter was determined using a powder flowability test instrument (Flodex Model WG-0110, Paul N. Gardner Company, Inc., Pompano Beach, FL, USA). Fifty grams of sample was allowed to flow through a stainless-steel funnel into a cylinder. The sample rested for 30 s in the cylinder, and then was evaluated based on the flow through an opening in a horizontal disc. The discs were 6 cm in diameter and the interior hole diameter ranged from 4 to 34 mm. A negative result was recorded when the sample did not flow through the opening in the disc or formed an off-center cylindrical tunnel or rathole. The disc hole size diameter was then sequentially increased by one-disc size until a positive result was observed. A positive result was recorded when the material flowed through the disc opening, forming an inverted cone shape. If a positive result was observed, the disc hole size diameter was decreased until a negative result was observed. Three positive results were used to determine the critical orifice diameter [8].

Compressibility was determined by measuring the initial and final tapped volume. A 100 g sample was poured into a 250 mL graduated cylinder and the initial volume was recorded. The cylinder was tapped until no further change in the volume was observed. The final volume was recorded and change in compressibility calculated. The change in compressibility, expressed as a percentage, was calculated by finding the difference between the initial and final volume, dividing by the initial volume, and multiplying by 100 [8].

Data were analyzed as 2×2 factorial using the PROC GLIMMIX procedure in SAS version 9.4 (SAS Institute Inc., Cary, NC, USA). Grinding run served as the experimental unit and each treatment was replicated three times. Results were considered significant if $p \leq 0.05$, and a trend if $0.05 < p \leq 0.10$. Tukey's test was used for comparisons between treatments.

3. Results

There was no evidence for a screen size $\times$ corn moisture interaction for MC, particle size, standard deviation, or flowability metrics (Table 1). Grinding corn using a 3 mm screen resulted in decreased ($p < 0.041$) MC compared to corn ground using the 6 mm screen. There was a decrease ($p < 0.031$) in particle size from the 6 mm screen to the 3 mm, but no evidence of difference was observed for the standard deviation of the mean particle sizes. There was a decrease ($p < 0.030$) in percent compressibility as screen size increased from 3 mm to 6 mm. Angle of repose tended to decrease ($p < 0.056$) when corn was ground using a 6 mm screen compared to a 3 mm screen. For the main effects of MC, 16.7% moisture corn had increased ($p < 0.001$) ground corn MC compared to 14.5% MC corn. The 14.5% moisture corn resulted in decreased ($p < 0.029$) particle size and an increase ($p < 0.038$) in standard deviation of the mean particle size compared to the 16.7%. The 16.7% moisture corn increased ($p < 0.038$) CFI and tended to decrease ($p < 0.100$) AOR and COD.

Table 1. Effect of initial corn moisture and screen size on physical characteristics of ground corn.

Item	3 mm		6 mm		SEM [1]	Probability, $p<$		
	14.5%	16.7%	14.5%	16.7%		Screen Size	Moisture	Interaction
Moisture, %	11.7	16.4	11.9	16.6	3.15	0.041	0.001	0.805
Particle size, d_{gw} μm [2]	348	401	438	563	44.58	0.031	0.029	0.240
Standard deviation, S_{gw}	2.49	2.39	2.75	2.41	0.03	0.240	0.038	0.189
Angle of repose, $^\circ$ [3]	52.07	47.21	47.24	46.00	1.35	0.056	0.055	0.219
Critical orifice diameter, mm [4]	30.0	26.0	27.3	23.3	1.79	0.175	0.056	1.00
Compressibility, % [5]	26.8	26.1	25.4	23.0	0.85	0.030	0.112	0.344
Composite flow index [6]	38.5	46.6	45.6	52.4	3.02	0.063	0.038	0.839

[1] SEM = standard error of the mean. [2] Particle size and standard deviation (Sgw) are determined according to ASABE 319.2 methods. [3] Angle of repose was determined by measuring the height and radius of the cone formed by the material and using the following equation Tan θ = height of cone (mm)/radius of cone (mm). [4] Critical orifice diameter was determined using a Flodex device to determine product mass flow characteristics through varying discharge outlet sizes. [5] The change in compressibility, expressed as a percentage, was calculated by finding the difference between the initial and final volume, dividing by the initial volume, and multiplying by 100. [6] The composite flow index is calculated by the following equation CFI = $(-0.667(\text{AoR Result}) + 50) + (-0.667(\%\text{C Result}) + 36.667) + (-1.111(\text{COD Result}) + 37.778)$.

4. Discussion

Corn particle size is a key quality measure when manufacturing feed. The experiment reported herein shows that corn of different initial MCs ground using two different screen sizes will result in different physical characteristics post grinding. Previous research [12] details the basic characteristics of grinding corn with a hammermill and the use of screens to control subsequent particle size. The decrease in particle size observed from the 6 mm screen to the 3 mm was expected. To achieve the increased particle size reduction, material may spend more time in the grinding chamber to reach the desired particle size [13]. The observed loss in MC with a larger reduction in particle size can be attributed to this increase in grind time, as more frictional heat is generated, causing more moisture and energy efficiency to be lost [14].

It has previously been described that increased MC influences the breaking behavior of material during the grinding process [15,16]. The results of the experiment reported herein suggest that as MC decreased, the Sgw of the ground material increased. The increase in MC and subsequent reduction in Sgw can be explained as a lower MC corresponds with a harder product to be ground. The hardness of the kernels impacts the shatter patterns of grain and, therefore, increases the production of fine particles as well as the overall variation in the ground material. Previous research evaluated corn as well as corn cobs at three different moisture contents, looking at grinding performance and the characteristics of the ground material [5]. An increase in post-grinding moisture loss as well as an increase in Sgw was observed as initial MC increased. A similar loss of moisture post grinding was seen in the experiment reported herein; however, corn with a 16.7% initial MC resulted in a smaller Sgw or a more uniform grind.

The particle size of the material as well as its distribution can cause segregation during handling and affects the flowability of materials [17]. Kalivoda et al. [8] reported that a reduction in particle size corresponded with poorer flowability characteristics, caused predominantly by fine particles or particles measuring less than 150 microns. The shape of these fine particles may be the main cause of their negative impacts on flowability [18]. Previous research details the characteristics of particles that may cause poor flowability, such as particle size, shape, density, and surface, among other factors [10]. Small or fine particles are a significant flowability concern due to the attractive forces between particles. Smaller particles result in a smaller distance between individual particles. Therefore, the cohesion forces between particles, primarily due to the Van der Waals attraction, is stronger and results in poorer flowability [19]. In the experiment reported herein, flowability was measured using angle of repose, percent compressibility, and critical orifice diameter. These analyses evaluate different characteristics of ground materials. Angle of repose corresponds to the inner-particulate friction or the resistance to movement between particles [20]. The percent compressibility is defined as an indication of the incremental, volumetric, structural and/or increase in external forces [21]. Critical orifice diameter is a method that employs a cylinder with a series of interchangeable base plate discs with different diameter orifices. The critical orifice diameter is the size of the smallest orifice in a base plate disc through which the powder in a cylinder will discharge, and is a direct measure of powder cohesiveness and arch strength [22]. These measurements were then compiled into a composite flow index using the equations [10]. A single value of flowability that considers all evaluations can be beneficial when evaluating flowability results, as it can demonstrate the overall rating as well as individual characteristics of ground material and where problems may arise. A previously reported scale [10] helps to further understand the results of the composite flow index. A score of less than 15 corresponds to very, very poor flowability, while a score over 85 results in an excellent rating. A powder with a composite flow index of 45 or greater is considered passable. In the experiment reported herein, all treatments met this standard, with the exception of 14.5% moisture corn that was ground using the 3 mm screen, which had a poor rating.

While energy consumption and animal performance were not evaluated in the experiment reported here in, there is evidence that MC can have an impact on mill efficiency and animal performance [23]. Tran et al. [24] found a linear relationship between MC and grinding energy that decreased with increased moisture content. This decrease in energy consumption is due to lower MCs, resulting in a harder product to be ground. However, Armstrong et al. [25] reported that changes in grinding energy consumption are more predictable in the region of from 10 to 13% moisture content. Commercial swine diets are fed on an as-is basis and corn makes up a large percentage of the complete diet. Therefore, increasing the moisture content of the corn dilutes the concentrations of nutrients provided by the corn. Further research is needed on the impact of moisture content in field conditions as well as changes in the grinding system on the particle size of corn, as well as the energy consumption and possible impacts on animal performance.

5. Conclusions

This experiment shows the different initial moisture content when corn samples are ground using two different screen sizes on a hammermill will lead to different physical characteristics post-grinding. Decreasing screen size decreased particle size and resulted in poorer flowability. A CFI of greater than 45 is considered passable and all treatments met this standard, except 14.5% moisture corn that was ground using the 3 mm screen. When corn was ground using the 3 mm screen, a 0.55% moisture loss was observed compared to the 6 mm screen, regardless of the initial corn moisture. Increasing the initial corn moisture resulted in increased particle size and an improved standard deviation, which created improved flow characteristics.

Author Contributions: Conceptualization, M.B., C.P., C.E. and C.S.; methodology, M.B., C.P., C.S.; formal analysis, M.B. and C.P.; investigation, M.B. and K.D.; resources, C.S.; data curation, M.B., C.E. and K.D.; writing—original draft preparation, M.B.; writing—review and editing, C.P., J.W. and C.S. All authors have read and agreed to the published version of the manuscript.

Funding: This research received no external funding.

Institutional Review Board Statement: Not applicable.

Informed Consent Statement: Not applicable.

Data Availability Statement: Not applicable.

Conflicts of Interest: The authors declare no conflict of interest.

References

1. Dabbour, M.I.; Bahnasawy, A.; Ali, S.; El-Haddad, Z. Grinding Parameters and their Effects on the Quality of Corn for Feed Processing. *J. Food Process. Technol.* **2015**, *6*, 482. [CrossRef]
2. Bitra, V.S.; Womac, A.; Chevanan, N.; Miu, P.I.; Igathinathane, C.; Sokhansanj, S.; Smith, D.R. Direct mechanical energy measures of hammer mill comminution of switchgrass, wheat straw, and corn stover and analysis of their particle size distributions. *Powder Technol.* **2009**, *193*, 32–45. [CrossRef]
3. Austin, L.G. A treatment of impact breakage of particles. *Powder Technol.* **2002**, *126*, 85–90. [CrossRef]
4. Mani, S.; Tabil, L.G., Jr.; Sokhansanj, S.; Oemler, J.A. Mechanical Properties of Corn Stover Grind. *Trans. ASAE* **2004**, *47*, 1983–1990. [CrossRef]
5. Probst, K.V.; Ambrose, R.P.K.; Pinto, R.L.; Bali, R.; Krishnakumar, P.; Ileleji, K.E. The Effect of Moisture Content on the Grinding Performance of Corn and Corncobs by Hammermilling. *Trans. ASABE* **2013**, *56*, 1025–1033. [CrossRef]
6. Fitzpatrick, J.; Barringer, S.; Iqbal, T. Flow property measurement of food powders and sensitivity of Jenike's hopper design methodology to the measured values. *J. Food Eng.* **2004**, *61*, 399–405. [CrossRef]
7. AOAC. *Official Methods of Analysis*, 15th ed.; Association of Official Analytical Chemists: Washington, DC, USA, 1990.
8. Kalivoda, J. Effect of Sieving Methodolgy on Determining Particle Size of Ground Corn, Sorghum, and Wheat by Sieving. Ph.D. Thesis, Kansas State University, Manhattan, KS, USA, 2016.
9. ASABE Standards. *S319.2: Method of Determining and Expressing Fineness of Feed Materials by Sieving*; ASABE: St. Joseph, MI, USA, 2008.
10. Horn, E. Development of a Composite Index for Pharmaceutical Powders. Ph.D. Thesis, North-West University, School for Pharmacy, Potchefstroom, South Africa, 2008.
11. Appel, W.B. *Physical Properties of Feed Ingredients*, 4th ed.; Feed Manufacturing Technology, American Feed Industry Association, Inc.: Arlington, VA, USA, 1994; p. 151.
12. Koch, K. *Hammermills and Roller Mills. 2002. MF-2048 Feed Manufacturing*; Department of Grain Science and Indus-try Kansas State University: Manhattan, KS, USA, 2002; pp. 1–4.
13. Martin, S. *Comparison of Hammermill and Roller Mill Grinding and the Effect of Grain Particle Size on Mixing and Pelleting*; Kansas State University: Kansas, MA, USA, 1981; pp. 7–12.
14. Guo, L.; Tabil, L.G.; Wang, D.; Wang, G. Influence of moisture content and hammer mill screen size on the physical quality of barley, oat, canola and wheat straw briquettes. *Biomass Bioenergy* **2016**, *94*, 201–208. [CrossRef]
15. Jindal, V.; Austin, L. The kinetics of hammer milling of maize. *Powder Technol.* **1976**, *14*, 35–39. [CrossRef]

16. Adapa, P.; Tabil, L.; Schoenau, G. Grinding performance and physical properties of non-treated and steam exploded barley, canola, oat and wheat straw. *Biomass Bioenergy* **2011**, *35*, 549–561. [CrossRef]
17. Barbosa-Canovas, G.V.; Ortega-Rivas, E.; Juliano, P.; Yan, H. *Food Powders: Physical Properties, Processing and Functionality*; Kluwer Academic/Plenum Publishers: New York, NY, USA, 2005.
18. Goodband, B.; Groesbeck, C.; Tokach, M.; Dritz, S.; DeRouchey, J.; Nelssen, J. Effects of diet particle size on pig growth performance, diet flow ability, and mixing characteristics. In *2006 Manitoba Swine Seminar*; University of Saskatchewan: Winnipeg, MB, Canada, 2006; pp. 6–19.
19. Yang, J.; Slíva, A.; Banerjee, A.; Dave, R.N.; Pfeffer, R. Dry particle coating for improving the flowability of cohesive powders. *Powder Technol.* **2005**, *158*, 21–33. [CrossRef]
20. United States Pharmacopoeial Convention. The United States pharmacopeia USP 30:NF 25. United States Pharmacopeial Convention: Rockville, MD, USA, 2007; p. 4134.
21. Harnby, N.; Hawkins, A.; Vandame, D. The use of bulk density determination as a means of typifying the flow characteristics of loosely compacted powders under conditions of variable relative humidity. *Chem. Eng. Sci.* **1987**, *42*, 879–888. [CrossRef]
22. Taylor, M.K.; Ginsburg, J.; Hickey, A.J.; Gheyas, F. Composite method to quantify powder flow as a screening method in early tablet or capsule formulation development. *AAPS PharmSciTech* **2000**, *1*, 20–30. [CrossRef] [PubMed]
23. Young, L.G. Moisture Content and Processing of Corn for Pigs. *Can. J. Anim. Sci.* **1970**, *50*, 705–709. [CrossRef]
24. Tran, T.; Deman, J.; Rasper, V. Measurement of Corn Kernel Hardness. *Can. Inst. Food Sci. Technol. J.* **1981**, *14*, 42–48. [CrossRef]
25. Armstrong, P.R.; Lingenfelser, J.E.; McKinney, L. The Effect of Moisture Content on Determining Corn Hardness from Grinding Time, Grinding Energy, and Near-Infrared Spectroscopy. *Appl. Eng. Agric.* **2007**, *23*, 793–799. [CrossRef]

Article

Evaluation of Hammermill Tip Speed, Air Assist, and Screen Hole Diameter on Ground Corn Characteristics

Michaela Braun [1], Haley Wecker [1], Kara Dunmire [1], Caitlin Evans [1], Michael W. Sodak [2], Maks Kapetanovich [2], Jerry Shepherd [2], Randy Fisher [2], Kyle Coble [2], Charles Stark [1] and Chad Paulk [1,*]

[1] Department of Grain Science & Industry, Kansas State University, Manhattan, KS 66503, USA; mbraun1@ksu.edu (M.B.); haley27@ksu.edu (H.W.); karadunmire@ksu.edu (K.D.); caitlinevans@ksu.edu (C.E.); crstark@ksu.edu (C.S.)
[2] JBS Foods, Greeley, CO 80634, USA; michaelw.sodak@jbssa.com (M.W.S.); maks.kapetanovich@jbssa.com (M.K.); jerry.shepherd@jbssa.com (J.S.); randal.fisher@jbssa.com (R.F.); kyle.coble@jbssa.com (K.C.)
* Correspondence: cpaulk@ksu.edu

Citation: Braun, M.; Wecker, H.; Dunmire, K.; Evans, C.; Sodak, M.W.; Kapetanovich, M.; Shepherd, J.; Fisher, R.; Coble, K.; Stark, C.; et al. Evaluation of Hammermill Tip Speed, Air Assist, and Screen Hole Diameter on Ground Corn Characteristics. *Processes* **2021**, *9*, 1768. https://doi.org/10.3390/pr9101768

Academic Editors: Yonghui Li and Xiaorong (Shawn) Wu

Received: 18 August 2021
Accepted: 12 September 2021
Published: 1 October 2021

Publisher's Note: MDPI stays neutral with regard to jurisdictional claims in published maps and institutional affiliations.

Abstract: This study was performed to evaluate hammermill tip speed, assistive airflow, and screen hole diameter on hammermill throughput and characteristics of ground corn. Corn was ground using two Andritz hammermills measuring 1 m in diameter each equipped with 72 hammers and 300 HP motors. Treatments were arranged in a $3 \times 3 \times 3$ factorial design with three tip speeds (3774, 4975, and 6176 m/min), three screen hole diameters (2.3, 3.9, and 6.3 mm), and three air flow rates (1062, 1416, and 1770 fan revolutions per minute). Corn was ground on three separate days to create three replications and treatments were randomized within day. Samples were collected and analyzed for moisture, particle size, and flowability characteristics. There was a 3-way interaction ($p = 0.029$) for standard deviation (S_{gw}). There was a screen hole diameter $\times$ hammer tip speed interaction ($p < 0.001$) for geometric mean particle size d_{gw} ($p < 0.001$) and composite flow index (CFI) ($p < 0.001$). When tip speed increased from 3774 to 6176 m/min, the rate of decrease in dgw was greater as screen hole diameter increased from 2.3 to 6.3 mm. For CFI, increasing tip speed decreased the CFI of ground corn when ground using the 3.9 and 6.3 mm screen. However, when grinding corn using the 2.3 mm screen, there was no evidence of difference in CFI when increasing tip speed. In conclusion, the air flow rate did not influence d_{gw} of corn, but hammer tip speed and screen size were altered and achieved a range of d_{gw} from 304 to 617 μm.

Keywords: corn; hammermill; moisture content; particle size

1. Introduction

Particle size reduction is one of the basic steps in processing grains [1]. Animal feed undergoes particle size reduction for many reasons such as expediting feed consumption, improving nutrient absorption, and reducing material handling and labor costs by facilitating easier transport of products [2,3]. As more information has become available on particle size and its influence, the knowledge of what is needed to optimize animal performance has also grown. This increase in understanding along with improved capabilities of grinding equipment has led to interest for targeting specific particle sizes for various species and growth stages. While this may seem a reasonable ask of the feed mill, there are limitations to what can be achieved.

Hammermills have become a cost-effective choice that offer the flexibility to create a wide range of particle sizes [4]. Hammermills consist of a rotor assembly within a screened chamber that houses hammers that rotate with the rotor assembly [5]. Particle size reduction is achieved by utilizing a combination of impact, shear, and compression forces exerted by the hammers in the grinding chamber with the largest proportion because of impact [6–8]. The most common method to alter the particle size when grinding with a hammermill would be to change the screens. The screen prevents ground material from leaving the chamber before it is properly sized to the size of the perforation holes of the screen. However, while screen changes are the most common, there are other options that can make smaller and more precise particle size adjustments without the added down time. Alternative solutions to controlling corn particle size and characteristics of the ground material are tip speed and air assist adjustments via a variable frequency drive (VFD). Adjusting the hammer tip speed allows for a range of particle sizes to be achieved with the same screen hole diameter being in place. Additionally, air assist systems are commonly installed in combination with hammermills to aid with removing sized particles from the grinding chamber. Adjusting the rate at which the air assist system is operating could also impact the final particle size by manipulating the time material spends in the grinding chamber [9]. As more air passes through the grinding chamber, sized particles will be removed from the chamber faster before more reduction occurs. All of these factors can potentially affect the particle size, standard deviation, and flowability characteristics of the resulting ground material. There is the potential to allow for a range of particle sizes to be achieved from one screen hole diameter, however with these changes a decrease in handling characteristics may be possible. Therefore, the objective of this study was to evaluate the effect of hammermill tip speed, assistive airflow rate, and screen hole diameter on hammermill throughput and characteristics of the ground material.

2. Materials and Methods

Whole yellow dent #2 corn was ground, and samples were collected at the JBS Live Pork LLC feed mill in Fremont, IA. Corn was ground using two 1-m Andritz hammermills (Model: 4330-6, Andritz Feed & Biofuel, Muncy, PA, USA). Both mills discharged to a shared plenum where samples were collected via a sample port. Each mill was equipped with 72 hammers and 300 HP motors on a VFD. Corn was ground on 3 separate days to create 3 replications per treatment and treatments were randomized within replication. A new lot of corn was used each day and corn moisture content was 15.4, 13.8, and 14.05% for day 1, 2, and 3, respectively. Treatments were arranged in a $3 \times 3 \times 3$ factorial design with 3 tip speeds (3774, 4975, and 6176 m/min), 3 screen hole diameters (2.3, 3.9, and 6.3 mm), and 3 air assist system fan RPM (1062, 1416, and 1770 fan RPM).

Motor load and outlet temperatures were recorded for both mills at three separate time points during each grinding run via the Repete operating system (Repete Corp., Sussex, WI, USA). Air flow was measured using a hot wire anemometer (PerfectPrime Model WD9829) and taken between the baghouse and grinders. Samples of each treatment were collected and analyzed for moisture, particle size, and flowability characteristics. Ground corn samples were analyzed for moisture according to the AOAC method [10].

Particle size analysis was conducted according to the ANSI/ASAE [11] standard particle size analysis method as described by Kalivoda [12]. A 100 ± 5 g sample was sieved with a 13-sieve stainless steel sieve stack containing sieve agitators with bristle sieve cleaners and rubber balls measuring 16 mm in diameter. Each sieve was individually weighed with the sieve agitators to obtain a tare weight. An additional 0.5 g of dispersing agent was mixed into the sample and then placed on the top sieve. The sieve stack was placed in a Ro-Tap machine (Model RX- 29, W. S. Tyler Industrial Group, Mentor, OH, USA) and tapped for 10 min. Once completed, each sieve was individually weighed with the sieve agitator(s) to obtain the weight of sample on each sieve. The amount of material on each sieve was used to calculate the d_{gw} and S_{gw} according to the equations described in ANSI/ASAE standard S319.2 [10]. The weight of the dispersing agent was not subtracted from the weight of the pan as specified in the ANSI/ASAE S319.2 [10]. Sieves were cleaned after each analysis with compressed air and a stiff bristle sieve cleaning brush.

The flowability characteristics of ground corn samples were evaluated using the results of percent compressibility, angle of repose, and critical orifice diameter which were then compiled into a composite flow index (CFI) using equations previously described [13].

$$CFI = y_1 + y_2 + y_3$$

where

Y_1 to y_3 are the transformed scores for test 1 to 3.

Critical orifice diameter value (COD; y_1) = $-1.111 \times$ COD + 37.778.

Compressibility (y_2) = $-0.667 \times$ Compresibility + 36.667/

Angle of repose (AoR; y_3) = $-0.667 \times$ AoR + 50.

Angle of repose was determined by allowing a sample to flow from a vibratory conveyor above a free-standing platform until it reached its maximum piling height. The angle between the free-standing platform of the sample pile and the height of the pile was calculated by taking the inverse tangent of the height of the pile divided by the platform radius [14]. The critical orifice diameter was determined using a powder flowability test instrument (Flodex Model WG-0110, Paul N. Gardner Company, Inc., Pompano Beach, FL, USA). Fifty grams of sample was allowed to flow through a stainless-steel funnel into a cylinder. The sample rested for 30 s in the cylinder, and it was then evaluated based on the flow through an opening in a horizontal disc. The discs were 6 cm in diameter and the interior hole diameter ranged from 4 to 34 mm. A negative result was recorded when the sample did not flow through the opening in the disc or formed an off-center cylindrical tunnel or rathole. The disc hole size diameter was then increased by one-disc size until a positive result was observed. A positive result was recorded when the material flowed through the disc opening forming an inverted cone shape. If a positive result was observed, the disc hole size diameter was decreased until a negative result was observed. Three positive results were used to determine the critical orifice diameter [12].

Compressibility was determined by measuring the initial and final tapped volume. A 100 g sample was poured into a 250 mL graduated cylinder and the initial volume was recorded. The cylinder was tapped until no further change in the volume was observed. The final volume was recorded and change in compressibility was calculated. The change in compressibility, expressed as a percentage, was calculated by finding the difference between the initial and final volume, dividing by the initial volume, and multiplying by 100 [12].

Data were analyzed as a $3 \times 3 \times 3$ factorial using the PROC GLIMMIX procedure of SAS (SAS Institute Inc., Cary, NC) with grinding run serving as the experimental unit and day of sample collection serving as the block. Contrast statements were used to separate treatment means with the comparison of the main effects screen (2.3 vs. 3.9 vs. 6.3), tip speed (3774 vs. 4975 vs. 6176), and air flow rate (1062 vs. 1416 vs. 1770). Linear and quadratic polynomials were used to test increasing parameters within each main effect. Results were considered significant if $p \leq 0.05$.

3. Results

There were no 3-way interactions for screen hole diameter $\times$ hammer tip speed $\times$ air flow for the d_{gw} or any flowability characteristics of ground corn (Table 1). However, there was a screen hole diameter $\times$ hammer tip speed $\times$ air flow interaction for S_{gw} ($p = 0.029$). When corn was ground using the 2.3 mm screen, increasing hammer tip speed decreased S_{gw} when the air assist setting was 1062 RPM. However, increasing tip speed did not influence S_{gw} when the air assist was set at 1416 or 1770 RPM. Furthermore, there was no evidence of difference in the S_{gw} when air assist was increased and corn was ground using hammer tip speeds of 3774, 4975, or 6176 m/min. When grinding with the 3.9 mm screen, increasing hammer tip speed reduced S_{gw}. However, the rate of S_{gw} reduction was greater when the air flow was increased. In addition, increasing the air flow rate from 1062 to 1770 RPM increased S_{gw} when corn was ground using a tip speed of 3774 m/min; however, there was no difference in air flow when a tip speed of 6176 was used. When corn was ground using the 6.3 mm screen, there was no evidence of difference in S_{gw} when increasing hammer tip speed when the air assist was set at 1062 RPM. Increasing hammer tip speed increased S_{gw} when the air assist was set at 1416 RPM, and increasing hammer tip decreased S_{gw} when the air assist motor was set at 1770 RPM. Furthermore, increasing air flow at hammer tip speeds of 3774 and 4975 m/min increased S_{gw} but no difference was observed at 6176 m/min.

There was a linear screen hole diameter $\times$ linear hammer tip speed interaction ($p = 0.001$) for d_{gw} (Table 2). When tip speed increased from 3774 to 6176 m/min the rate of decrease in d_{gw} was greater as screen hole diameter increased from 2.3 to 6.3 mm resulting in a 67, 111, and 254 µm decrease in d_{gw} for corn ground using the 2.3, 3.9, and 6.3 mm screen hole diameter, respectively. There was a linear screen hole diameter $\times$ linear hammer tip speed interaction was also observed for COD ($p = 0.018$). When grinding using a hammer tip speed of 3774 m/min a decrease in COD was observed as screen hole diameter increased from 2.3 mm to 6.3 mm, but as tip speed increased to 4975 and 6176 m/min no differences in COD were observed with increasing screen hole diameter. Additionally, an interaction of screen hole diameter and hammer tip speed was observed for percent compressibility (Quadratic $\times$ Linear, $p = 0.015$). Increasing screen hole diameter had a quadratic effect on percent compressibility and increasing hammer tip speed decreased percent compressibility when using the 2.3 mm screen but increased with the 3.9 mm and 6.3 mm screens. Furthermore, an interaction of screen hole diameter and hammer tip speed was also observed for the composite flow index (Linear $\times$ Linear, $p = 0.040$). Composite flow index results increased with increasing screen hole diameter when corn was ground using a hammer tip speed of 3774 m/min but no differences were observed as tip speed increased to 4975 and 6176 m/min. An interaction of screen hole diameter and hammer tip speed was observed for mill motor load (Quadratic $\times$ Quadratic, $p = 0.001$). Mill motor load was decreased as screen hole diameter increased from 2.3 mm to 6.3 mm but increased as hammer tip speed was increased with the most significant reductions being observed as tip speed was increased from 3774 m/min to 4975 m/min on the 2.3 mm screen. Last, an interaction of screen hole diameter and hammer tip speed was observed for mill outlet temp (Linear $\times$ Linear, $p < 0.036$), where mill outlet temperature decreased as screen hole diameter was increased. However, as hammer tip speed was increased on the 2.3 mm screen mill outlet temperature decreased, where on the 3.9 mm and 6.3 mm screens increasing hammer tip speed resulted in increased outlet temperatures.

A significant interaction of screen hole diameter and air flow was observed for the compressibility (Quadratic $\times$ Linear, $p < 0.046$) and composite flow index (Linear $\times$ Linear, $p < 0.026$) results (Table 3). Compressibility results increased as air flow rate was increased on the 2.3 mm and 6.3 mm screens but decreased as air flow was increased on the 3.9 mm screen. Furthermore, screen hole diameter increased percent compressibility in a quadratic fashion with the highest measurements resulting from the 3.9 mm screen. The CFI increased as screen hole diameter was increased and increased as air flow was increased on the 2.3 mm and the 3.9 mm but decreased with increasing air flow on the 6.3 mm screen. There were no hammer tip speed by air flow interactions.

Table 1. Influence of 3-way interaction of screen hole diameter × hammer tip speed × air flow on the particle size and standard deviation of hammermilled corn [1].

Screen Hole Diameter, mm [2]	2.3			3.9			6.3				Probability, $p <$ [6]
Hammer Tip Speed, m/min [3]	3774	4975	6176	3774	4975	6176	3774	4975	6176	SEM	
Particle size, µm [5]											
Air flow [4]											
1062	344	341	296	443	395	334	580	437	380		
1416	361	314	273	477	390	336	652	437	357	25.26	0.227
1770	408	330	342	433	389	349	620	452	351		
Standard deviation, S_{gw} [5]											
Air flow											
1062	3.07	2.97	2.86	3.25	3.03	2.93	3.10	3.21	3.23		
1416	2.97	2.82	2.89	3.05	2.91	2.87	2.89	3.07	3.24	0.086	0.029
1770	3.13	2.97	2.97	3.47	3.13	2.97	3.54	3.46	3.27		

[1] Treatments were arranged in a 3 × 3 × 3 factorial design with main effects of tip speed, screen hole diameter, and air flow rate. Each treatment was replicated 3 times. [2] Corn was ground using screen hole diameters of 2.3, 3.9 or 6.3 mm. [3] Corn was ground using three motor speeds: 1100, 1450, or 1800 rpm. Hammer tip speed was then calculated by multiplying π by the hammermill diameter (m) and motor speed (rpm). [4] Corn was ground using three air flow settings of 1062, 1416, or 1770 fan RPM. [5] Particle size and standard deviation (S_{gw}) are determined according to ASABE 319.2 methods. [6] For the standard deviation result, a linear screen hole diameter × linear tip speed × linear air flow response was observed.

Table 2. Influence of 2-way interaction of screen hole diameter × hammer tip speed on the energy consumption, particle size, and flowability of hammermilled corn [1].

Screen Hole Diameter, mm [2]	2.3			3.9			6.3				Probability, $p <$
Tip Speed, m/min [3]	3774	4975	6176	3774	4975	6176	3774	4975	6176	SEM	
Physical Analysis											
Particle size, µm	371 [cd]	328 [ef]	304 [f]	451 [b]	391 [c]	340 [def]	617 [a]	442 [b]	363 [cde]	15.73	0.001 *
Standard deviation, S_{gw}	3.05 [b]	2.92 [a]	2.90 [c]	3.25 [c]	3.02 [bc]	2.92 [c]	3.17 [a]	3.24 [a]	3.24 [a]	0.064	0.002 *[†]
Critical orifice diamete r [4]	32.4 [ab]	31.7 [ab]	32.8 [a]	31.1 [b]	30.8 [b]	31.7 [ab]	29.1 [c]	31.5 [ab]	32.2 [ab]	0.576	0.018 *
Compressibility, % [5]	18.34 [abc]	19.01 [ab]	17.25 [bc]	16.96 [c]	19.75 [a]	19.33 [a]	18.96 [ab]	18.52 [abc]	19.49 [a]	0.641	0.015 [†]
Composite flow index [6]	46.14 [c]	45.91 [c]	45.75 [c]	49.34 [ab]	46.91 [bc]	45.43 [c]	51.53 [a]	47.58 [bc]	46.43 [bc]	1.264	0.040 *
Energy											
Motor Load, kW	147.07 [a]	102.41 [b]	99.39 [b]	89.58 [c]	88.96 [c]	89.73 [c]	76.01 [d]	76.16 [d]	78.21 [d]	2.677	0.001 [‡]
Mill Temp, °C	29.81 [a]	27.58 [cd]	29.00 [abc]	27.46 [d]	28.06 [bcd]	29.67 [a]	27.01 [d]	28.34 [abcd]	29.21 [ab]	0.960	0.036 *[¶]

[1] Treatments were arranged in a 3 × 3 × 3 factorial design with main effects of tip speed, screen hole diameter, and air flow rate. Each treatment was replicated 3 times. [2] Corn was ground using screen hole diameters of 2.3, 3.9 or 6.3. [3] Corn was ground using three motor speeds: 1100, 1450, or 1800 rpm. Hammer tip speed was then calculated by multiplying π by the hammermill diameter (mm) and motor speed (rpm). [4] Critical orifice diameter was determined using a Flodex device to determine product mass flow characteristics through varying discharge outlet sizes. [5] Percent compressibility is calculated by using the Hausner ratio (PTapped/PBulk). [6] The composite flow index is calculated by the following equation CFI = (−0.667(AoR Result) + 50) + (−0.667(%C Result) + 36.667) + (−1.778(COD Result) + 37.778). * Denotes Linear Screen hole diameter × Linear Tip Speed response. ¶ Denotes Linear Screen hole diameter × Quadratic Tip Speed response [†] Denotes Quadratic Screen hole diameter × Linear Tip Speed response [‡] Denotes Quadratic Screen hole diameter × Quadratic Tip Speed response. [abcdef] Within a row, means without a common superscript differ ($p < 0.05$).

Table 3. Influence of 2-way interaction of screen hole diameter $\times$ air flow on the compressibility and composite flow index of hammermilled corn [1].

Screen Hole Diameter, mm [2]	2.3			3.9			6.3				Probability, $p <$
Air flow, RPM [3]	1062	1416	1770	1062	1416	1770	1062	1416	1770	SEM	
Physical Characteristic											
Compressibility, % [4]	17.52 [b]	18.37 [b]	18.71 [ab]	19.03 [ab]	18.65 [ab]	18.36 [b]	18.15 [b]	18.44 [b]	20.39 [a]	0.641	0.046 [†]
Composite flow index [5]	45.48 [cd]	44.62 [d]	47.69 [abc]	46.12 [bcd]	46.51 [bcd]	49.05 [ab]	50.07 [a]	47.72 [abc]	47.76 [abc]	1.264	0.026 [*]

[1] Treatments were arranged in a $3 \times 3 \times 3$ factorial design with main effects of tip speed, screen hole diameter, and air flow rate. Each treatment was replicated 3 times. [2] Corn was ground using screen hole diameters of 2.3, 3.9, or 6.3 mm. [3] Corn was ground using three air flow settings of 1062, 1416, or 1770 fan RPM. [4] Percent compressibility is calculated by using the Hausner ratio (PTapped/PBulk). [5] The composite flow index is calculated by the following equation CFI = $(-0.667(\text{AoR Result}) + 50) + (-0.667(\%\text{C Result}) + 36.667) + (-1.778(\text{COD Result}) + 37.778)$. [*] Denotes Linear Screen hole diameter $\times$ Linear Air Flow response. [†] Denotes Quadratic Screen hole diameter $\times$ Linear Air Flow response. [abcd] Within a row, means without a common superscript differ ($p < 0.05$).

4. Discussion

Corn particle size, or geometric mean diameter (d_{gw}), and geometric standard deviation (S_{gw}) are key quality measures when manufacturing feed. The experiment reported herein shows that varying equipment and settings used during the grinding process will produce different physical characteristics post grinding. Previous research details the basic characteristics of grinding with a hammermill and the use of screens to control grind size [4]. The decrease in particle size observed from the 6.3 mm to the 2.3 mm screen is to be expected as decreasing screen hole diameter will increase the time material spends in the grinding chamber [15]. This decrease in screen hole diameter also increased energy consumption. This was also expected as grinding to obtain smaller particles will increase the amount of time spent in the grinding chamber and therefore energy consumption [16,17]. As the time spent in the grinding chamber increases friction becomes increasingly significant as particles contact the hammers as well as the surface of the screen which leads to increased fragmentation of particles [18]. Increased time spent in the grinding chamber can be a result of many factors. Decreased screen hole diameter as previously mentioned, as well as hammer tip speed and the rate of assistive air flowing through the grinding chamber can all effect the grinding time. As hammer tip speed is increased, the impact forces exerted by the hammers is increased and causes a more severe shatter pattern of the grain [19]. Research also found that increasing hammer tip speed and screen hole diameter increased hammermill through put [20]. While an interaction of the two factors was not evaluated in that study, an interaction of screen hole diameter and hammer tip speed on the energy used by the grinders was observed in the experiment reported herein. Furthermore, increasing screen hole diameter was shown to increase d_{gw} and S_{gw} as well as impact flow characteristics.

Increasing in grinding time and therefore fragmentation of particles results in a greater proportion of fine, flour-like particles. There is evidence that an increase in fine particles negatively impacts the flowability of ground material. Kalivoda et al. [12] reported a reduction in particle size that corresponded with poorer flowability characteristics caused predominantly by fine particles. A similar reduction in flow properties was observed with changes in screen hole diameter, hammer tip speed, and assistive air flow rate. According to a scale developed by Horn et al. [13], flowability decreases as CFI and angle of repose increases. In the experiment reported herein increasing screen hole diameter or air flow decreased the angle of repose while increasing tip speed increased AoR (Table 4). The particle size of material as well as its distribution can cause segregation during handling and affects the flowability of materials [21]. Haque [22] also suggested that flow properties are impacted more so by the physical characteristics of ground material rather than any chemical properties. This includes the d_{gw}, S_{gw}, particle shape, and electrostatic charge [22].

In the experiment reported herein, a three-way interaction of screen hole diameter, hammer tip speed, and assistive air flow rate was observed for the S_{gw} of corn ground using a hammermill. There are few published data evaluating screen hole diameter, hammer tip speed, and air assist simultaneously. There is a particular lack of understanding of the impact of air assist on ground material characteristics. Previous research determined that applying air flow through a hammer mill aids to improve the capacity of the mill as well as achieve a more uniform grind, or lower S_{gw} [9]. This was demonstrated in the experiment reported herein as a quadratic response of the main effect of air flow where the S_{gw} of corn ground using the median air flow setting resulted in the lowest S_{gw} value followed by the low and then high setting respectively. This result was unexpected as increased assistive air flow rate should aid to remove appropriately sized particles from the grinding chamber faster and therefore reduce the amount of time particles are subject to grinding forces. The magnitude of the response however was dependent on the screen hole diameter and hammer tip speed. Along with the response of S_{gw} to different air flows, a significant decrease in angle of repose was seen with an increase to the maximum air flow setting as well as an interaction of screen hole diameter and air flow on the CFI of ground corn. It can be hypothesized that an increase in fine particles created from increased time spent in the grinding chamber influenced the responses of S_{gw} and flowability characteristics as a result of assistive air flow rate.

Table 4. Influence of main effects of screen hole diameter, tip speed, and air flow on energy consumption, particle size, and flowability of hammermilled corn [1,2,3,4].

	Screen Hole Diameter, mm			Tip Speed, m/min			Air Flow, RPM				*p*-Value		
	2.3	3.9	6.3	3774	4975	6176	1062	1416	1770	SEM	Screen	Tip	AS
Physical Analysis													
Particle size d_{gw}, µm [5]	334 c	394 b	474 a	480 a	381 b	335 c	394	400	408	10.83	0.001 *	0.001 *	0.477
Standard deviation, S_{gw}	2.96 c	3.06 b	3.22 a	3.16 a	3.06 b	3.02 b	3.07 b	2.96 c	3.21 a	0.033	0.001 *	0.001 *	0.001 †
Angle of repose, ° [6]	45.57 a	45.01 a	43.26 b	43.55 b	44.98 a	45.30 a	44.84 a	45.75 a	43.25 b	1.034	0.002 *	0.025 *	0.004 †
Critical orifice diameter [7]	32.37 a	31.25 b	30.96 b	30.88 b	31.40 ab	32.29 a	31.62	31.77	31.18	0.337	0.010 *	0.014 *	0.424
Compressibility, % [8]	18.20	18.68	18.99	18.09	19.09	18.69	18.23	18.49	19.15	0.370	0.322	0.163	0.200
Composite flow index [9]	45.93 a	47.23 ab	48.52 a	49.00 a	46.80 a	45.87 b	47.23	46.28	48.17	0.912	0.018 *	0.002 *	0.109
Moisture, %	13.9	13.7	16.6	13.9	13.5	16.7	13.8	13.6	16.8	0.686	0.439	0.383	0.368
Energy													
Motor Load, kW	116.29 a	89.43 b	76.05 c	104.21 a	89.18 b	88.37 b	92.64	93.74	95.41	1.969	0.001 *	0.001 *	0.311
Mill Temp, °C	28.80	28.40	28.19	28.09 b	27.99 b	29.30 a	28.27	28.53	28.57	0.538	0.377	0.007 *	0.754
Fan Speed, m/min	17.39	19.10	17.45	17.72	18.45	17.78	13.87 c	17.79 b	22.27 a	1.309	0.380	0.842	0.001 *

[1] Treatments were arranged in a $3 \times 3 \times 3$ factorial design with main effects of tip speed, screen hole diameter, and air flow rate. Each treatment was replicated 3 times. [2] Corn was ground using screen hole diameters of 2.3, 3.9, or 6.3. [3] Corn was ground using three motor speeds: 1100, 1450, or 1800 rpm. Hammer tip speed was then calculated by multiplying π by the hammermill diameter (mm) and motor speed (rpm). [4] Corn was ground using three air flow settings of 1062, 1416, or 1770 fan RPM. [5] Particle size and standard deviation (Sgw) are determined according to ASABE 319.2 methods. [6] Angle of repose was determined by measuring the height and radius of the cone formed by the material and using the following equation Tan θ = height of cone (mm)/radius of cone (mm). [7] Critical orifice diameter was determined using a Flodex device to determine product mass flow characteristics through varying discharge outlet sizes. [8] Percent compressibility is calculated by using the Hausner ratio (PTapped/PBulk). [9] The composite flow index is calculated by the following equation CFI = (−0.667(AoR Result) + 50) + (−0.667(%C Result) + 36.667) + (−1.778(COD Result) + 37.778). * Denotes Linear Screen hole diameter × Linear Air Flow response. † Denotes Quadratic Screen hole diameter × Linear Air Flow response. [abc] Within a row and main effect, means without a common superscript differ ($p < 0.05$).

While no nutritional values were evaluated in this experiment, it is a significant consideration for optimum animal performance. The positive impacts reduced particle size have on swine has been widely reported. Healey et al. [23] observed improvement in the performance of pigs when corn was reduced from 1000 to 500 micron. Callan et al. [24] demonstrated improved feed conversion when finishing pigs were fed a complete diet ground through a 3 mm screen compared to a 6mm hammermill screen. For poultry, decreasing the d_{gw} of diets showed no effect when fed to broilers or turkey poults [25,26]. However, increasing the S_{gw} to include a larger portion of large particles was shown to improve broiler performance [27]. This improvement is driven by the larger particles stimulating gizzard development [28].

5. Conclusions

In summary, the results of the experiment reported herein show that hammer tip speed and air flow rate are viable options for adjusting ground material characteristics when grinding using a hammermill alongside the traditional screen variations. This experiment showed when using a 2.3, 3.9, and 6.3 mm screens at hammer tip speeds of 3774, 4975, and 6176 m/min as well as air flow settings of 1062, 1416, and 1770 that a wide range of particle sizes can be achieved. Along with the range of particle sizes capable of being produced, an increased level of accuracy can also be achieved with hammer tip speed and air flow adjustments with minimizing the down time necessary for screen changes. However, while increasing hammer tip speeds gives added flexibility, there are negatives that should be considered. Increasing the hammer tip speed with a VFD will increase the energy usage, as motor load will be increased especially on screens will smaller hole diameters. Furthermore, results of this study showed that when grinding using a 3.9 or 6.3 mm screen hole diameter, increasing hammer tip speed decreased the flowability of ground corn. However, increasing the air assist rate helped to improve the flowability characteristics.

Author Contributions: Conceptualization, M.B., C.P., C.E., and C.S.; methodology, M.B., C.P., and C.S.; formal analysis, M.B. and C.P.; investigation, M.B., H.W., M.W.S., M.K., J.S., R.F., K.C., C.E., and K.D.; resources, C.S.; data curation, M.B., H.W., M.W.S., M.K., J.S., R.F., K.C., C.E., and K.D.; writing—original draft preparation, M.B.; writing—review and editing, C.P. and C.S. All authors have read and agreed to the published version of the manuscript.

Funding: This research received no external funding.

Institutional Review Board Statement: Contribution no. 22-080-J from the Kansas Agric. Exp. Stn., Manhattan, KS 55606-0210.

Informed Consent Statement: Not applicable.

Data Availability Statement: The data presented in this study are available on request from the corresponding author. The data are not publicly available due to the privacy of the location where the data was collected.

Conflicts of Interest: The authors declare no conflict of interest.

References

1. Dziki, D. The crushing of wheat kernels and its consequence on the grinding process. *Powder Technol.* **2008**, *185*, 181–186. [CrossRef]
2. Berk, Z. Chapter 6: Size reduction. In *Food Process Engineering and Technology*, 2nd ed.; Academic Press: San Diego, CA, USA, 2013; pp. 167–191.
3. Wennerstrum, S.; Kendrick, T.; Tomaka, J.; Cain, J. Size reduction solutions for hard-to-reduce materials. *Powder Bulk Eng.* **2002**, *16*, 43–49.
4. Koch, K. Hammermills and Roller Mills. In *MF-2048 Feed Manufacturing*; Department of Grain Science and Industry Kansas State University: Manhattan, KS, USA, 2002; pp. 1–4.
5. Heiman, M. Particle Size Reduction, Chapter 8. In *Feed Manufacturing Technology V*; Schofield, E.K., Ed.; American Feed Industry Association: Arlington, VA, USA, 2005; pp. 108–126.
6. Austin, L.G. A treatment of impact breakage of particles. *Powder Technol.* **2002**, *126*, 85–90. [CrossRef]

7.   Probst, K.V.; Ambrose, R.P.K.; Pinto, R.L.; Bali, R.; Krishnakumar, P.; Ileleji, K.E. The Effect of Moisture Content on the Grinding Performance of Corn and Corncobs by Hammermilling. *Trans. ASABE* **2013**, 1025–1033. [CrossRef]
8.   Saravacos, G.D.; Kostaropoulos, A.E. *Handbook of Food Processing Equipment*; Springer: New York, NY, USA, 2002.
9.   Lyu, F.; Thomas, M.; Hendriks, W.H.; Van der Poel, A.F.B. Size reduction in feed technology and methods for determining, expressing and predicting particle size: A review. *Anim. Feed Sci. Technol.* **2020**, *261*, 114347. [CrossRef]
10. AOAC. *Official Methods of Analysis*, 15th ed.; AOAC: Arlington, VA, USA, 1990.
11. ASABE Standards. *S319.2: Method of Determining and Expressing Fineness of Feed Materials by Sieving*; ASABE: St. Joseph, MI, USA, 2008.
12. Kalivoda, J. *Effect of Sieving Ethodology on Determining Particle Size of Ground Corn, Sorghum, and Wheat by Sieving*; Kansas State University, Degree Granting Institution: Manhattan, KS, USA, 2016.
13. Horn, E. Development of a Composite Index for Pharmaceutical Powders. Ph.D. Thesis, North-West University, School for Pharmacy, Potchefstroom, South Africa, 2008.
14. Appel, W.B. *Physical Properties of Feed Ingredients*, 4th ed.; Feed Manufacturing Technology; American Feed Industry Association, Inc.: Arlington, VA:, USA, 1994; p. 151.
15. Martin, S. *Comparison of Hammermill and Roller Mill Grinding and the Effect of Grain Particle Size on Mixing and Pelleting*; Kansas State University: Manhattan, KS, USA, 1981; pp. 7–12.
16. Alrabadi, G.J. Influence of hammer mill screen size on processing parameters and starch enrichment in milled sorghum. *Cereal Res. Commun.* **2013**, *41*, 493–499. [CrossRef]
17. Miao, Z.; Grift, T.; Hansen, A.; Ting, K. Energy requirement for comminution of biomass in relation to particle physical properties. *Ind. Crop. Prod.* **2011**, *33*, 504–513. [CrossRef]
18. Milanovic, S. Literature review on the influence of million and pelleting on nutirional quality, physical characteristics, and production cost of pelleted poultry feed. 2018. Available online: http://hdl.handle.net/11250/2574133 (accessed on 20 September 2021).
19. Tran, T.; Deman, J.; Rasper, V. Measurement of Corn Kernel Hardness. *Can. Inst. Food Sci. Technol. J.* **1981**, *14*, 42–48. [CrossRef]
20. Bochat, A.; Wesolowski, L.; Zastempowski, M. A Comparative Study of New and Traditional Designs of a Hammer Mill. *Trans. ASABE* **2015**, *58*, 585–596. [CrossRef]
21. Barbosa-Canovas, G.V.; Ortega-Rivas, E.; Juliano, P.; Yan, H. *Food Powders: Physical Properties, Processing and Functionality*; Kluwer Academic/Plenum Publishers: New York, NY, USA, 2005.
22. Haque, M. Variation of Flow Property of Different Set of Formulas of Excipients against Variable Ratio of Different Diluents. Ph.D. Thesis, East West University, Department of Pharmacy, Dhaka, Bangladesh, 2010.
23. Healy, B.J.; Hancock, J.D.; Kennedy, G.A.; Bramel-Cox, P.J.; Behnke, K.C.; Hines, R.H. Optimum particle size of corn and hard and soft sorghum for nursery pigs. *J. Anim. Sci.* **1994**, *72*, 2227–2236. [CrossRef] [PubMed]
24. Callan, J.; Garry, B.; O'Doherty, J. The effect of expander processing and screen size on nutrient digestibility, growth performance, selected faecal microbial populations and faecal volatile fatty acid concentrations in grower–finisher pigs. *Anim. Feed. Sci. Technol.* **2007**, *134*, 223–234. [CrossRef]
25. Deaton, J.; Lott, B.; Simmons, J. Hammer Mill Versus Roller Mill Grinding of Corn for Commercial Egg Layers. *Poult. Sci.* **1989**, *68*, 1342–1344. [CrossRef]
26. Charbeneau, R.A.; Roberson, K.D. Effects of Corn and Soybean Meal Particle Size on Phosphorus Use in Turkey Poults. *J. Appl. Poult. Res.* **2004**, *13*, 302–310. [CrossRef]
27. Hetland, H.; Svihus, B.; Olaisen, V. Effect of feeding whole cereals on performance, starch digestibility and duodenal particle size distribution in broiler chickens. *Br. Poult. Sci.* **2002**, *43*, 416–423. [CrossRef] [PubMed]
28. Svihus, B. The gizzard: Function, influence of diet structure and effects on nutrient availability. *World's Poult. Sci. J.* **2011**, *67*, 207–224. [CrossRef]

processes

Article

Food Powder Flow in Extrusion: Role of Particle Size and Composition

Cameron McGuire [1], Kaliramesh Siliveru [1,*], Kingsly Ambrose [2] and Sajid Alavi [1,*]

[1] Department of Grain Science & Industry, Kansas State University, Manhattan, KS 66506, USA; camomcg@ksu.edu

[2] Department of Agricultural & Biological Engineering, Purdue University, West Lafayette, IN 47907, USA; rambrose@purdue.edu

* Correspondence: kaliramesh@ksu.edu (K.S.); salavi@ksu.edu (S.A.)

Citation: McGuire, C.; Siliveru, K.; Ambrose, K.; Alavi, S. Food Powder Flow in Extrusion: Role of Particle Size and Composition. *Processes* **2022**, *10*, 178. https://doi.org/10.3390/ pr10010178

Academic Editor: Shawn/Xiaorong Wu

Received: 31 October 2021
Accepted: 3 December 2021
Published: 17 January 2022

Publisher's Note: MDPI stays neutral with regard to jurisdictional claims in published maps and institutional affiliations.

Abstract: Innovations in food extrusion technology are enabling its rapid expansion and applicability in diverse areas related to bioprocessing and value addition. This study relates raw material particulate rheology to the granular flow in a single screw food extruder. Raw materials based on corn (i.e., meal, flour, and starch), wheat (i.e., farina, flour, and starch), and sucrose (i.e., granulated, superfine, and powdered) were used as model particulate systems for the study. Various particulate-scale characteristics and flow parameters of these nine materials were determined using a powder rheometer, a promising new offline tool. Properties such as basic flow energy, specific energy, cohesion, stability index, flow function, and effective angle of internal friction were good indicators of flowability in an extruder. Corn meal exhibited lower energy requirements and a higher propensity for flow than corn flour (6.7 mJ/g versus 10.7 mJ/g, and "free-flowing" versus "cohesive," according to Flow Function classifications), with wheat farina showing similar results when compared to wheat flour (5.8 mJ/g versus 7.9 mJ/g, and "highly free-flowing versus "cohesive," according to Flow Function classifications), although both wheat systems showed comparatively lower energy requirements than their comparable corn systems. Sugar, being of a different base material and particle shape, behaved differently than these starch-based materials—flow energy decreased and propensity to flow increased (51.7 mJ/g versus 8.0 mJ/g, and "free-flowing" versus "highly free-flowing"). This large energy requirement for coarse sugar particles may be attributed more to particle shape than composition, as the sharp edges of sugar can interlock and increase restriction to movement through the sample. The starch-based results were validated in a particulate flow study involving the above model systems (corn meal, corn flour, wheat farina, and wheat flour) in a pilot-scale single screw extruder. Visualization data, obtained using a transparent plexiglass window during extrusion, confirmed that the flours exhibited higher flow energy requirements and a lower flow factor when compared to the coarser-particle size corn meal during extrusion, seen by the increased peak heights and barrel fill.

Keywords: granular flow; particulate flow; extrusion; food powders; powder rheology; particle size; composition; corn; wheat; sugar

1. Introduction

In extrusion processing, material flow starts in a hopper and is fed through a feeder screw, through a preconditioning system and finally into the extruder. All of this flow takes place as a granular material. After entering into the kneading and cooking zones (in the extruder), the granular material undergoes pressure and temperature changes and begins the transition into a fluidized mass before exiting the die at the end of the barrel.

While this is a simplified illustration of an extrusion system, each target product has different optimal processing parameters: moisture, thermal energy, screw profile, barrel temperature, and physiochemical changes that occur in these zones. The analysis of flow and cook patterns changes with each zone, as the material transitions from individual granules to a compacted solid to a viscous melt, but the understanding of these patterns leads to improved developments of screw configuration, screw/barrel metallurgy, and operational parameters. In the feeding zone, specifically, understanding the behavior of particle flow as a function of size and composition is the key to these developments. As extrusion occurs in an enclosed system with forward movement aided by a rotating screw, particle-surface interactions (i.e., the interactions between particles and the screw and barrel) play a significant role in flow, in addition to interparticle forces. Flowability is a result of physical, chemical, and environmental variables. Determining the flow properties (dynamic, bulk, and shear) of these granular materials is necessary to predict how efficiently these materials are conveyed from the start of the barrel into the kneading and cooking zones [1]. If a material will not flow well upon entering the feeding zone, the entire process can back up, choking the extruder and resulting in time lost due to clearing the blockage [2]. Even if a material flows, a blended mixture may have components that behave differently: (i) one element being more adhesive to the barrel surface, and (ii) another agglomerating to like particles more. As such, it is important to understand how the individual raw materials will behave during the extrusion process to predict the flow problems in the extruder barrel as well as to estimate the system performance.

Beyond the particle-wall interactive forces present in this system, additional factors impact the effectiveness of material flow, such as particle shape. Yamane et al., [3] performed discrete element modeling on the dynamic angle of repose of non-spherical mustard seeds compared to spherical particles, with the non-spherical particles showing a greater angle of repose at any given rotational speed of a drum. These observations were similar to the observations of Dury et al., [4], where two species of mustard seeds were compared against spherical glass beads rotated in a drum, and an increased coefficient of friction was observed for both varieties of non-spherical seeds over the spherical beads. Additionally, particles that are irregularly shaped, with sharp corners or other non-rounded sides/edges, have even higher angles of internal friction than lenticular or ellipsoidal particles due to their ability to interlock and subsequently resist flow action [5].

The composition of the material also contributes to the differences in flow patterns. The starch-based, protein-based, and sucrose-based powders all have different roles in extrusion processing for achieving the desired end product. All these materials have different flow properties and contribute to the flow problems in the extruder. Fitzpatrick et al., [6] analyzed flow functionality of one dozen food powders which have similar particle sizes (fine) and found that the flow index of these powders was influenced by composition (including equilibrium moisture). However, there were no other extensive studies on comparing the flow functionality of food powders in the extrusion process. This study was undertaken to characterize the flow behavior of commonly used food powders (corn, wheat, and sucrose) in the extrusion process. The specific objectives of this study are to:

1. Explore the significance of particle size on the flow functionality and energy requirements of corn, wheat, and sucrose powders.
2. Explore the correlation between composition of food powders and their flow behavior patterns in the extruder.

2. Materials and Methods

2.1. Materials

The materials used for this experiment were as follows: Corn Starch (Argo, Engelwood Cliffs, NJ, USA), corn flour (Bunge, Atchison, KS, USA), corn meal (Aunt Jemima, Chicago, IL, USA), wheat starch (MGP Ingredients, Atchison, KS, USA), wheat flour (Gold Medal, Minneapolis, MN), wheat farina (Hal Ross Mill, Kansas State University, Manhattan, KS, USA), powdered sugar (C&H, Yonkers, NY, USA), superfine sugar (C&H, Yonkers, NY, USA), and granulated sugar (C&H, Yonkers, NY, USA). The moisture content of the food powders was determined using AOAC Standard 925.10 [7] by drying 2–3 g of sample in a hot air oven at 130 °C for 60 min.

2.2. Particle Size Measurement

Preliminary particle size was measured via Rotap sieve shaker (W.S. Tyler, Mentor. OH, USA). The particle size measurements were carried out on representative 100 g samples using the ASABE Standard S319.4 method [8]. The average particle size of the food powders was calculated using Equation (1) as follows:

$$\text{Average Particle Size} = 10^{\frac{\sum \left((mass\ on\ screen) * \log (screen\ opening) \right)}{\sum (mass\ on\ screen)}} \tag{1}$$

2.3. Flow Properties

The Freeman Technology FT4 Powder Rheometer (FT4, Freeman Technologies, Tewkesbury, UK) was used to evaluate the flow properties of the powders. The detailed description of this equipment and its use in powder flow characterization can be found in Freeman [1]. The usage of equipment and description is described briefly here for enhanced readability and completeness of this manuscript.

2.3.1. Dynamic Flow Properties of Food Powders

The basic Flowability Energy (BFE) is the total energy required to establish a specific flow pattern in the food powders when they are confined in a storage container [1]. The higher this value, the more energy is required to establish the flow. The BFE was calculated using Equation (2).

$$\text{BFE (mJ)} = \int_0^{\Delta x} \left(T\dot\theta + Fv_x \right) v_x^{-1} dx \tag{2}$$

where T = rotation resistance or torque experienced by the blade (N·m); F = vertical resistance or force experienced by the blade (N); $\dot\theta$ = angular speed of the blade (rad/s); v_x = vertical speed of the blade (m/s); and Δx = vertical distance traversed by the blade.

Specific Basic Flow Energy (SBFE) is BFE divided by total mass of product in the cylinder to give Joules/gram, which allows for a more uniform comparison across products with different densities, as formulations for products are mixed on a per-mass basis, not per-volume basis.

In contrast to BFE, Specific Energy (SE) represents the energy taken to move from the base of the cylinder to the top, representing unconfined flow, and is calculated by dividing with the mass of the sample to give a per-unit-mass value (Equation (3)):

$$\text{SE (mJ/g)} = \frac{(FE_6 + FE_7)/2}{m} \tag{3}$$

where FE_6 = upward flow energy required during test cycle 6 (mJ); FE_7 = upward flow energy required during test cycle 7 (mJ); and m = mass of sample (g).

Stability Index (SI) shows whether a powder expands, compacts, or remains at the same volume through the test cycles. A value near 1.00 indicated the powder maintains its volume, while a value greater or less than 1.00 indicated the powder had a tendency to compact or expand, respectively [1]. The SI is computed using Equation (4) below.

$$\text{SI} = \frac{BFE_7}{BFE_1} \tag{4}$$

where BFE_1 = Flow energy required during test cycle 1; and BFE_7 = Flow energy required during test cycle 7.

The rate at which the flours are handled in the extrusion system varies from time to time. To account for these variations and to understand the process of conveyance in the extrusion system, the Flow Rate Index (FRI) is computed using Equation (5) as follows:

$$\mathrm{FRI} = \frac{\mathrm{BFE}_{11}}{\mathrm{BFE}_8} \tag{5}$$

where BFE_8 = Flow energy required during test cycle 8; and BFE_{11} = Flow energy required during test cycle 11.

Eleven different blade tip speeds were used in computing the FRI, to account for the variations in the handling and conveyance of the food powders.

2.3.2. Compressibility

The compressibility tests accounts for the changes in density of the food powders as a result of mechanical compaction that occurs during the pre-conditioning and conditioning steps of the extrusion process. For the compressibility test, a 48 mm helical blade was used to condition the sample, with a glass cylinder of 85 mm × 50 mm used for the base. After the conditioning cycle was completed, the blade was replaced by a 48 mm-diameter vented piston. The top cylinder was split to remove excess powder, leaving a standard volume of product, and then the piston was lowered at increasing force levels: 0.5, 1, 2, 4, 6, 8, 10, 12, and 15 kPa. The percentage compression of the powder was recorded at each force interval. Then, compressibility is calculated as a percentage change in volume.

2.3.3. Shear Flow Properties of Food Powders

Testing of shear properties allows for further understanding of the inter-particulate forces that powders are subjected to during handling and processing, such as the yield point of powder flow initiation. Preparation for the shear cell test involved using the 48 mm helical blade, followed by the 48 mm diameter vented piston, and then splitting the two 85 mm × 50 mm glass cylinders, leaving a compacted volume of sample for the test. The shear cell attachment, with the same radius as the vented piston but with small blades on the underside, was used to carry out the test by inducing rotational and vertical stress. Once the powder bed in the cylinder yielded to the stress applied by the shear head, the stress value was recorded.

These results utilized Mohr Circle analysis to calculate values such as cohesion, major principle stress, unconfined yield strength, and flow factor, as illustrated in Figure 1. Test points were plotted along a graph to determine cohesion factor (y-intercept value of yield locus—line through data points) and effective angle of internal friction (angle of line drawn between farthest test point and origin compared to *x*-axis). Unconfined Yield Strength (UYS) was developed by drawing a half-circle from the origin, tangent to the yield locus, and the point the half-circle crossed the *x*-axis was labeled as the UYS. Major Principal Stress (MPS) was calculated in a similar matter, with a semi-circle drawn between the farthest test point and the pre-shear point, with the higher end of the semi-circle labeled as the MPS.

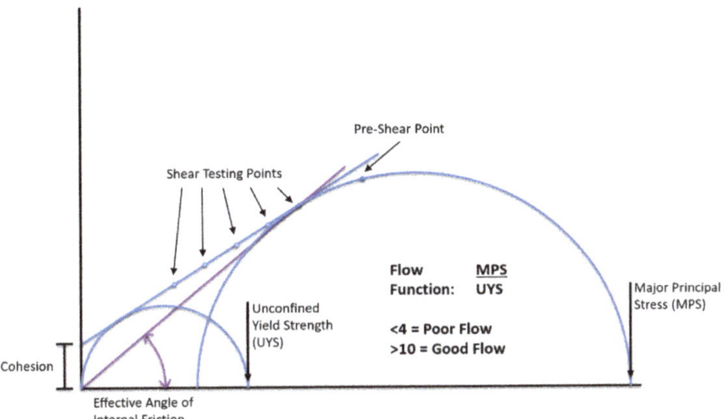

Figure 1. Typical Mohr Circle plot.

Flow Function (FF) was calculated using Equation (6), which directly indicates how easily a powder will flow, with a higher value indicating a greater propensity to flow and a lower number indicating a resistance to flow [9]. These values have been elaborated by Thomas and Schubert [10] and divided into the categories seen in Table 1.

$$FF = \frac{MPS}{UYS} \tag{6}$$

Table 1. Classification of particulate flowability based on flow function.

Type of Flow	Flow Function Value
Not Flowing	FF < 1
Very Cohesive	1 < FF < 2
Cohesive	2 < FF < 4
Easy-Flowing	4 < FF < 10
Free-Flowing	10 < FF

2.3.4. Wall Friction

This test measures the resistance of flow of powders in relation to the process equipment surface by using a friction disc head that applies both vertical and rotational stress on a powder at rest to determine the torque necessary to overcome the resistance of the powder bed. Preparation for the Wall Friction test involved using the 48 mm helical blade, followed by the 48 mm diameter vented piston, and then splitting the two 85 mm × 50 mm glass cylinders, leaving a compacted volume of sample for the test. A Wall Friction disc, with a friction coefficient value of 0.05 (low friction interference), was used for the test.

The torque required to maintain the rotational momentum of the disc was measured and used to calculate a 'steady-state' shear stress. The normal stress was maintained at a constant value throughout the measurement. From the relationship between normal stress (σw) and shear stress (τw), the wall friction angle (Φ), is calculated using Equation (7) as follows:

$$\Phi = \tan^{-1}\left(\frac{\tau_w}{\sigma_w}\right) \tag{7}$$

2.4. Extrusion Visualization

A pilot-scale X-20 37.3 kW single-screw extruder (Wenger Manufacturing, Sabetha, KS, USA) was used for in-line powder flow visualization trials with a plexiglass window along one-third of the circumference of the barrel. The screw diameter was 82.1 mm and the L:D ratio 8:1. Four regularly spaced steam locks were used dispersed along the length of the extruder screw, which in a typical extrusion process provide resistance to flow, increase material fill, and lead to the desired mechanical energy input. In the visualization trials, the steam locks developed peaks of flowing material along the length of the barrel, which were subjectively evaluated for determining flowability of corn meal, corn flour, wheat farina, and wheat flour. There was no die or any other restriction at the end of the screw. Material feed rate was set at 80 kg/hr. An extruder screw speed of 125 rpm was used for corn meal and wheat farina and 250 rpm for corn flour and wheat flour.

2.5. Data Analysis

All flow property tests were performed in triplicate and new food powder samples were used for each test. The flow property values were expressed as mean (standard deviation). Mean flow property values were analyzed using the GLIMMIX procedure with mean comparisons done using the Tukey's HSD procedure ($p \leq 0.05$). All analyses were conducted using SAS 9.3 software (SAS Institute, Cary, NC, USA).

3. Results and Discussion

3.1. Particle Size

Particle size of food powders, measured as average particle size diameter, is a critical factor in the determining the powder's usefulness and application in the extrusion process. The average particle size for each material was determined to be: corn starch (12 μm), corn flour (154 μm), corn meal (622 μm); wheat starch (23 μm), wheat flour (72 μm), wheat farina (410 μm); powdered sugar (12 μm), superfine sugar (150 μm), and granulated sugar (450 μm).

Since food particles are a variety of shapes, there is error associated with the assumption that all particles are spherical. The diversity of shapes also creates error in sieve measurements, as a long skinny particle may or may not pass through a screen opening, depending on its orientation. If it passes through an opening when it orients vertically, it is counted as being smaller than if it had not bounced to that orientation and thus remained on the larger sieve screen.

3.2. Moisture Content

Moisture content for each material was found to be as follows: corn starch, 10.32; corn flour, 11.04; corn meal, 13.13; wheat starch, 9.40; wheat flour, 12.28; wheat farina, 13.73; powdered sugar, 0.38; superfine sugar, 0.09; and granulated sugar, 0.06 (%wet basis).

3.3. Dynamic Flow Properties

Stability and variable flow rate testing shows that, as particle size increases (for corn and wheat powders), the energy requirements for confined and unconfined flow decreases (Table 2). The increase in energy with smaller particles can be attributed to an increase in surface area that increased the interparticle friction resulting in higher resistance to flow [1]. This higher flow resistance with decreasing particle size translated to poor granular flow in extrusion as could be seen from the visualization trials on the pilot-scale single screw extruder (Figure 2). Higher barrel fill for corn flour versus corn meal, and similarly higher barrel fill for wheat flour versus wheat farina, was a direct result of poor flowability of the finer particle size material in each case. Trials for corn and wheat starches on the extruder were not done, but it is reasonable to extrapolate from the above results that they will exhibit poorest flow, as has also been noted anecdotally. The SI results shows that corn flour and corn starch both compact during the testing, while corn meal slightly expands as the blade rotates through the sample (Table 2). The FRI shows that corn starch is more sensitive to changes in blade speed when compared to corn flour and corn meal. The FRI also shows that wheat starch was more sensitive to changes in blade speed when compared to wheat flour or wheat farina (Table 2). This could be due to the shape of the particles, as the SI of 1.0 indicates that farina tends to neither compact nor expand throughout the testing process. Contrarily, the starch and flour powders settled and compacted during the conditioning of the test and required noticeably more energy when blade speed was reduced for FRI testing. With a FRI value of < 1.0 for farina (Table 2), it can be inferred that a slower blade speed is more energy efficient for moving through the particles. From this test, it can also be concluded that for mixing or conditioning the wheat farina samples, slower mixing blade speeds have to be employed.

Table 2. Dynamic flow properties of food powders.

Sample	SBFE (mJ/g)	SE (mJ/g)	SI	FRI
Corn starch	11.01 ± 0.08 [A]	12.93 ± 1.19 [A]	1.18 ± 0.03 [A]	1.61 ± 0.04 [A]
Corn flour	10.71 ± 0.05 [B]	8.36 ± 0.09 [B]	1.07 ± 0.04 [B]	1.38 ± 0.01 [B]
Corn meal	6.70 ± 0.00 [C]	3.37 ± 0.10 [C]	0.97 ± 0.01 [B]	1.40 ± 0.02 [B]
Wheat starch	14.24 ± 0.11 [A]	8.64 ± 0.26 [A]	1.07 ± 0.03 [A B]	1.63 ± 0.01 [A]
Wheat flour	7.94 ± 0.08 [B]	6.50 ± 0.85 [B]	1.12 ± 0.03 [A]	1.21 ± 0.01 [B]
Wheat farina	5.81 ± 0.09 [C]	2.66 ± 0.01 [C]	1.00 ± 0.00 [B]	0.95 ± 0.01 [C]
Powdered sugar	9.30 ± 0.12 [C]	9.27 ± 0.02 [B]	1.05 ± 0.02 [A]	1.72 ± 0.00 [A]
Superfine sugar	8.02 ± 0.49 [B]	4.17 ± 0.49 [C]	1.11 ± 0.09 [A]	1.02 ± 0.02 [B]
Granulated sugar	51.73 ± 5.05 [A]	10.90 ± 0.40 [A]	1.07 ± 0.03 [A]	0.91 ± 0.04 [C]

Values followed by the same upper case letters indicate no significant difference among the particle sizes for a particular food powder material as well as for a particular dynamic flow property test ($p < 0.05$).

(a)

(b)

(c)

Figure 2. *Cont.*

(**d**)

Figure 2. In-line visualization of powder flow in a pilot-scale single screw extruder using (**a**) corn meal, (**b**) corn flour, (**c**) wheat farina, and (**d**) wheat flour.

Sugar, being the only non-starch-based powder tested, showed different results with increments in particle size (Table 2). While powdered and superfine sugar resulted in similar trends (as that of corn and wheat powders) for SBFE and SE tests, the granulated sugar required over four times the energy for both the tests (Table 2). This could be partly due to the shape of the powder, as powdered and superfine sugars are both more rounded particles, while granulated sugar is longer with asymmetrical edges [11]. This shape may cause granules to interlock and create a more difficult matrix for the blade to traverse, in both confined and unconfined flow. Conversely, as particle size increased, sensitivity to blade tip speed was reduced, denoted by the decreasing FRI value. Powdered sugar was the most sensitive food powders (among the tested powders), to a ten-fold decrease in blade speed, while superfine sugar was mostly unaffected. Granulated sugar, with a SI value of < 1.0 (Table 2), indicates that a slower blade speed was more efficient for traversing through the powder. Particle geometry may again explain this, as a slower tip speed could gently disrupt the interlocking particles smoother than the faster blade speed (similar to how non-Newtonian fluids behave as solids when acted upon by high forces, but flow freely when forces below the threshold for behaving as a solid are applied).

3.4. Shear Flow Properties

The shear testing results provide further understanding on whether the food powders will flow through the extrusion process or whether bridging, caking, and choking in the extruder are likely. According to the shear test results and classification of powders by Jenike [9], corn starch, wheat starch, and super fine sugar are free flowing powders as the FF values of these powders are very close to or greater than 10 (Table 3). The cohesion values of these powders (corn starch, wheat starch and super fine sugar, and also wheat farina) are lesser than the other powders (Table 3). This could be due to the spherical shape and smother texture of the starch particles [12], which might have resulted slipping movement of the particles. However, this result could also be attributed to the constant stress the powders are under during testing, once corn starch (or any material) is under enough constant pressure, it fluidizes and results in the inflated values that reflect a free-flowing liquid. Marston et al., [13] found that decreasing particle size of materials resulted in behavior more similar to water when struck with a solid object at a constant velocity, although specific values for fluidization were not found. This does explain the deviation from dynamic flow properties discussed above. Corn starch, other starches, and similar fine particulate materials are known for not flowing well in an extruder, especially during unconfined, granular flow in the feeding section, which is predicted reasonably from the dynamic flow tests parameters such as specific energy. The high FF of corn starch appears contrary, but this type of particulate fluidization for very fine granulation materials is not observed in extrusion feeding zone due to the low pressure and compaction regime. The lower FF and cohesive resistance of corn flour as compared to corn meal is, however, consistent with the higher specific energy observed for the former in dynamic testing and its lower propensity to flow in an extruder.

Table 3. Shear flow properties of food powders.

Sample	Cohesion (kPa)	MPS (kPa)	UYS (kPa)	FF	AIF (°)	Φ (°)
Corn Starch	0.440 ± 0.099 [C]	23.20 ± 0.15 [A]	1.93 ± 0.04 [C]	12.03 ± 0.26 [A]	34.1 ± 0.2 [C]	13.1 ± 0.5 [A]
Corn Flour	1.797 ± 0.073 [A]	20.43 ± 0.09 [B]	8.67 ± 0.30 [A]	2.36 ± 0.07 [C]	58.5 ± 0.3 [A]	7.7 ± 0.2 [B]
Corn Meal	0.843 ± 0.051 [B]	19.73 ± 0.09 [C]	4.02 ± 0.22 [B]	4.94 ± 0.27 [B]	49.0 ± 0.1 [B]	5.8 ± 0.1 [C]
Wheat Starch	0.468 ± 0.048 [B]	16.93 ± 0.09 [C]	1.82 ± 0.10 [B]	9.36 ± 0.53 [B]	38.2 ± 0.2 [B]	17.6 ± 0.2 [A]
Wheat Flour	1.647 ± 0.078 [A]	24.70 ± 0.15 [B]	7.53 ± 0.32 [A]	3.29 ± 0.13 [C]	49.9 ± 0.5 [A]	4.2 ± 0.3 [C]
Wheat Farina	0.395 ± 0.048 [B]	28.37 ± 0.41 [A]	1.54 ± 0.09 [B]	18.4 ± 0.77 [A]	34.0 ± 0.1 [C]	5.6 ± 0.0 [B]
Powdered Sugar	2.180 ± 0.130 [A]	15.10 ± 0.15 [B]	10.83 ± 0.50 [A]	1.40 ± 0.06 [C]	65.9 ± 1.0 [A]	26.8 ± 1.0 [A]
Superfine Sugar	0.191 ± 0.039 [C]	15.13 ± 0.03 [B]	0.62 ± 0.06 [C]	24.73 ± 2.31 [A]	37.8 ± 0.3 [C]	12.8 ± 0.5 [B]
Granulated Sugar	0.807 ± 0.092 [B]	27.17 ± 0.93 [A]	3.28 ± 0.34 [B]	8.53 ± 1.17 [B]	40.5 ± 0.4 [B]	9.5 ± 0.2 [C]

Values followed by the same upper case letters indicate no significant difference among the particle sizes for a particular food powder material as well as for a particular shear flow property test ($p < 0.05$).

Shear flow results for wheat starch, flour, and farina followed the same trend as corn-based powders. The FF for superfine sugar was the highest of all powders (Table 3), while powdered sugar was lowest of all, which is confirmed by the cohesion values being the lowest and highest, respectively. The shear flow property results of granulated and powdered sugars were in agreement with the results reported by Stasiak and Molenda [14].

Cohesion values (Table 3) tended to relate to yield stress, flow function, and angle of internal friction. The greater the cohesion value a powder has, the more the particles interact with each other, which results in a higher yield strength, lower flow function, and greater angle of internal friction. The corn flour, wheat flour, and powdered sugar have higher cohesion values (Table 3). This could be due to the smaller size of the particles. As in the production smaller particles, more forces are applied in the production process which results in unusual rough texture of the particles. This surface roughness combined with the higher surface area would tend to promote mechanical bridging and result in higher cohesion. Corn and wheat starches are an exception as they are not produced via isolation or separation techniques rather than size reduction.

The angle of internal friction or AIF indicates the interparticle friction as powder starts to slide. The powders with higher AIF values are more resistant to flow than the powders with lower AIF values. The flour powders (corn and wheat) and powdered sugar have higher AIF values, indicating that these powders are more resistant to flow than the other powders.

3.5. Compressibility

The compressibility tests results showed a correlation with the cohesion results for corn flour, wheat flour, as they had the highest cohesion value and was compressed the most (Figure 3). However, corn meal was shown to be more cohesive but less compressible than corn starch, despite having a higher cohesion value. Like corn and wheat powders, sugar showed a correlation between cohesion and compressibility (Figure 3c). Powdered sugar had the highest cohesion of any of the nine powders tested and was the most compressible, as well. Both superfine and granulated sugars had low cohesion values, which was reflected in a low compressibility (comparatively lower compressibility for superfine than granulated sugar, to match with the comparatively lower cohesion value). These results contribute to the hypothesis that the constant rotational stress in the previous test fluidized corn starch and yielded artificially lowered results. Cohesive powders have interparticle forces that create bridges and void spaces in a given volume, whereas non-cohesive powders tend to flow freely to occupy as much of a given volume as possible. The latter results in very little compressibility due to the lack of void space available for particles to nestle into when force is applied. The former, however, has much more space (further increasing as cohesiveness increases) that allows for particles to compact, resulting in increasing compressibility in tandem with increasing cohesive properties [15].

3.6. Wall Friction Angle (WFA)

The wall friction testing showed that the wall friction angle decreases with increment in particle size for corn and wheat powders (Table 3). A greater decrease in the wall friction angle is observed between corn starch and corn flour than from corn flour to corn meal, similar to the trends cohesion values (Table 3). This indicates a relationship between cohesion and wall friction: an increase in interparticle forces results in a decrease in the impact of external forces, such as friction from a wall. Due to the differences in particle sizes between wheat flour and farina was much less than corn flour and meal, the values for wheat were much closer in this test. Additional forces, such as interparticle friction or cohesion, may play a role in these values as well; a greater internal influence may negate or lessen the effect of external forces applied to a powder.

The wall friction testing for sugar powders yielded results that appear to run contrary to the corn and wheat powders (Table 3). The cohesion values had an inverse relationship with wall friction angle of corn and wheat; particle size appeared to have an inverse correlation with wall friction angle for sugar. This could be due to the differences in surface chemical composition of the sugar powders when compared to that of wheat and corn powders.

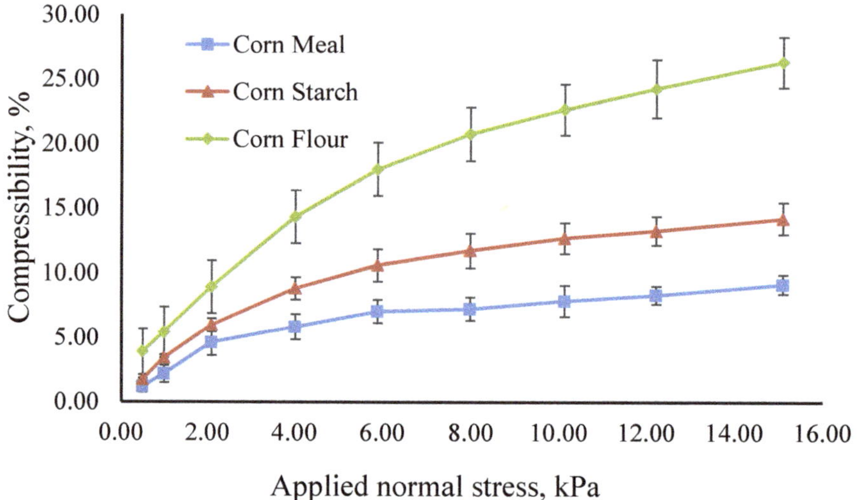

(a)

Figure 3. *Cont.*

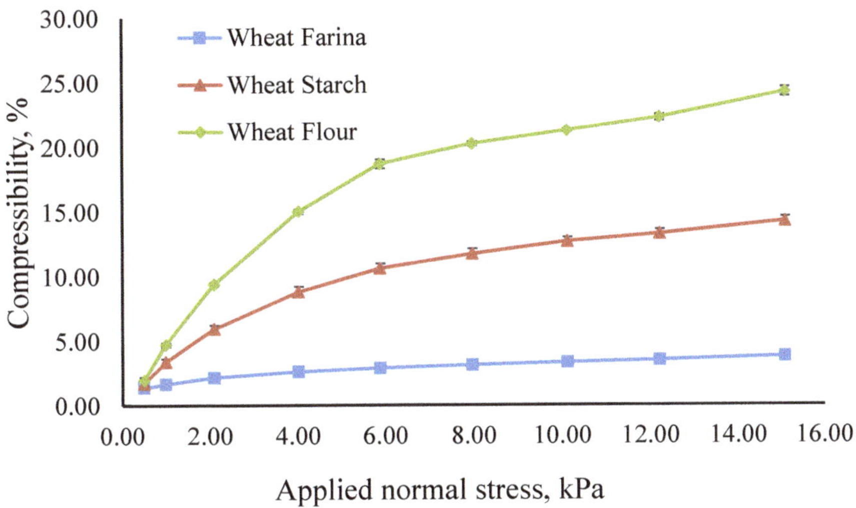

(b)

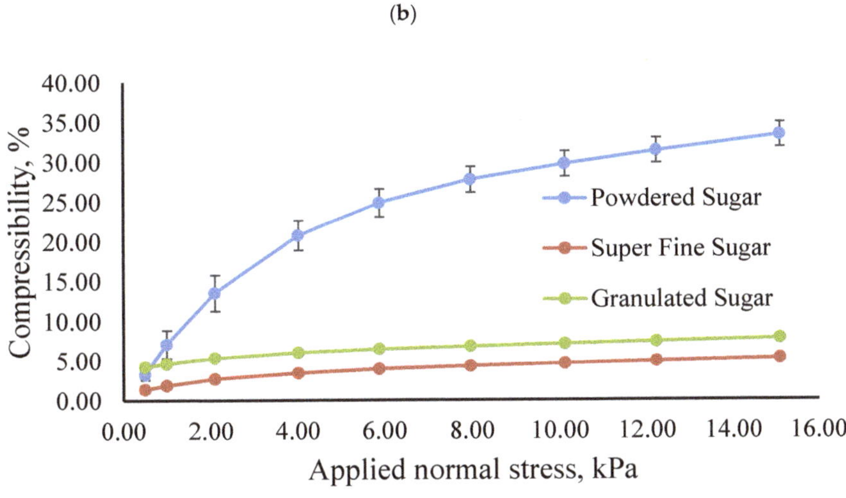

(c)

Figure 3. Effect of applied normal stress on compressibility of (**a**) corn (**b**) wheat, and (**c**) sugar powders.

4. Conclusions

For starch-based powders, energy per unit mass tends to decrease as particle size increases. The larger particles also flow more readily, which makes them ideal for systems such as extruders where flowability is an important factor that controls formulation flexibility and production rates, while granulated sugar was shown to be much higher in energy consumption in confined and unconfined flow. All these experiments were run at ambient moisture contents, which are very much different from that of the extruder moisture contents, and the effect of moisture will be studied in future research. As extruder barrels completely encase the food powders, the wall friction values determined in this study will be useful in predicting the flow of the tested food powders. The determined shear and dynamic flow property values also correspond to granular flow in an extruder. Thus, the efficiency of the extruder screw to move different formulations or mixes from the beginning of the barrel to the subsequent zones can be predicted and potentially altered based on the type of powder, to achieve maximum efficiency.

The stability and flow rate indices illustrate the impact that changing the screw speed would have on these food powders, as well as the impact of using a gravimetric feed system into the process-with powders that have an SI smaller or larger than 1 (essentially every tested powder, to varying degrees of severity), volume will change to be greater or lesser, respectively, than the initial volume, and could lead to under- or over-feeding the extruder. Cohesion results, in tandem with SBFE and FF results provide an insight to the working principle of the extruder's feeding zone, as understanding the impendence or propensity to flow of a powder can allow proactive modifications to be made to a system. Changes to recipes, increasing or reducing screw speed at start-up (or process throughout) to ensure material is sufficiently conveyed forward, or adding water to agglomerate or reduce the intensity of interparticle forces and thus aid in forward conveying are just some of the potential solutions for a more efficient extrusion process.

Author Contributions: Conceptualization, S.A.; methodology, C.M., R.P.K.A. and S.A.; formal analysis, C.M. and K.S.; investigation, C.M., S.A. and K.S.; resources, R.P.K.A. and S.A.; data curation, C.M. and K.S.; writing—original draft preparation, C.M.; writing—review and editing, K.S. and S.A.; visualization, C.M.; supervision, S.A.; project administration, S.A.; funding acquisition, S.A. All authors have read and agreed to the published version of the manuscript.

Funding: This research received no external funding.

Institutional Review Board Statement: Not applicable.

Informed Consent Statement: Not applicable.

Data Availability Statement: Data is contained within the article.

Acknowledgments: The authors would like to acknowledge Wenger Manufacturing, Inc. (Sabetha, KS) for their support to the Kansas State University Extrusion Lab and Eric Maichel for his assistance with the extrusion visualization studies.

Conflicts of Interest: The authors declare no conflict of interest.

References

1. Freeman, R.E. The Flowability of Powder-an Empirical Approach. In *Powder to Bulk/International Conference on Powder and Bulk Solids Handling International Conference on Powder and Bulk Solids Handling*; Wiley: New York, NY, USA, 2000; pp. 545–556.
2. Alavi, S.; Ambrose, R.P.K. Particulate Flow and Agglomeration in Food Extrusion. In *Production, Handling and Characterization of Particulate Materials*; Springer: Cham, Germany, 2016; pp. 257–289.
3. Yamane, K.; Nakagawa, M.; Altobelli, S.A.; Tanaka, T.; Tsuji, Y. Steady Particulate Flows in a Horizontal Rotating Cylinder. *Phys. Fluids.* **1998**, *10*, 1419–1427. [CrossRef]
4. Dury, C.M.; Ristow, G.H.; Moss, J.L.; Nakagawa, M. Boundary Effects on the Angle of Repose in Rotating Cylinders. *Phys. Rev. E.* **1998**, *57*, 4491–4497. [CrossRef]
5. Juliano, P.; Muhunthan, B.; Barbosa-Cánovas, G.V. Flow and Shear Descriptors of Preconsolidated Food Powders. *J. Food Eng.* **2006**, *72*, 157–166. [CrossRef]
6. Fitzpatrick, J.J.; Barringer, S.A.; Iqbal, T. Flow Property Measurement of Food Powders and Sensitivity of Jenike's Hopper Design Methodology to the Measured Values. *J. Food Eng.* **2004**, *61*, 399–405. [CrossRef]
7. AOAC International. *Official Methods of Analysis*, 17th ed.; AOAC International: Gaithersburg, MD, USA, 2006.
8. ASABE Standards. *S3194.4: Method of Determining and Expressing Fineness of Feed Materials by Sieving*; ASABE Standards: St. Joseph, MI, USA, 2008.
9. Jenike, A.W. *Gravity Flow of Bulk Solids*; Bulletin No. 108; The University of Utah: Salt Lake City, UT, USA, 1961.
10. Thomas, J.; Schubert, H. Particle Characterization. *Proc. Partec.* **1979**, *79*, 301–319.
11. Rogé, B.; Mathlouthi, M. Caking of Sucrose Crystals: Effect of Water Content and Crystal Size. *Zuckerindudtrie* **2000**, *125*, 336–340.

12. Siliveru, K.; Kwek, J.W.; Lau, G.M.L.; Ambrose, R.P.K. An Image Analysis Approach to Understand the Differences in Flour Particle Surface and Shape Characteristics. *Cereal Chem.* **2016**, *93*, 234–241. [CrossRef]
13. Marston, J.O.; Li, E.Q.; Thoroddsen, S.T. Evolution of Fluid-like Granular Ejecta Generated by Sphere Impact. *J. Fluid Mech.* **2012**, *704*, 5–36. [CrossRef]
14. Stasiak, M.; Molenda, M. Direct Shear Testing of Flowability of Food Powders. *Res. Agric. Eng.* **2004**, *50*, 6–10. [CrossRef]
15. Peleg, M.; Mannheim, C.H.; Passy, N. Flow Properties of Some Food Powders. *J. Food Sci.* **1973**, *38*, 959–964. [CrossRef]

Article

Changes in Hydrophobic Interactions among Gluten Proteins during Dough Formation

Sonoo Iwaki [1,2,*], Katsuyuki Hayakawa [1], Bin-Xiao Fu [3] and Chikako Otobe [2,4]

[1]   Cereal Science Research Center of Tsukuba, Nisshin Flour Milling Inc., 13 Ohkubo Tsukuba, Ibaraki 300-2611, Japan; hayakawa.katsuyuki@nisshin.com
[2]   Degree Programs in Life and Earth Science, Graduate School of Science and Technology, University of Tsukuba, 1-1-1 Tennodai Tsukuba, Ibaraki 305-8577, Japan; ochika@affrc.go.jp
[3]   Grain Research Laboratory, Canadian Grain Commission, 303 Main Street, Winnipeg, MB R3C 3G8, Canada; binxiao.fu@grainscanada.gc.ca
[4]   Institute of Crop Science, National Agriculture and Food Research Organization, 2-1-2 Kannondai Tsukuba, Ibaraki 305-8518, Japan
*   Correspondence: iwaki.sonoo@nisshin.com; Tel.: +81-298-651-177; Fax: +81-298-651-237

Abstract: In this study, changes in hydrophobic interactions among gluten proteins were analyzed during dough mixing. Size-exclusion high-performance chromatography and two-dimensional fluorescence difference gel electrophoresis were performed on proteins extracted with 1-propanol by weakening the hydrophobic interaction. The amount of proteins extracted with 30% 1-propanol increased from the start of mixing to peak consistency, suggesting that the hydrophobic interactions among the strongly aggregated proteins weakened and resulted in disaggregation. The amount of proteins extracted with 10% 1-propanol decreased during hydration, indicating that these proteins aggregated through relatively weak hydrophobic interactions. The proteins that extractability decreased were mainly low molecular weight glutenin, α-gliadin, and γ-gliadin. The amount of monomeric proteins extracted with 30% 1-propanol decreased after peak consistency. The decreased protein was mainly ω-gliadin, indicating that ω-gliadin aggregated with other proteins through hydrophobic interactions. A front-face fluorescence analysis was performed on the dough with the addition of 8-anilino-1-naphthalenesulfonic acid or thioflavin T. The fluorescence intensity increased as a result of exposure to the hydrophobic groups of the gluten proteins and the formation of protein aggregates during dough mixing. These results indicate the importance of hydrophobic interactions in dough formation.

Keywords: wheat; flour; dough; mixing; hydrophobic interaction; aggregation

Citation: Iwaki, S.; Hayakawa, K.; Fu, B.-X.; Otobe, C. Changes in Hydrophobic Interactions among Gluten Proteins during Dough Formation. *Processes* **2021**, *9*, 1244. https://doi.org/10.3390/pr9071244

Academic Editors: Yonghui Li and Shawn/Xiaorong Wu

Received: 10 June 2021
Accepted: 16 July 2021
Published: 19 July 2021

1. Introduction

Bread-making involves various processes, such as mixing, fermentation, proofing, and baking. Dough mixing is an important process that can affect the quality of the bread substantially, as viscoelastic gluten is formed during the mixing process. Many studies have been conducted to elucidate the mechanisms underlying gluten formation and structure [1–3]. Some early models emphasized the importance of intermolecular disulfide bonds [4–6]. However, since the late 1990s, most proposed gluten structures, such as the loop-train model [7] and hyper-aggregation model [8], have focused on noncovalent bonds. A recent study proposed a model in which different bonds or interactions play a role depending on the ratio of glutenin to gliadin during mixing [9]. Much debate remains in regard to the structure and formation of the gluten network [1,10–12]. This research theme has been one of the major issues that have continued until the present.

In terms of changes in protein during dough mixing, a decrease in glutenin macropolymer [13,14], a decrease in unextractable polymeric protein (UPP) by sodium dodecyl sulfate (SDS), and a decrease in the molecular weight of UPP have been reported [15].

These results have been attributed to protein depolymerization [13–15]. Additionally, a decrease [16] or increase [17,18] in free sulfhydryl groups during mixing has been reported. In addition, recent studies have reported data showing that SS binding does not continue to increase or decrease during mixing but increases in the early stage and then decreases significantly [9,19] or slightly [20] as the mixing intensity increases and then decreases. Thus, there were many reports on changes in SS bonds during dough formation. However, as no method has been established to measure noncovalent bonds directly, little remains known about changes in noncovalent bonds during mixing.

Noncovalent bonds include electrostatic interactions, hydrophobic interactions, and hydrogen bonds. In our previous investigation [21], we reported that noncovalent bonds weaken and the molecular size of polymeric proteins decrease during dough mixing, while ω-gliadin aggregates with other proteins through noncovalent bonds. However, we did not elucidate the behavior of each type of noncovalent bond, so which noncovalent bond was responsible for the aggregation of ω-gliadin during mixing remained unclear. Therefore, in this study, we focused on hydrophobic interactions, which is a kind of noncovalent bond, and studied changes in hydrophobic interactions during dough formation.

Wheat gluten proteins contain few ionizing groups but more than 35% hydrophobic amino acids [2] and, thus, promote hydrophobic interactions. The impact of salt on the properties of dough is often explained by gluten aggregation resulting from hydrophobic interactions [22]. Kinsella et al. [23] showed that, in the presence of F^- or Cl^-, the hydrophobic interaction becomes stronger, and the proteins remain aggregated, making hydration difficult. Melnyk et al. [24] demonstrated that kosmotropes, such as NaCl, reduce the water absorption of gluten and strengthen hydrophobic interactions.

Assessing the strength of hydrophobic interactions among proteins remains very challenging. Chen et al. [25] used the SDS-binding capacity [26] as an indicator for evaluating hydrophobic interactions and examined the effect of a NaCl addition but did not find a significant difference. Hydrophobic interactions have also been evaluated based on data from dough surface hydrophobicity generated by the front-face fluorescence method [27,28]. Bonomi et al. [27] pointed out that the variation in front-face fluorescence is large and discussed surface hydrophobicity without evaluating any significant differences, but Jazaeri et al. [28] discussed the data at a significance level of 10% ($p < 0.1$). In our previous report [21], we evaluated the strength of noncovalent bonds by extracting proteins with a series of SDS solutions, which weaken noncovalent bonds, at various concentrations. In this study, we analyzed hydrophobic interactions by extracting proteins in different concentrations of 1-proponol solutions, which weaken hydrophobic interactions. Furthermore, we used the front-face fluorescence method to examine the changes in the hydrophobic interactions during dough formation.

2. Materials and Methods

2.1. Wheat Flour

In the present study, we used the same two commercial flour samples (Nisshin Flour Milling Inc., Tokyo, Japan) as those used in a previous report [21]: one milled from hard wheat with a protein content of 14.5% (referred to as "HF" (high-protein flour)) and the other from semi-hard wheat with a protein content of 11.7% (referred to as "LF" (low-protein flour)) (dry basis).

2.2. Dough Sampling during Mixing

The dough was sampled as described in a previous report [21]. First, flour (200 g), distilled water (HF: 128.8 mL, LF: 117.4 mL; mean water absorption for the farinograph), and NaCl (4 g) were mixed at a constant temperature of 27 °C using a Swanson mixer (National Mfg. Co., Lincoln, NE, USA). All dough samples were mixed at 120 rpm (Association for Cereal Chemistry International standard mixing method 54-70.01) [29] for 20 min to reach overmixing conditions. A data logger (AF-1700; ATTO Co., Tokyo, Japan) was used to monitor the change in electric power during mixing. Figure 1 shows the mixing curves.

Dough samples were collected at the following time points: dough buildup (HF: 4 min, LF: 6 min), peak consistency (HF: 8 min, LF: 11 min), dough breakdown (HF: 12 min, LF: 16 min), and overmixing (20 min).

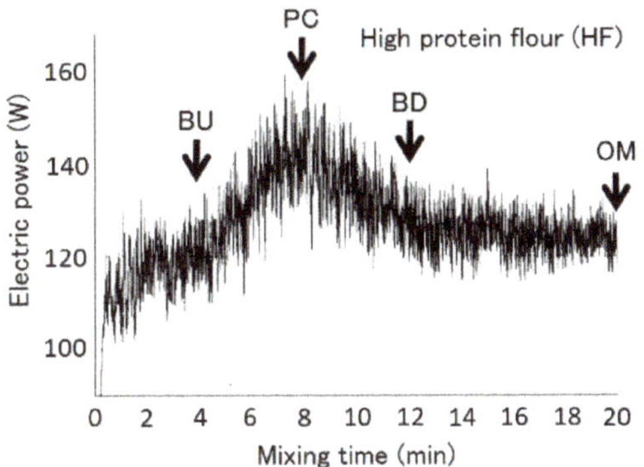

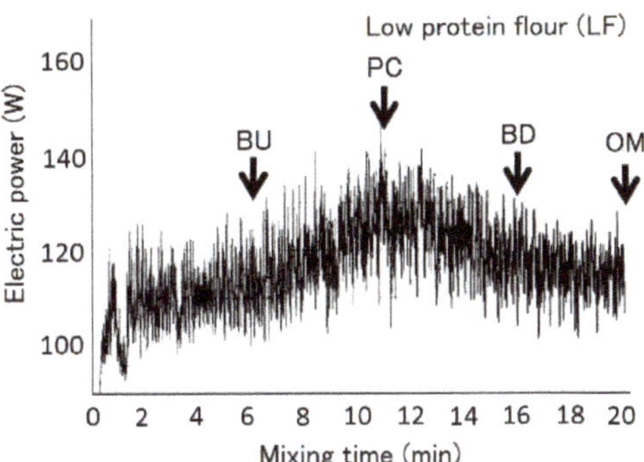

Figure 1. Dough mixing curves for high-protein flour (**left**) and low-protein flour (**right**). Arrows indicate sampling points. BU, PC, BD and OM indicate buildup, peak consistency, breakdown, and overmixing, respectively. Reprinted from Iwaki et al. 2020 [21].

2.3. Protein Extraction

Regarding the protein extraction, 30% or 10% 1-propanol was used as a solvent to weaken the relatively strong or weak hydrophobic interaction, and the proteins were extracted as described in a previous report [21]. Immediately after sampling, the dough sample (1 g) was homogenized with 20 mL of 30% or 10% 1-propanol (hereafter "extraction solvent") at 10,000 rpm for 5 min (Ace-AM10; Nihonseiki Kaisha Ltd., Tokyo, Japan). After centrifugation at $5000\times g$ for 10 min, the supernatant was collected, and extraction solvent (1 mL) was added to the residue. The suspension was centrifuged after stirring, and the supernatant was collected. The two supernatants were then combined and mixed, and an automated protein analyzer (Kjeltech 8400; FOSS, Hillerød, Denmark) was used to determine the amount of protein (N × 5.7). Each extraction procedure was repeated six times.

2.4. Analysis of Protein Size Distribution by Size-Exclusion High-Performance Liquid Chromatography (SE-HPLC)

Size-exclusion high-performance liquid chromatography (SE-HPLC) was performed in accordance with a previous report [21]. The extracted protein solutions were made up to 100 mL with extraction solvent and then diluted twice with extraction solvent and passed through a 0.45-μm filter. An SE-HPLC system composed of an HPLC system (Chromaster; Hitachi High-tech Science, Tokyo, Japan) consisting of an online degasser, gradient pump, auto injector, column oven, and ultraviolet (UV) detector was used to analyze the molecular size distribution of the proteins. A size-exclusion column (300 × 7.8 mm i.d., 3 μm, Yarra SEC-4000; Phenomenex, Torrance, CA, USA) was used for separation. Phosphate buffer (50 mM, pH 7.0) containing 0.5% SDS was used as the mobile phase. All samples (50 μL) were then eluted at 35 °C at a flow rate of 0.5 mL/min.

Open LAB software (Hitachi High-tech Science) was used to integrate the UV signal (214 nm). We then multiplied the amount of extracted protein by the area ratio in the HPLC chromatogram to calculate the amount of protein in each fraction.

2.5. Analysis of Protein Composition by SDS-Polyacrylamide Gel Electrophoresis (SDS-PAGE)

The individual protein fractions separated by SE-HPLC in four independent runs were then combined, and trichloroacetic acid was added to make a final concentration of

20%, followed by vortexing. After 30 min of immersion in ice, the sample was centrifuged at $12,000 \times g$ for 5 min to precipitate proteins. After, the supernatant was discarded, and 2 mL of cold acetone was added. Then, the mixture was vortexed and centrifuged at $12,000 \times g$ for 5 min to remove nonprotein contaminants. This operation was repeated. After discarding the supernatant, the precipitated protein was dissolved in the sample buffer (20 µL; 1% SDS, 0.1-M Tris HCl, 20% glycerol, and an appropriate amount of bromophenol blue) for SDS-PAGE. Both with and without 2-mercaptoethanol (3 µL), the samples were heated at 95 °C for 3 min, cooled at room temperature, and then analyzed by SDS-PAGE on a Mini Protean Tetra System (Bio-Rad Laboratories, Hercules, CA, USA) at 200 A for 30 min using a mini protean 7.5% TGX gel and electrophoresis buffer (25-mM Tris-HCl, 192-mM glycine, and 0.1% SDS).

The gel was removed from the plate after electrophoresis, dyed with Oriole Fluorescent Gel Stain (Bio-Rad Laboratories) for 90 min, and scanned (ImageQuant LAS 4000; Cytiva, Tokyo, Japan).

2.6. Protein Identification by Two-Dimensional Fluorescence Difference Gel Electrophoresis (2D-DIGE)

Two-dimensional fluorescence difference gel electrophoresis (2D-DIGE) was performed as described in a previous report [21]. The mixtures of protein aggregates (A) or monomeric proteins (B) in Figure 2 collected twice were concentrated to 200 µL using a 10-kDa or 3-kDa cutoff ultrafiltration column (Vivaspin 20; Sartorius Lab Instruments, Göttingen, Germany). Next, 1.8 mL of cold acetone containing 10% trichloroacetic acid was added to the concentrated protein solution, which was then stored at −20 °C for 4 h. Following centrifugation at $14,000 \times g$ for 8 min to precipitate proteins, the supernatant was discarded. Then, 1 mL of cold acetone was added and stored at −20 °C for 10 min. Following centrifugation at $14,000 \times g$ for 8 min, the supernatant was discarded, and the residue was air-dried for 5 min. Next, 15 µL of swelling solution in the Glycine system reagent set for Auto2D (Merck, Tokyo, Japan) was added, followed by vortexing for 10 min. Then, 1 µL of 200 pmol fluorescent dye (IC3-Osu; Dojindo Laboratories, Kumamoto, Japan) or 1 µL of 200 pmol fluorescent dye (IC5-Osu; Dojindo Laboratories) dissolved in N,N-dimethylformamide, and 0.5 µL of 1.5-M Tris-HCl buffer (pH 8.8) was added to 10 µL of each sample at each mixing time. As an internal standard, 2 µL of each sample at each mixing time was mixed, followed by the addition of 1 µL of 200 pmol cyDye DIGE Fluor Cy2 minimal dye (Cytiva) and 0.5 µL of 1.5-M Tris- hydrochloric acid buffer (pH 8.8). Each sample and the internal standard were stored in the dark at 4 °C for 1 h for labeling; after which, 1 µL of 10-mM lysine was added to each sample and the labeled internal standard, and the mixture was stored at 4 °C for 10 min. Then, 2 µL of internal standard labeled with Cy2, 2 µL of sample labeled with IC3, 2 µL of sample labeled with IC5, and 7 µL of swelling fluid (swelling fluid:1-M dithiothreitol:ampholyte = 113.4:6:0.6) were mixed. The prepared mixture was then loaded for 2D-DIGE using Auto-2D plus (Merck). The electrophoresis conditions were set based on the desalting mode recommended by the manufacturer. An isoelectric focusing chip (pH 3–10) and a 10% PAGE chip (Merck) were used for the Auto-2D electrophoresis. Next, the gel was removed from the plate and scanned using Amersham Typhoon (Cytiva), and then, an Ettan DIGE analysis was carried out using Melanie 9 software (Cytiva).

2.7. Front-Face Fluorescence Method

The water was replaced with 400-µM 8-anilinonaphthalene sulfonic acid (8-ANS) and 50-µM thioflavin T to analyze the dough surface hydrophobicity and protein aggregation, respectively. The dough used for the front-face fluorescence analysis was prepared as described in Section 2.2. HF and LF samples were collected from the start of the mixing after 2, 4, 6, 8, 10, 12, 16, and 20 min. Immediately after sampling, the dough was covered with plastic wrap, and an optical fiber from the fluorescence spectrometer (F-7000; Hitachi High-tech Science) was pushed into the samples to measure the fluorescence intensity. Next, 8-ANS and thioflavin T fluorescence were monitored by the emission spectra from

350 to 600 nm, with excitation at 384 nm and 435 nm, respectively, and the height of the peaks in the emission spectra was assumed to correspond to the fluorescence intensity, which was then compared between samples.

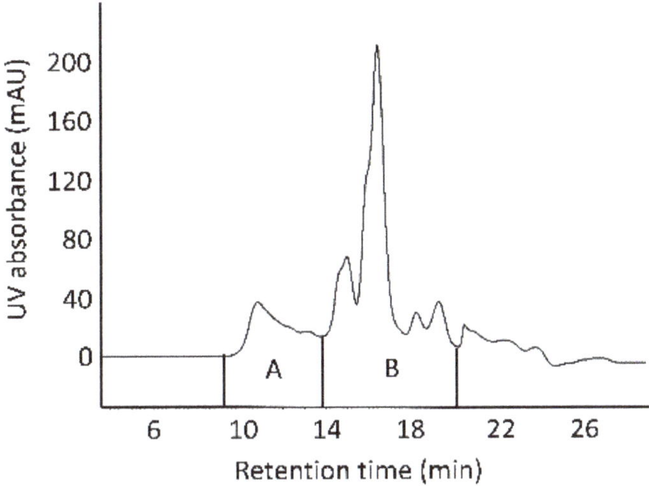

Figure 2. Size-exclusion high-performance liquid chromatography (SE-HPLC) chromatogram of the protein extracted with 30% 1-propanol (pH 7.0) from high-protein flour (HF). "A" represents protein aggregates and "B" monomeric proteins.

2.8. Statistics

Tukey's method was applied to determine significant differences in the SE-HPLC ($p < 0.05$, $n = 6$) and front-face fluorescence analyses ($p < 0.1$, $n = 6$) using JMP software (SAS Institute Inc., Cary, NC, USA), and analysis of variance was applied to determine significant differences ($p < 0.05$, $n = 4$) in the 2D-DIGE analysis using Melanie 9 software (Cytiva).

3. Results and Discussion

3.1. Definition of Protein Fractions

Figure 2 shows an SE-HPLC chromatogram of the protein extracted with 30% 1-propanol, with two fractions (A and B) collected for analysis.

Each of the fractions was concentrated, and SDS-PAGE was performed. The results are shown in Figure 3. Under nonreducing conditions (A and B), some of proteins in fraction A were stacked in the end part of the gel, and others appeared to be smeared in the high molecular region, because they existed as "polymers" and/or "aggregates", whereas the proteins in fraction B existed only as "monomers" because no stacked or smeared proteins were observed. Fraction B did not show any high molecular glutenin subunits under the reducing condition, which indicates that this fraction did not contain any polymeric proteins. Fraction A was defined and discussed as "polymeric proteins" in previous reports [30,31]. However, a small amount of ω-gliadins appeared to be present in this polymeric protein fraction. Since ω-gliadins do not have cysteines and cannot form disulfide bonds, they are most likely bonded to the polymeric glutenin proteins through strong, noncovalent bonds. Fraction A consisted largely of polymeric proteins of glutenin subunits linked through disulfide bonds, with some protein aggregates of monomeric proteins linked by noncovalent bonds. As this report focuses on hydrophobic interactions, fraction A is defined as "protein aggregates".

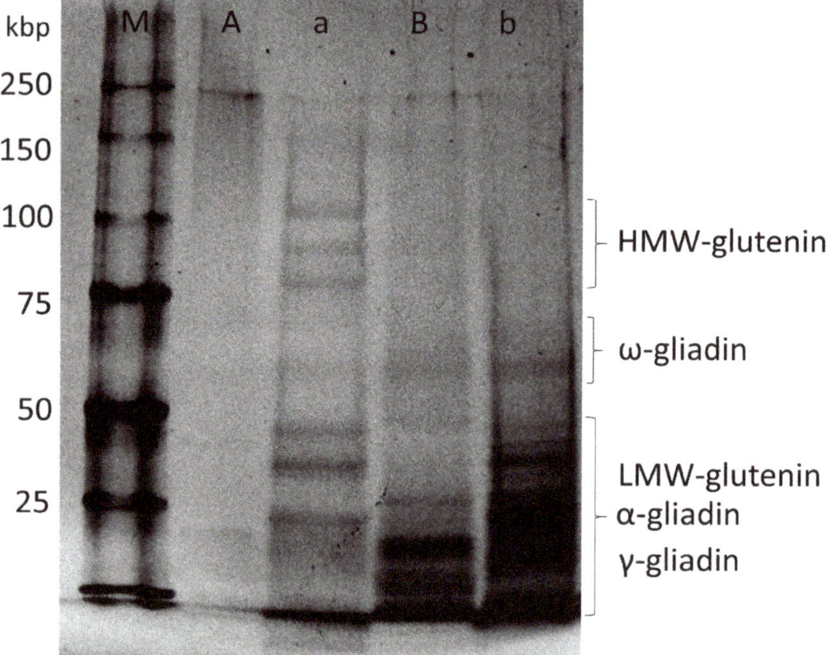

Figure 3. Electrophoretic images of SDS-PAGE for fractions A and B under nonreducing (A, B) and reducing (a, b) conditions. Fractions A and B are shown in Figure 2. The M lane is a size marker (from the top: 250, 150, 100, 75, 50, and 25 kbp).

3.2. Changes in Hydrophobic Interactions during Mixing

The changes in the amounts of monomeric proteins and protein aggregates—respectively, "A" and "B", as defined in Figure 2—extracted with 30% 1-propanol are shown in Figure 4. The solvent 30% 1-propanol weakens relatively strong hydrophobic interactions. In Figure 4, the protein aggregates and monomeric proteins extracted with 30% propanol are shown in orange and blue, respectively. The amount of proteins extracted with 30% 1-propanol was subtracted from the total amount of proteins and is shown in gray as unextractable proteins. The hydrophobic interactions in the unextractable proteins are so strong that they cannot be extracted with 30% 1-propanol. The monomeric proteins and protein aggregates increased and unextractable proteins decreased until reaching the peak consistency in HF and LF. These results indicate that the hydrophobic interactions were weakened up to the peak consistency, so the unextractable proteins disaggregated and became extractable. After reaching the peak consistency, the monomeric proteins decreased, protein aggregates increased, and the amount of unextractable proteins did not change in HF or LF. Thus, the aggregation of the monomeric proteins due to hydrophobic interactions was stronger than the disaggregation after reaching the peak consistency.

The changes in the amount of monomeric proteins and protein aggregates extracted with 10% 1-propanol are shown in Figure 5. The solvent 10% 1-propanol weakens the relatively weak hydrophobic interactions. The monomeric proteins and protein aggregates extracted with 10% 1-propanol decreased significantly when water was added (from flour to buildup) in HF, whereas only the monomeric proteins decreased in LF. These data suggested that monomeric proteins aggregated through relatively weak hydrophobic interactions during hydration and could not be extracted with 10% 1-propanol.

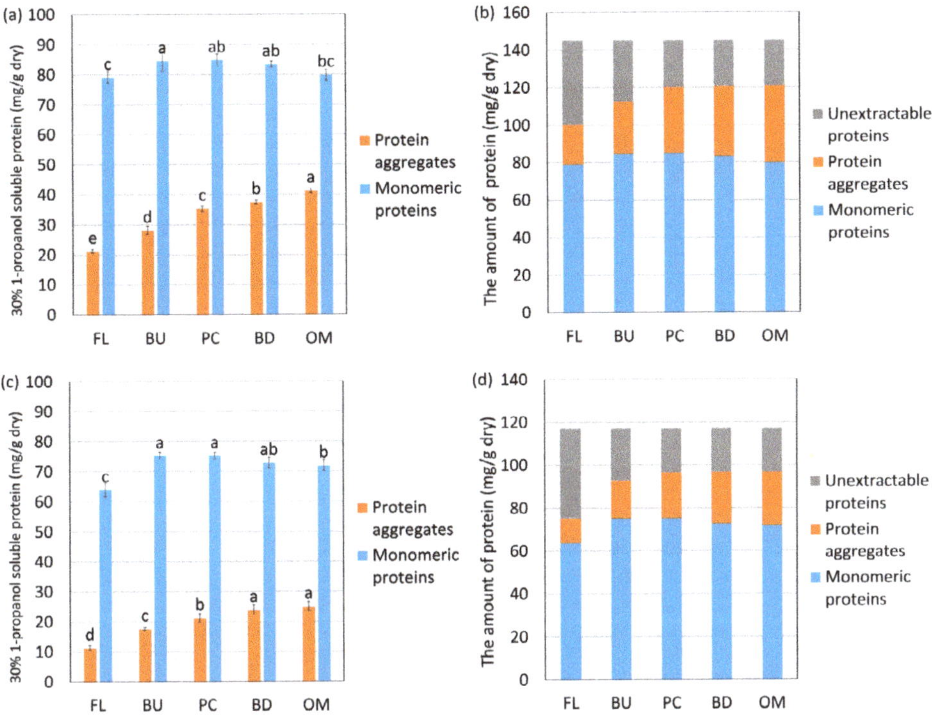

Figure 4. Changes in the amount of protein aggregates, monomeric proteins extracted with 30% 1-propanol, and unextractable proteins during mixing from high-protein flour (HF) (**a,b**) and low-protein flour (LF) (**c,d**). FL, BU, PC, BD and OM indicate flour (before mixing), buildup, peak consistency, breakdown, and overmixing, respectively. Protein aggregates mean "A", and monomeric proteins mean "B" in Figure 2. Unextractable proteins mean proteins not extracted with 30% 1-propanol. The error bars show standard deviations ($n = 6$). Different letters on the graph indicate significant differences ($p < 0.05$).

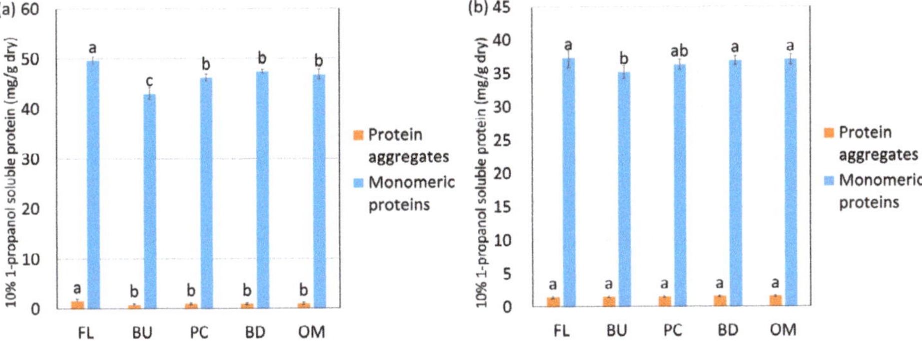

Figure 5. Changes in the amount of protein aggregates and monomeric proteins extracted with 10% 1-propanol during mixing from high-protein flour (HF) (**a**) and low-protein flour (LF) (**b**). FL, BU, PC, BD and OM indicate flour (before mixing), buildup, peak consistency, breakdown, and overmixing, respectively. Protein aggregates mean "A", and monomeric proteins mean "B" in Figure 2. The error bars show standard deviations ($n = 6$). Different letters on the graph indicate significant differences ($p < 0.05$).

Gluten proteins in dough are aggregated through noncovalent bonds and are hardly soluble in water but soluble in organic solvents [32–34]. The reason why proteins are solubilized in organic solvents is that the hydrophobic groups of the organic solvent weaken the hydrophobic interactions of the proteins. High concentrations of organic solvents are highly hydrophobic and, therefore, weaken almost all hydrophobic interactions of proteins, including relatively strong ones, whereas low concentrations of organic solvents weaken only relatively weak ones. It is possible to evaluate the hydrophobic interactions of different strengths by quantifying the proteins extracted in each concentration of organic solvent.

A solvent with high hydrophobicity should be effective in weakening strong hydrophobic interactions; however, it is difficult to solubilize highly hydrophobic solvents, such as hexane, in an aqueous environment. This is because highly hydrophobic solvents repel the water in the dough, making it difficult for the solvent to enter. Therefore, lower alcohols with an appropriate degree of hydrophilicity are advantageous for the evaluation of hydrophobic interactions. Bean et al. [35] reported that the amount of protein extracted with 30–50% 1-propanol was the highest among various concentrations of lower alcohols. In the present study, we defined hydrophobic interactions that could be weakened by 10% propanol as "relatively weak" and those that could be weakened by 30% but not 10% 1-propanol as "relatively strong".

3.3. Identification of Monomeric Proteins Incorporated in Protein Aggregates during Mixing

Figure 6 shows electrophoretic images of monomeric proteins and protein aggregates extracted with 30% 1-propanol. The blue circles indicate decreased protein, and the red circles indicate increased protein after the dough buildup stage ($n = 4$, $p < 0.05$). Each protein was identified based on Dupont et al. [36]. The results are shown in Tables 1 and 2 (HF) and in Tables 3 and 4 (LF).

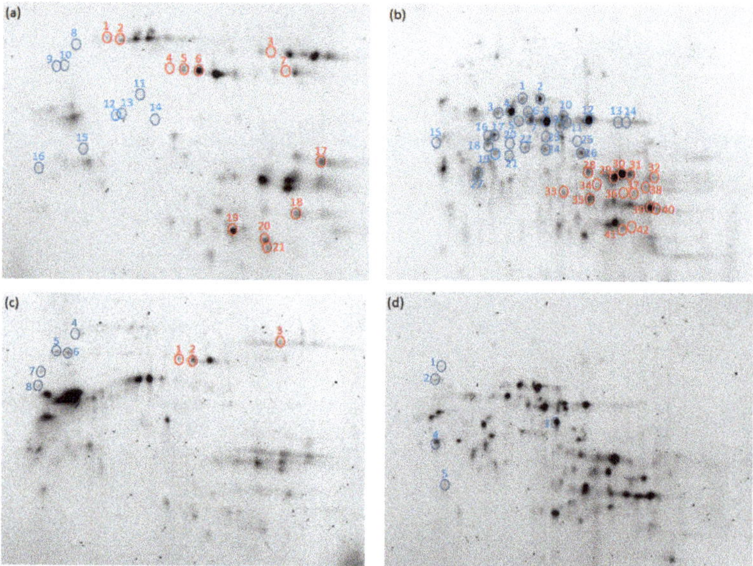

Figure 6. Electrophoretic images of the protein aggregates (**a,c**) and monomeric proteins (**b,d**) in high-protein flour (HF) (**a,b**) and low-protein flour (LF) (**c,d**) extracted with 30% 1-propanol. Blue circles show proteins with decreased intensities after buildup ($n = 4$, $p < 0.05$), and red circles show proteins with increased intensities after buildup ($n = 4$, $p < 0.05$). The numbers correspond to those in Tables 1–4.

Table 1. The increased or decreased protein in protein aggregates extracted with 30% 1-propanol in HF [a] during mixing.

Spot No. [b]	Identification [c]	Spot Ratio (%) [d]					Increase (↑) or Decrease (↓)
		Flour	Build Up	Peak Top	Break Down	Overmixing	
1	HMW-glutenin	0.66 ± 0.29	0.83 ± 0.15	0.99 ± 0.14	1.03 ± 0.15	1.09 ± 0.07	↑
2	HMW-glutenin	0.96 ± 0.23	0.93 ± 0.19	1.48 ± 0.49	1.47 ± 0.16	1.39 ± 0.32	↑
3	HMW-glutenin	0.65 ± 0.22	0.8 ± 0.22	1.1 ± 0.47	1.14 ± 0.15	1.15 ± 0.22	↑
4	HMW-glutenin	0.58 ± 0.21	0.72 ± 0.14	0.81 ± 0.28	1.13 ± 0.14	1.14 ± 0.21	↑
5	HMW-glutenin	0.55 ± 0.14	0.83 ± 0.22	1.35 ± 0.33	1.48 ± 0.07	1.61 ± 0.32	↑
6	HMW-glutenin	0.68 ± 0.18	0.75 ± 0.2	1.35 ± 0.41	1.51 ± 0.52	1.44 ± 0.71	↑
7	HMW-glutenin	0.7 ± 0.25	0.92 ± 0.15	1.01 ± 0.46	1.22 ± 0.18	1.27 ± 0.13	↑
8	Enzymes	1.38 ± 0.89	0.75 ± 0.39	0.26 ± 0.12	0.4 ± 0.1	0.41 ± 0.13	↓
9	Enzymes	2.1 ± 1.01	0.66 ± 0.34	0.29 ± 0.1	0.35 ± 0.16	0.28 ± 0.15	↓
10	Enzymes	2.01 ± 1.21	0.71 ± 0.35	0.26 ± 0.08	0.4 ± 0.2	0.28 ± 0.1	↓
11	ω-gliadin	0.66 ± 0.19	1.81 ± 9.69	0.36 ± 0.18	0.31 ± 0.09	0.38 ± 0.16	↓
12	ω-gliadin	1.77 ± 0.39	1.35 ± 0.19	1.25 ± 0.25	1.07 ± 0.24	0.99 ± 0.09	↓
13	ω-gliadin	0.96 ± 0.41	1.25 ± 0.28	0.87 ± 0.34	0.53 ± 0.12	0.51 ± 0.2	↓
14	ω-gliadin	1.11 ± 0.53	1.43 ± 0.16	1.17 ± 0.35	0.77 ± 0.16	0.71 ± 0.26	↓
15	ω-gliadin	1.53 ± 0.21	1.29 ± 0.25	1.13 ± 0.17	1.18 ± 0.17	1.02 ± 0.2	↓
16	ω-gliadin	1.68 ± 0.86	1.12 ± 1.28	0.58 ± 0.19	0.59 ± 0.31	0.46 ± 0.18	↓
17	LMW-glutenin	1.41 ± 0.15	1.19 ± 0.17	1.84 ± 0.44	1.52 ± 0.08	1.52 ± 0.23	↑
18	γ-gliadin	1.47 ± 0.31	1.16 ± 0.04	1.96 ± 0.58	1.61 ± 0.27	1.45 ± 0.22	↑
19	α-gliadin	1.24 ± 0.08	1.13 ± 0.35	1.73 ± 0.34	1.67 ± 0.47	1.68 ± 0.32	↑
20	α-gliadin	1.61 ± 0.5	1.21 ± 0.33	2.15 ± 0.26	2.01 ± 0.7	1.99 ± 0.35	↑
21	α-gliadin	1.88 ± 0.43	1.21 ± 0.3	2.28 ± 0.43	1.86 ± 0.55	1.99 ± 0.4	↑

[a] Spots were detected by 2D-DIGE; significant differences ($p < 0.05$, $n = 4$) among any of the groups were determined by an ANOVA test. [b] These spot numbers correspond to the spot numbers in Figure 6a. [c] The identification of each protein was estimated from the report of Dupont et al. [36]. [d] Values are the means ± standard deviation.

Table 2. The increased or decreased protein in monomeric protein extracted with 30% 1-propanol in HF [a] during mixing.

Spot No. [b]	Identification [c]	Spot Ratio (%) [d]					Increase (↑) or Decrease (↓)
		Flour	Build Up	Peak Top	Break Down	Overmixing	
1	ω-gliadin	0.93 ± 0.05	1.97 ± 0.51	1.05 ± 0.18	0.88 ± 0.29	0.88 ± 0.17	↓
2	ω-gliadin	0.96 ± 0.1	2.02 ± 0.44	1.2 ± 0.2	0.97 ± 0.17	0.91 ± 0.11	↓
3	ω-gliadin	0.88 ± 0.11	1.79 ± 0.51	1.01 ± 0.14	0.86 ± 0.24	0.83 ± 0.14	↓
4	ω-gliadin	0.95 ± 0.15	2 ± 0.66	1.21 ± 0.25	0.94 ± 0.18	1.03 ± 0.2	↓
5	ω-gliadin	0.94 ± 0.31	2.01 ± 0.76	1.24 ± 0.22	0.89 ± 0.22	1.05 ± 0.21	↓
6	ω-gliadin	0.91 ± 0.17	1.78 ± 0.6	1.21 ± 0.25	0.88 ± 0.33	1.02 ± 0.19	↓
7	ω-gliadin	0.99 ± 0.13	1.89 ± 0.55	1.08 ± 0.2	0.82 ± 0.21	0.9 ± 0.1	↓
8	ω-gliadin	0.91 ± 0.1	1.8 ± 0.49	1.08 ± 0.18	0.97 ± 0.21	0.96 ± 0.14	↓
9	ω-gliadin	0.87 ± 0.02	2.26 ± 0.54	1.16 ± 0.19	0.92 ± 0.2	1 ± 0.16	↓
10	ω-gliadin	0.99 ± 0.22	1.93 ± 0.82	1.05 ± 0.19	0.86 ± 0.26	0.85 ± 0.16	↓
11	ω-gliadin	0.87 ± 0.16	1.65 ± 0.56	0.93 ± 0.07	0.81 ± 0.23	0.72 ± 0.14	↓
12	ω-gliadin	0.92 ± 0.08	1.68 ± 0.59	0.98 ± 0.11	0.86 ± 0.14	0.9 ± 0.12	↓
13	ω-gliadin	0.81 ± 0.17	1.67 ± 0.71	0.96 ± 0.06	0.75 ± 0.25	0.72 ± 0.13	↓
14	ω-gliadin	0.91 ± 0.13	1.77 ± 0.58	1.12 ± 0.19	0.85 ± 0.18	0.94 ± 0.15	↓
15	ω-gliadin	0.86 ± 0.13	1.82 ± 0.76	1.05 ± 0.05	0.89 ± 0.23	0.94 ± 0.19	↓
16	ω-gliadin	0.9 ± 0.26	2.3 ± 0.77	1.39 ± 0.1	0.96 ± 0.31	1 ± 0.32	↓
17	ω-gliadin	1.15 ± 0.24	2.39 ± 0.76	1.07 ± 0.22	1.06 ± 0.21	0.96 ± 0.14	↓
18	ω-gliadin	0.85 ± 0.08	1.79 ± 0.65	1 ± 0.14	0.87 ± 0.17	0.77 ± 0.17	↓
19	ω-gliadin	0.9 ± 0.18	1.72 ± 0.57	1.1 ± 0.1	0.92 ± 0.19	0.9 ± 0.17	↓
20	ω-gliadin	0.73 ± 0.16	1.68 ± 0.75	0.95 ± 0.17	0.77 ± 0.15	0.87 ± 0.08	↓
21	ω-gliadin	0.85 ± 0.12	1.75 ± 0.6	1.02 ± 0.15	0.96 ± 0.22	0.93 ± 0.17	↓
22	ω-gliadin	0.86 ± 0.24	2.14 ± 0.61	1.12 ± 0.16	0.99 ± 0.25	0.9 ± 0.2	↓
23	ω-gliadin	0.87 ± 0.12	1.77 ± 0.67	1.06 ± 0.13	0.95 ± 0.08	0.86 ± 0.1	↓
24	ω-gliadin	0.89 ± 0.09	1.98 ± 0.62	1.18 ± 0.36	0.96 ± 0.14	1.13 ± 0.26	↓
25	ω-gliadin	0.92 ± 0.15	1.61 ± 0.47	0.99 ± 0.08	0.76 ± 0.32	0.8 ± 0.12	↓
26	ω-gliadin	0.91 ± 0.16	1.78 ± 0.49	1.18 ± 0.14	0.88 ± 0.08	0.97 ± 0.15	↓
27	ω-gliadin	1.25 ± 0.3	1.31 ± 1.76	0.51 ± 0.18	0.57 ± 0.33	0.2 ± 0.15	↓
28	LMW-glutenin	0.93 ± 0.33	0.33 ± 0.34	1.1 ± 0.11	0.93 ± 0.18	0.86 ± 0.13	↑
29	LMW-glutenin	1.25 ± 0.31	0.4 ± 0.44	1.54 ± 0.46	1.31 ± 0.15	1.33 ± 0.13	↑
30	LMW-glutenin	1.19 ± 0.32	0.4 ± 0.53	1.49 ± 0.24	1.43 ± 0.36	1.33 ± 0.23	↑
31	LMW-glutenin	1.06 ± 0.29	0.41 ± 0.62	1.66 ± 0.34	1.43 ± 0.47	1.39 ± 0.25	↑
32	LMW-glutenin	1.06 ± 0.29	0.44 ± 0.35	1.36 ± 0.17	1.23 ± 0.15	1.24 ± 0.29	↑
33	α-gliadin	1.1 ± 0.26	0.25 ± 0.39	1 ± 0.66	1.15 ± 0.37	1.22 ± 0.2	↑

Table 2. *Cont.*

Spot No. [b]	Identification [c]	Spot Ratio (%) [d]					Increase (↑) or Decrease (↓)
34	γ-gliadin	1.07 ± 0.2	0.49 ± 0.62	1.67 ± 0.43	1.26 ± 0.2	1.44 ± 0.26	↑
35	α-gliadin	0.94 ± 0.5	0.24 ± 0.43	1.03 ± 0.66	1.32 ± 0.45	1.28 ± 0.21	↑
36	γ-gliadin	1.33 ± 0.71	0.32 ± 0.27	1.41 ± 0.35	1.42 ± 0.37	1.58 ± 0.3	↑
37	LMW-glutenin	0.96 ± 0.61	0.34 ± 0.43	1.56 ± 0.56	1.66 ± 0.56	1.73 ± 0.4	↑
38	LMW-glutenin	1.2 ± 0.35	0.51 ± 0.23	1.23 ± 0.37	1.17 ± 0.21	1.06 ± 0.4	↑
39	γ-gliadin	1.15 ± 0.54	0.3 ± 0.43	1.07 ± 0.65	1.53 ± 0.46	1.28 ± 0.26	↑
40	γ-gliadin	1.23 ± 0.41	0.48 ± 0.4	1.01 ± 0.35	1.43 ± 0.37	1.47 ± 0.15	↑
41	α-gliadin	1.28 ± 0.41	0.57 ± 0.21	1 ± 0.13	0.99 ± 0.2	0.96 ± 0.06	↑
42	α-gliadin	1.26 ± 0.23	0.71 ± 0.18	1.09 ± 0.21	1.2 ± 0.31	1.32 ± 0.24	↑

[a] Spots were detected by 2D-DIGE; significant differences ($p < 0.05$, $n = 4$) among any of the groups were determined by an ANOVA test. [b] These spot numbers correspond to the spot numbers in Figure 6b. [c] The identification of each protein was estimated from the report of Dupont et al. [36]. [d] Values are the means ± standard deviation.

Table 3. The increased or decreased protein in protein aggregates extracted with 30% 1-propanol in LF [a] during mixing.

Spot No. [b]	Identification [c]	Spot Ratio (%) [d]					Increase (↑) or Decrease (↓)
		Flour	Build Up	Peak Top	Break Down	Overmixing	
1	HMW-glutenin	0.62 ± 0.28	0.71 ± 0.33	0.86 ± 0.13	0.96 ± 0.16	1.61 ± 0.21	↑
2	HMW-glutenin	0.81 ± 0.36	0.88 ± 0.43	1.28 ± 0.49	1.44 ± 0.35	2.33 ± 0.42	↑
3	HMW-glutenin	0.66 ± 0.14	0.75 ± 0.24	0.91 ± 0.11	1.05 ± 0.43	1.22 ± 0.19	↑
4	Enzymes	2.65 ± 2.85	0.54 ± 0.26	0.76 ± 0.19	0.9 ± 0.26	0.49 ± 0.65	↓
5	Enzymes	2.21 ± 0.67	0.79 ± 0.3	0.85 ± 0.14	0.92 ± 0.51	0.42 ± 0.37	↓
6	Enzymes	3.09 ± 4.13	0.79 ± 0.11	0.78 ± 0.3	0.97 ± 0.73	0.48 ± 0.71	↓
7	ω-gliadin	1.13 ± 0.67	0.51 ± 0.25	0.87 ± 0.25	0.61 ± 0.29	0.29 ± 0.23	↓
8	ω-gliadin	1.2 ± 0.24	0.83 ± 0.35	0.73 ± 0.34	1.4 ± 0.79	0.35 ± 0.2	↓

[a] Spots were detected by 2D-DIGE; significant differences ($p < 0.05$, $n = 4$) among any of the groups were determined by an ANOVA test. [b] These spot numbers correspond to the spot numbers in Figure 6c. [c] The identification of each protein was estimated from the report of Dupont et al. [36]. [d] Values are the means ± standard deviation.

Table 4. The increased or decreased protein in monomeric protein extracted with 30% 1-propanol in LF [a] during mixing.

Spot No. [b]	Identification [c]	Spot Ratio (%) [d]					Increase (↑) or Decrease (↓)
		Flour	Build Up	Peak Top	Break Down	Overmixing	
1	Enzymes	1.51 ± 0.61	0.81 ± 0.31	0.48 ± 0.29	0.61 ± 0.32	0.44 ± 0.34	↓
2	Enzymes	2.35 ± 0.61	1.06 ± 0.85	0.86 ± 0.12	0.95 ± 0.18	0.75 ± 0.34	↓
3	ω-gliadin	1.88 ± 0.87	2.61 ± 1.64	1.9 ± 0.25	1.7 ± 0.67	0.8 ± 0.44	↓
4	ω-gliadin	3.73 ± 2.13	2 ± 1.33	1.27 ± 0.36	1.11 ± 0.29	0.76 ± 0.52	↓
5	Serpins	1.88 ± 0.87	2.61 ± 1.64	1.9 ± 0.25	1.7 ± 0.67	0.8 ± 0.44	↓

[a] Spots were detected by 2D-DIGE; significant differences ($p < 0.05$, $n = 4$) among any of the groups were determined by an ANOVA test. [b] These spot numbers correspond to the spot numbers in Figure 6d. [c] The identification of each protein was estimated from the report of Dupont et al. [36]. [d] Values are the means ± standard deviation.

Omega-gliadin in the monomeric proteins in HF increased when hydrated (from flour to buildup) and decreased after buildup (Figure 6b and Table 2). Omega-gliadin in the protein aggregates in HF decreased during mixing (Figure 6a and Table 1). These results suggest that the hydrophobic interactions weakened and ω-gliadin disaggregated during hydration; after which, the hydrophobic interactions between ω-gliadin and the other proteins became stronger and aggregated, and then, ω-gliadin in the protein aggregates aggregated further and insolubilized. Omega-gliadin in the monomeric proteins (Figure 6d and Table 4) and protein aggregates (Figure 6c and Table 3) in LF decreased during mixing, suggesting that ω-gliadin aggregates with other proteins.

HMW-glutenin, LMW-glutenin, α-gliadin, and γ-gliadin in the protein aggregates in HF (Figure 6a and Table 1) and HMW-glutenin in the protein aggregates in LF (Figure 6c and Table 3) all increased during mixing. These data indicate that the hydrophobic interactions of unextractable proteins weakened during mixing. LMW-glutenin, α-gliadin, and γ-gliadin in the monomeric proteins in HF (Figure 6b and Table 2) increased after buildup, indicating that these proteins disaggregate to release monomeric proteins.

Figure 7 shows an electrophoretic image of the monomeric proteins and protein aggregates extracted with 10% 1-propanol. The blue circles indicate spots whose intensities decreased during hydration (from flour to buildup), and the red circles show spots whose intensities increased during hydration. Each protein was identified based on Dupont et al. [36], and these are shown in Table 5 (HF) and Table 6 (LF).

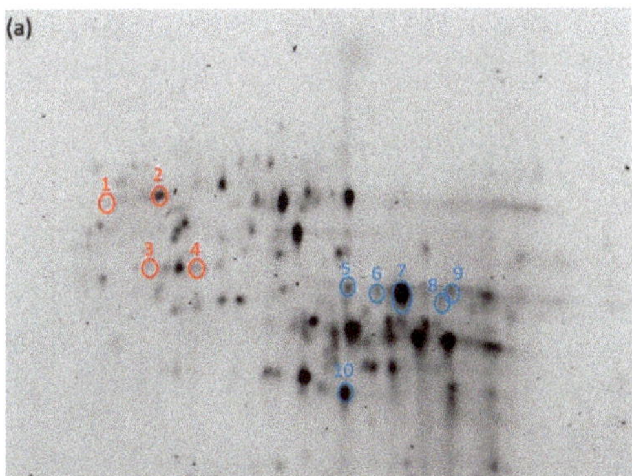
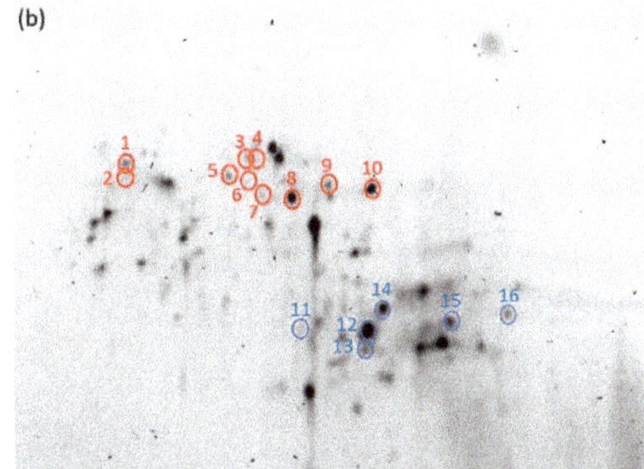

Figure 7. Electrophoretic images of monomeric proteins (**a,b**) in high-protein (HF) and low-protein flour (LF) extracted with 10% 1-propanol. Blue circles show proteins whose intensities decreased from flour to buildup ($n = 4$, $p < 0.05$). Red circles show proteins whose intensities increased from flour to buildup ($n = 4$, $p < 0.05$). The numbers correspond to those in Tables 5 and 6.

Table 5. The increased or decreased proteins in the monomeric protein extracted with 10% 1-propanol in HF [a] during mixing.

Spot No. [b]	Identification [c]	Spot Ratio (%) [d]		Increase (↑) or Decrease (↓)
		Flour	**Build Up**	
1	ω-gliadin	0.64 ± 0.23	1.4 ± 0.69	↑
2	ω-gliadin	1.07 ± 0.44	2.1 ± 0.34	↑
3	ω-gliadin	0.65 ± 0.12	1.55 ± 0.44	↑
4	ω-gliadin	0.73 ± 0.27	1.38 ± 0.39	↑
5	LMW-glutenin	1.31 ± 0.24	0.68 ± 0.19	↓
6	LMW-glutenin	1.43 ± 0.52	0.73 ± 0.17	↓
7	LMW-glutenin	2.11 ± 0.53	0.84 ± 0.21	↓
8	LMW-glutenin	1.16 ± 0.2	0.86 ± 0.13	↓
9	LMW-glutenin	1.19 ± 0.27	0.74 ± 0.09	↓
10	α-gliadin	1.55 ± 0.36	0.9 ± 0.28	↓

[a] Spots were detected by 2D-DIGE; significant differences ($p < 0.05$, $n = 4$) among any of the groups were determined by an ANOVA test. [b] These spot numbers correspond to the spot numbers in Figure 7a. [c] The identification of each protein was estimated from the report of Dupont et al. [36]. [d] Values are the means ± standard deviation.

LMW-glutenin and α-gliadin in HF (Figure 7a and Table 5) and LMW-glutenin, α-gliadin, and γ-gliadin in LF (Figure 7b and Table 6) decreased during hydration. These data indicate that these proteins aggregate through relatively weak hydrophobic interactions. It was also found that LMW-glutenin, α-gliadin, and γ-gliadin aggregated during hydration (Figure 5). On the other hand, ω-gliadin in the monomeric protein fraction increased during hydration (Figure 7 and Tables 5 and 6). This is likely due to the disaggregation of ω-gliadin extracted with 30% 1-propanol in HF during hydration (Figure 6 and Table 5).

While gliadins are thought to act as a plasticizer or diluent [37], they are also thought to play a role in extensibility and viscosity [38]. Gliadins are also considered to be involved in dough formation and swelling during baking, but their detailed role remains unclear [37]. Considering the fact that ω-gliadin is incorporated into protein aggregates through hydrophobic interactions, they may act as an adhesive between glutenin and other proteins, thereby contributing to the viscosity of the dough. On the other hand, it is interesting that α-gliadin and γ-gliadin disaggregate during mixing. Thus, the distance between α-gliadin or γ-gliadin and the other molecules may determine the extensibility of the dough.

Table 6. The increased or decreased proteins in the monomeric protein extracted with 10% 1-propanol in LF [a] during mixing.

Spot No. [b]	Identification [c]	Spot Ratio (%) [d]		Increase (↑) or Decrease (↓)
		Flour	Build Up	
1	ω-gliadin	0.35 ± 0.17	1.61 ± 0.26	↑
2	ω-gliadin	0.61 ± 0.12	1.37 ± 0.38	↑
3	ω-gliadin	1.17 ± 0.24	1.87 ± 0.28	↑
4	ω-gliadin	1.19 ± 0.05	1.6 ± 0.37	↑
5	ω-gliadin	1 ± 0.24	2.1 ± 0.43	↑
6	ω-gliadin	0.83 ± 0.28	1.35 ± 0.24	↑
7	ω-gliadin	0.59 ± 0.05	0.77 ± 0.11	↑
8	ω-gliadin	1.53 ± 0.43	2.41 ± 0.27	↑
9	ω-gliadin	1.45 ± 0.3	2.14 ± 0.18	↑
10	ω-gliadin	1.47 ± 0.35	2.08 ± 0.28	↑
11	α-gliadin	0.88 ± 0.2	0.51 ± 0.13	↓
12	α-gliadin	2.46 ± 0.9	0.85 ± 0.64	↓
13	α-gliadin	1.53 ± 0.65	0.45 ± 0.56	↓
14	γ-gliadin	2.08 ± 0.48	1.44 ± 0.31	↓
15	LMW-glutenin	2.8 ± 1.09	0.9 ± 0.99	↓
16	LMW-glutenin	2.09 ± 0.53	0.96 ± 0.65	↓

[a] Spots were detected by 2D-DIGE; significant differences ($p < 0.05$, $n = 4$) among any of the groups were determined by an ANOVA test. [b] These spot numbers correspond to the spot numbers in Figure 7b. [c] The identification of each protein was estimated from the report of Dupont et al. [36]. [d] Values are the means $\pm$ standard deviation.

3.4. Changes in the Hydrophobicity of the Dough Surface and Aggregation during Mixing

The results of a front-face fluorescence analysis using 8-ANS as a fluorescent reagent are shown in Figure 8a (HF) and Figure 8b (LF). The changes in fluorescence intensity indicate the changes in the hydrophobicity [27,28,39] of the dough surface. The hydrophobicity increased with the mixing time to the peak consistency in HF, which indicates that the structure of the proteins was loosened, and the hydrophobic groups were exposed. The fluorescence intensity in HF tended to decrease slightly after reaching the peak consistency. It is considered that the hydrophobic groups were bonded to each other and that the fluorescence intensity slightly decreased. This result complements that shown in Figure 4, which demonstrates the hydrophobic interactions of ω-gliadin. No significant changes were seen in the hydrophobicity of the dough surface in LF.

The results of a front-face-fluorescence analysis using thioflavin T as a fluorescent reagent are shown in Figure 8c (HF) and Figure 8d (LF). The changes in fluorescence intensity indicate the changes in aggregates [39,40] present on the dough surface. The aggregates increased slightly with mixing and then decreased. It was considered that there was a time lag between the exposure of the hydrophobic groups and the aggregation through the hydrophobic interactions. No significant changes were seen in the aggregation of the dough surface in LF.

Bonomi et al. [27] and Jazaeri et al. [28] reported data on the changes in the surface hydrophobicity, but their results were not conclusive, depending on the type of sample used. Bonomi et al. [27] measured the changes in fluorescence due to a tryptophan (a type of hydrophobic amino acid) residue and surface hydrophobicity during 15 min of farinograph mixing using commercial wheat flour and semolina flour as samples without evaluating the presence of significant differences. In semolina flour, the fluorescence intensity by tryptophan residues increased to the peak consistency and then became constant, suggesting that tryptophan exposure was associated with network formation and the completion of changes in dough consistency. However, they reported finding no changes in the wheat flour. Jazaeri et al. [28] measured the surface hydrophobicity by farinograph mixing for 10 min using hard and soft flours as samples. They claimed that the hard flour behaved similarly to the semolina flour of Bonomi et al., although it was not significantly changed ($p > 0.1$). The results of the present study showed that the hydrophobic groups were significantly exposed during mixing, even with the hard wheat flour.

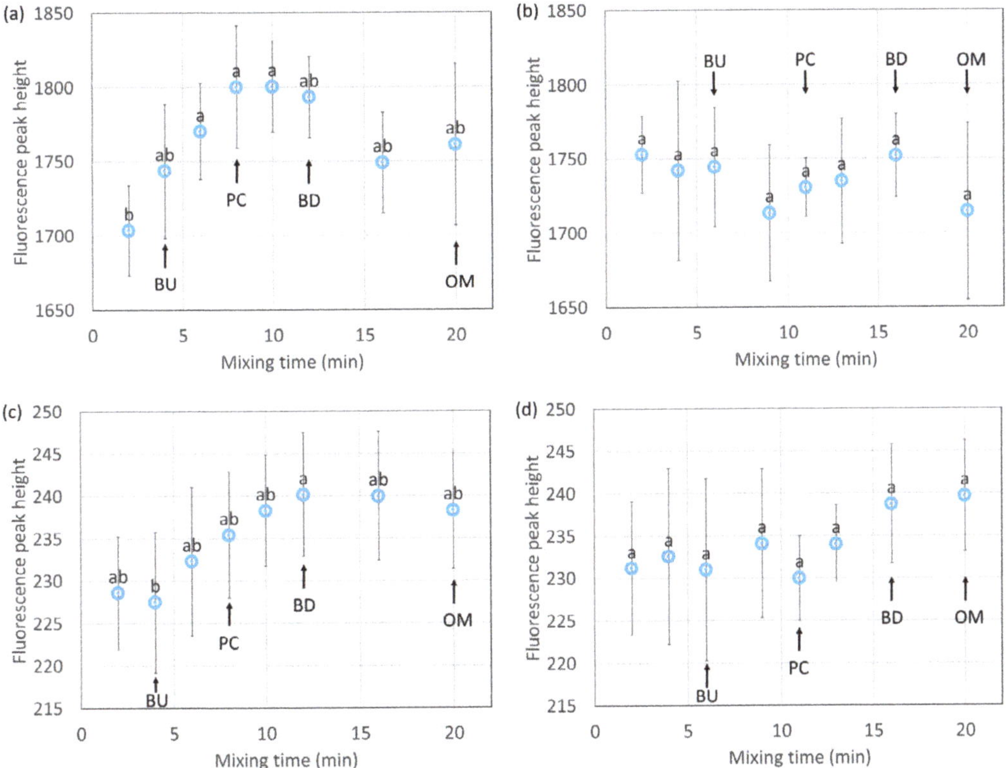

Figure 8. Changes in the fluorescence intensity of the dough with the addition of 8-anilino-1-naphthalene-sulfonic acid (**a,b**) or thioflavin T (**c,d**) during mixing from high-protein flour (HF) (**a,c**) and low-protein flour (LF) (**b,d**). BU, PC, BD and OM indicate buildup, peak consistency, breakdown, and overmixing, respectively. The error bars show standard deviations ($n = 6$). Different letters on the graph indicate significant differences ($p < 0.1$).

The changes in the hydrophobic interactions and aggregation were very small in LF during mixing. The extraction rate of the proteins with 30% 1-propanol was lower in LF than in HF, and the proteins in LF aggregated through relatively strong hydrophobic interactions. The time to reach the peak consistency was longer in LF than in HF, indicating that its relatively strong aggregation made hydration difficult. In LF, hydrophobic changes may be less likely to occur because of the relatively strong aggregation of the protein. In our previous report [21], LF also showed a small change in molecular size.

Based on the results of this report, it is considered that the hydrophobic interactions of proteins change during dough formation as follows. First, the hydrophobic interactions of proteins that originally aggregate through relatively strong hydrophobic interactions weaken up to the peak consistency, and the protein aggregates disaggregate to monomers (mainly ω-gliadin). At the same time, the hydrophobic group is exposed, and the surface hydrophobicity increases. Some monomeric proteins (e.g., α-gliadin, ω-gliadin, and LMW-glutenin) aggregate through relatively weak hydrophobic interactions during the addition of water. After the peak consistency, the exposed hydrophobic proteins interact with each other and the ω-gliadin aggregates. The relatively strong hydrophobic interactions weaken while the proteins aggregate, so the hydrophobic interactions are expected to converge at a certain level. Khan and Bushuk [41] proposed a structure in which glutenin polymerizes with disulfide bonds while glutenin polymers bond together with hydrogen bonds and hydrophobic interactions. Weegels et al. [42] proposed a model in which hydrophobic gliadin aggregates at the C- and N-terminals of glutenin through hydrophobic interactions during mixing; the glutenins line up side-by-side, and then, the glutenins bond to other glutenin with disulfide bonds; after which, the disulfide bonds strengthen the protein structure. Since we did not analyze the behavior of the disulfide or hydrogen bonds in this study, we could not verify either model. In a recent study, Liu et al. proposed a model in which proteins aggregate by hydrophobic interactions until optimum mixing and then are destroyed [9]. This behavior is different from our data. This may be due to the different experimental conditions in each report, such as the samples, extraction methods, and the method of analysis. We could analyze the changes in hydrophobic interactions during dough formation in detail, thereby showing the importance of hydrophobic interactions.

4. Conclusions

This study showed changes in the hydrophobic interactions among proteins during dough formation. The hydrophobic interactions among proteins weaken, and a part of the protein aggregates disaggregate to monomers (mainly ω-gliadin) up to the peak consistency. After the peak consistency, the exposed hydrophobic groups in the protein interact with each other, and ω-gliadin reaggregates.

In a previous study [21], it was reported that the molecular size of proteins decreases, because noncovalent bonds weaken during dough formation, while ω-gliadin aggregates with other proteins. In the present study, we confirmed the same behavior in the hydrophobic interactions as in the noncovalent interactions. This suggests that the effect of noncovalent bonds during dough formation is mainly due to hydrophobic interactions.

These findings suggest that, to gain a better understanding of the mechanism underlying gluten formation, it is important to analyze disulfide and other noncovalent bonds to obtain a comprehensive picture of dough formation.

Author Contributions: K.H. and S.I. designed the research experiments. S.I. performed the experiments. S.I. and K.H. analyzed the data. S.I. wrote the paper. S.I., K.H., B.-X.F., and C.O. reviewed and edited the paper. The manuscript was critically revised and approved by all the authors. All authors have read and agreed to the published version of the manuscript.

Funding: This research received no external funding.

Institutional Review Board Statement: Not applicable.

Informed Consent Statement: Not applicable.

Data Availability Statement: All data are presented in this article in the form of figures and tables.

Conflicts of Interest: The authors declare no conflict of interest.

References

1. Delcour, J.A.; Joye, I.J.; Pareyt, B.; Wilderjans, E.; Brijs, K.; Lagrain, B. Wheat Gluten Functionality as a Quality Determinant in Cereal-Based Food Products. *Annu. Rev. Food Sci. Technol.* **2012**, *3*, 469–492. [CrossRef]
2. Bock, J.; Seetharaman, K. Unfolding gluten. *Cereal Foods World* **2012**, *57*, 209–214. [CrossRef]
3. MacRitchie, F. Theories of glutenin/dough systems. *J. Cereal Sci.* **2014**, *60*, 4–6. [CrossRef]
4. Graveland, A.; Bosveld, P.; Lichtendonk, W.; Marseille, J.; Moonen, J.; Scheepstra, A. A model for the molecular structure of the glutenins from wheat flour. *J. Cereal Sci.* **1985**, *3*, 1–16. [CrossRef]

5. Kasarda, D.D. Glutenin structure in relation to wheat quality. In *Wheat is Unique*; Pomeranz, Y., Ed.; Am. Assoc. Cereal Chem.: St. Paul, MN, USA, 1989; pp. 277–302.

6. Wrigley, C.W. Giant proteins with flour power. *Nat. Cell Biol.* **1996**, *381*, 738–739. [CrossRef] [PubMed]

7. Belton, P. Mini Review: On the Elasticity of Wheat Gluten. *J. Cereal Sci.* **1999**, *29*, 103–107. [CrossRef]

8. Hamer, R.J.; van Vliet, T. Understanding the structure and properties of gluten: An overview. In *Wheat Gluten*; Shewry, P.R., Tatham, A.S., Eds.; Royal Society of Chemistry: Cambridge, UK, 2000; pp. 125–131.

9. Li, M.; Yue, Q.; Liu, C.; Zheng, X.; Hong, J.; Wang, N.; Bian, K. Interaction between gliadin/glutenin and starch granules in dough during mixing. *LWT* **2021**, *148*, 111624. [CrossRef]

10. MacRitchie, F. Letters to the editor. *J. Cereal Sci.* **2007**, *60*, 96–97. [CrossRef]

11. Belton, P. Letters to the editor. *J. Cereal Sci.* **2007**, *60*, 97–98. [CrossRef]

12. Van Vliet, T.; Hamer, R.J. Letters to the editor. *J. Cereal Sci.* **2007**, *60*, 98–99. [CrossRef]

13. Skerritt, J.H.; Hac, L.; Bekes, F. Depolymerization of the glutenin macropolymer during dogh mixing: I. Changes in levels, molecular weight distribution, and overall composition. *Cereal Chem.* **1999**, *76*, 395–401. [CrossRef]

14. Skerritt, J.H.; Hac, L.; Lindsay, M.P.; Bekes, F. Depolymerization of the glutenin macropolymer during dogh mixing: II. Differences in relation of specific glutenin subunits. *Cereal Chem.* **1999**, *76*, 402–409. [CrossRef]

15. Aussenac, T.; Carceller, J.L.; Kleiber, D. Changes in SDS solubility of gluten polymers during dough mixing and resting. *Cereal Chem.* **2001**, *78*, 39–45. [CrossRef]

16. Okada, K.; Negishi, Y.; Nagao, S. Factors affecting dough breakdown during overmixing. *Cereal Chem.* **1987**, *64*, 428–434.

17. Lee, L.; Ng, P.K.W.; Steffe, J.F. Biochemical Studies of Proteins in Nondeveloped, Partially Developed, and Developed Doughs. *Cereal Chem. J.* **2002**, *79*, 654–661. [CrossRef]

18. Morel, M.-H.; Redl, A.; Guilbert, S. Mechanism of Heat and Shear Mediated Aggregation of Wheat Gluten Protein upon Mixing. *Biomacromolecules* **2002**, *3*, 488–497. [CrossRef]

19. Lancelot, E.; Fontaine, J.; Grua-Priol, J.; Assaf, A.; Thouand, G.; Le-Bail, A. Study of structural changes of gluten proteins during bread dough mixing by Raman spectroscopy. *Food Chem.* **2021**, *358*, 129916. [CrossRef] [PubMed]

20. Feng, Y.; Zhang, H.; Wang, J.; Chen, H. Dynamic changesin glutenin macropolymer during different dough mixing and resting prosses. *Molecules* **2021**, *26*, 541. [CrossRef]

21. Iwaki, S.; Aono, S.; Hayakawa, K.; Fu, B.X.; Otobe, C. Changes in Protein Non-Covalent Bonds and Aggregate Size during Dough Formation. *Foods* **2020**, *9*, 1643. [CrossRef]

22. Miller, R.A.; Hoseney, R.C. Role of salt in baking. *Cereal Foods World* **2008**, *53*, 4–6. [CrossRef]

23. Kinsella, J.E.; Hale, M.L. Hydrophobic associations and gluten consistency: Effects of specific anions. *J. Agric. Food. Chem.* **1984**, *32*, 1054–1056. [CrossRef]

24. Melnyk, J.P.; Dreisoerner, J.; Bonomi, F.; Marcone, M.F.; Seetharaman, K. Effect of the Hofmeister series on gluten aggregation measured using a high shear-based technique. *Food Res. Int.* **2011**, *44*, 893–896. [CrossRef]

25. Chen, G.; Ehmke, L.; Sharma, C.; Miller, R.; Faa, P.; Smith, G.; Li, Y. Physicochemical properties and gluten structures of hard wheat flour doughs as affected by salt. *Food Chem.* **2019**, *275*, 569–576. [CrossRef] [PubMed]

26. Kato, A.; Matsuda, T.; Matsudomi, N.; Kobayashi, K. Determination of protein hydrophobicity using sodium dodecyl sul-fate binding method. *J. Agric. Food. Chem.* **1984**, *32*, 284–288. [CrossRef]

27. Bonomi, F.; Mora, G.; Pagani, M.A.; Iametti, S. Probing structural features of water-insoluble proteins by front-face fluorescence. *Anal. Biochem.* **2004**, *329*, 104–111. [CrossRef] [PubMed]

28. Jazaeri, S.; Bock, J.E.; Bagagli, M.P.; Iametti, S.; Bonomi, F.; Seetharaman, K. Structural Modifications of Gluten Proteins in Strong and Weak Wheat Dough During Mixing. *Cereal Chem. J.* **2015**, *92*, 105–113. [CrossRef]

29. Ogilvie, O.; Roberts, S.; Sutton, K.; Gerrard, J.; Larsen, N.; Domigan, L. The effect of dough mixing speed and work input on the structure, digestibility and celiac immunogenicity of the gluten macropolymer within bread. *Food Chem.* **2021**, *359*, 129841. [CrossRef]

30. Kuktaite, R.; Larsson, H.; Johansson, E. Variation in protein composition of wheat flour and its relationship to dough mixing behavior. *J. Cereal Sci.* **2004**, *40*, 31–39. [CrossRef]

31. La Gatta, B.; Rutigliano, M.; Rusco, G.; Petrella, G.; Di Luccia, A. Evidence for different supramolecular arrangements in pasta from durum wheat (Triticum durum) and einkorn (Triticum monococcum) flours. *J. Cereal Sci.* **2017**, *73*, 76–83. [CrossRef]

32. Fu, B.X.; Sapirstein, H.D. Procedure for isolating monomeric proteins and polymeric glutenin of wheat flour. *Cereal Chem.* **1996**, *73*, 143–152.

33. Sapirstein, H.D.; Fu, B.X.; Sapirstein, H.D.; Fu, B.X. Intercultivar Variation in the Quantity of Monomeric Proteins, Soluble and Insoluble Glutenin, and Residue Protein in Wheat Flour and Relationships to Breadmaking Quality. *Cereal Chem. J.* **1998**, *75*, 500–507. [CrossRef]

34. Cinco-Moroyoqui, F.J.; MacRitchie, F. Quantitation of LMW-GS to HMW-GS Ratio in Wheat Flours. *Cereal Chem. J.* **2008**, *85*, 824–829. [CrossRef]

35. Bean, S.R.; Lyne, R.K.; Tilley, K.A.; Chung, O.K.; Lookhart, G.L. A Rapid Method for Quantitation of Insoluble Polymeric Proteins in Flour. *Cereal Chem. J.* **1998**, *75*, 374–379. [CrossRef]

36. Dupont, F.M.; Vensel, W.H.; Tanaka, C.K.; Hurkman, W.J.; Altenbach, S.B. Deciphering the complexities of the wheat flour proteome using quantitative two-dimensional electrophoresis, three proteases and tandem mass spectrometry. *Proteome Sci.* **2011**, *9*, 1–29. [CrossRef] [PubMed]

37. Branlard, G.P.; Metakovsky, E.V. Some Gli alleles related to common wheat dough quality. In *Gliadin and Glutenin—The Unique Balance of Wheat Quality*; Wrigley, C.W., Bekes, F., Bushuk, W., Eds.; Am. Assoc. Cereal Chem.: St. Paul, MN, USA, 2006; pp. 115–137.

38. Mills, E.N.C.; Burgess, S.R.; Tatham, A.S.; Shewry, P.R.; Chan, H.W.S.; Morgan, M.R.A. Characterization of a panel of mono-clonal anti-gliadin antibodies. *J. Cereal Sci.* **1990**, *11*, 89–101. [CrossRef]

39. Huschka, B.; Bonomi, F.; Marengo, M.; Miriani, M.; Seetharaman, K. Comparison of lipid effects on structural features of hard and soft wheat flour proteins assessed by front-face fluorescence. *Food Chem.* **2012**, *133*, 1011–1016. [CrossRef]

40. Rasmussen, P.; Barbiroli, A.; Bonomi, F.; Faoro, F.; Ferranti, P.; Iriti, M.; Picariello, G.; Iametti, S. Formation of structured polymers upon controlled denaturation of β-Lactoglobilin with different chaotropes. *Biopolymers* **2007**, *86*, 57–72. [CrossRef] [PubMed]

41. Khan, K.; Bushuk, W. Studies of glutenin. XII. Comparison by sodium dodecyl sulfate-polyacrylamide gel electrophoresis of unreduced and reduced glutenin from various isolation and purification procedures. *Cereal Chem.* **1979**, *56*, 63–68.

42. Weegels, P.L.; Marseille, J.P.; Jager, A.M.; Hamer, R.J. Structure-function relationships of gluten proteins. In *Gluten Proteins*; Bushuk, W., Tkachuk, R., Eds.; AACC International Press: St. Paul, MN, USA, 1990; pp. 98–111.

processes

Article

Effect of Pulse Type and Substitution Level on Dough Rheology and Bread Quality of Whole Wheat-Based Composite Flours

Yiqin Zhang [1], Ruijia Hu [1], Michael Tilley [2], Kaliramesh Siliveru [1] and Yonghui Li [1,*]

[1] Department of Grain Science and Industry, Kansas State University, Manhattan, KS 66506, USA; cicy0202@ksu.edu (Y.Z.); ruijia@ksu.edu (R.H.); kaliramesh@ksu.edu (K.S.)

[2] Center for Grain and Animal Health Research, Agricultural Research Service, USDA, 1515 College Ave, Manhattan, KS 66502, USA; michael.tilley@usda.gov

* Correspondence: yonghui@ksu.edu; Tel.: +1-785-532-4061

Abstract: Pulse flours are commonly added to food products to improve the functional properties, nutritional profiles, product quality and health benefits. This study aimed at assessing the effects of the partial replacement (0–25%) of whole wheat flour with diversified whole pulse flours (yellow pea, green pea, red lentil, and chickpea) on dough properties and bread quality. The pulse flours had higher protein contents and ash, but lower moisture content and larger average particle size, compared to whole wheat flour. Increasing the substitution level of pulse flours decreased dough viscosity, stability, development time and bread volume, and accelerated bread retrogradation. The incorporation of 5% yellow pea flour led to a similar bread quality as that with only whole wheat flour. Among all the tested pulse flours, the composite flour containing yellow pea flour or chickpea flour had overall better potential for bread making by providing good dough handling properties and product quality. This study will benefit the development of more nutritious food products by combining cereal and pulse ingredients.

Keywords: whole grain bread; pulse; yellow pea; green pea; lentil; chickpea; Mixolab; dough rheology; bread texture

Citation: Zhang, Y.; Hu, R.; Tilley, M.; Siliveru, K.; Li, Y. Effect of Pulse Type and Substitution Level on Dough Rheology and Bread Quality of Whole Wheat-Based Composite Flours. *Processes* **2021**, *9*, 1687. https://doi.org/10.3390/pr9091687

Academic Editor: Péter Sipos

Received: 24 August 2021
Accepted: 16 September 2021
Published: 21 September 2021

Publisher's Note: MDPI stays neutral with regard to jurisdictional claims in published maps and institutional affiliations.

1. Introduction

Pulses, such as peas, lentils, chickpeas, and dry beans, are widely consumed as a staple food in many countries, due to their high nutritional values [1]. Pulses contain a high amount of dietary fiber, proteins, vitamins, minerals, and phytochemical antioxidants (e.g., phenolic acids, flavonoids, and isoflavones), which are beneficial to human health [2–4]. Pulses generally contain about 15 to 30% of protein with a high level of lysine, which is a limiting amino acid in cereals [2,5]. The phytonutrients in pulse, such as tannins, flavonoids, and phenolic acids, have high potential for antioxidant, anti-inflammatory, and antimicrobial properties [5]. For example, a study indicated that lentil can provide 0.167 mg thiamin, 0.072 mg riboflavin, 1.049 mg niacin, 0.632 mg pantothenic acid, and 0.176 mg pyridoxine per $\frac{1}{2}$ cup of dry seed [5]. Fully cooked pulses can function as low-glycemic foods that inhibit appetite and glycemia in the short term [5]. Consuming pulses may reduce the risk of cardiovascular disease, gastrointestinal disorders, obesity, type-2 diabetes, and cancer as well as lower cholesterol levels [4,6–9].

Nowadays, consumers are becoming more informed and aware of health and wellness needs [10]. The tendency of including pulses to improve the nutritional value of foods has become more popular. Whole grain foods are considered healthier, as they contain all the original nutrients present in bran, germ, and endosperm. Whole wheat flour contains better nutritional profiles and more health benefits than refined wheat flour, especially because it is rich in vitamins, minerals, fibers, antioxidants, and phytochemicals [11]. The consumption of whole wheat foods can reduce the risk of cardiovascular disease, obesity, cancer, and diabetes, as well as maintain body weight [12]. However, baking with whole wheat flour always leads to bread with a smaller size, bitter taste, and coarser and harder texture, compared to white bread, which is less appealing to consumers. Moreover, wheat contains relatively low concentrations of protein (8–15%), and it is an incomplete protein source, due to the lower amount of the essential amino acid, lysine. Pulses, on the other hand, generally have a protein content of approximately 15–30%, and are rich in lysine (about 64 ± 10 mg/g of protein). Incorporating pulse flours into wheat breads can increase the lysine content and amino acid score [13].

Marchini et al. [14] found that 10% flour substitution with red lentil flour provided the best baking properties, and larger particle size fraction (>200 μm) generated better properties than the finer fractions. Pulse flour blends showed higher water absorption than common wheat flour, due to the high amount of polysaccharides and protein content [15]. A study on white flour/pea flour blends showed that the bread properties (such as specific volume, crumb texture, and density) were positively related to dough rheological properties, and the bread specific volume decreased as the amount of pea flour was increased [16]. Additionally, several studies demonstrated that incorporating pulse flour significantly improved nutritive values of wheat-based bakery products [1,17]. Compared with the continuous network and unique viscoelasticity of the wheat dough protein matrix, the protein matrix of pulse flour was less desirable for bread making. Pulses may also have some negative effects on food products, such as having a strong beany flavor and intense aroma and introducing anti-nutritional compounds. A study showed that the higher fiber content in wheat flour and chickpea flour resulted in lower wheat bread volume, due to the interaction between its hydroxyl groups and water through hydrogen bonding, and posed a negative influence on dough stability [15]. Including pulse flours into bread dough dilutes the gluten protein and affects both gluten development and starch–protein complexes, which are important to the dough rheology and quality of bread [16,18]. Pea flour also interrupted the starch–gluten matrix, resulting in weaker and less elastic dough [16]. Different pretreatments of pulses, such as germination, extrusion, and fermentation, were used to improve the quality of foods containing pulses [19].

So far, there is limited literature on incorporating whole pulse flours into whole wheat bread products. We hypothesized that different pulse flours would have different techno-functions and influences on dough and bread properties. This study aimed to determine the effects of different types and amount of whole pulse flours (e.g., yellow pea, green pea, red lentil, and chickpea) on whole wheat dough properties and bread quality. This research will benefit grain scientists in developing more nutritious and palatable whole grain food products.

2. Materials and Methods

2.1. Materials

Whole wheat flour (protein content 15.1%, moisture content 12.9%) was supplied by Mennel Milling Company (Fostoria, OH, USA). Commercial whole yellow pea flour was obtained from Harvest Innovations (Indianola, IA, USA). Dried whole yellow pea and whole green pea grains were purchased from Rani (Houston, TX, USA). Whole red lentil grain was provided by Food to Live (Brooklyn, NY, USA). Whole chickpea grain was from Palouse Brand (Palouse, WA, USA). Active dry instant yeast, sucrose, salt, and shortening were purchased from a local grocery store.

2.2. Flour Characterization

Pulse grains were ground with a Wiley Mill (Thomas Scientific, Swedesboro, NJ, USA) equipped with a 0.1 mm sieve. The protein, moisture, and ash of the flours were measured following AACC approved methods [20–22]. The lipid content was measured according to AACC approved method 30-10.01 with some modification [23]. Briefly, flour and ethyl ether were mixed at a ratio of 1:20 (w:v) in a conical flask for 30 min, then centrifuged at $3000 \times g$ for 10 min, and the extract was collected. The extraction process was repeated twice. The extract was then pooled and left in a fume hood overnight to evaporate the solvent, and then the weight of the lipids was measured. Total carbohydrate was calculated based on the percentage of ash, total fat, moisture, and protein content, using Equation (1):

$$Carbohydrates\ (\%) = 100 - Ash - Total\ Fat - Moisture - Protein \tag{1}$$

The particle size of the flours was measured, using a Ro-Tap sieve shaker (Model B, W.S. Tyler Mentor, OH, USA) with sieve sizes of 53 to 3360 µm. Approximately 100 g of the sample was weighed, transferred to the top of the shaker, and then shaken for 10 min. The weight of the flour on each sieve and the pan was measured. The average particle size was calculated by Equation (2):

$$Avg\ particle\ size\ (\mu m) = \Sigma \left[\frac{Wi}{Wt} \times di \right] \tag{2}$$

where Wi is the weight of flour remains in each sieve; Wt is the weight of total flour; di is the diameter of the ith sieve in the stack.

2.3. Mixolab Analysis of Whole Wheat/Pulse Composite Flours

Dough mixing and pasting properties of blends of whole wheat flour with different types (commercial yellow pea flour, yellow pea flour, green pea flour, red lentil flour, and chickpea flour) and amounts (5, 15, and 25%, based on total composite flour weight) of pulse flours were analyzed, using Mixolab (Chopin Technologies, Villeneuve-la-Garenne, France), according to AACC Method 54-60.01 [24]. The substitution level of 5–25% was decided based on our preliminary studies so as to maximize the pulse additions but maintain a processible dough system. Whole wheat flour was also measured for comparison. The flours were pre-mixed to achieve a homogenous composite system in an external mixer before further testing.

2.4. Bread Baking of Whole Wheat/Pulse Blends

Breads from different whole wheat/pulse flour blends were prepared following AACC Method 10-10.03 [25]. The formulation included 100 g flour, 2 g instant dry yeast, 3 g shortening, 6 g sucrose, 1.5 g salt, 0.2 g malt flour, and an optimal amount of water. The amount of water and dough mixing time were determined based on the Mixograph analysis. The bread making was conducted in duplicate for each formulation.

2.5. Bread Quality Analysis

Bread volume and specific volume were determined following AACC Method 10-05.01 [26]. The breads were then cut into slices, approximately 15 mm thick, and stored in foil bags with oxygen absorbers for further analysis. The bread crust color was measured, using a CIE-LAB color system (XITIAN machine equipment Co., Ltd., Huizhou, China) to obtain the Hunter L*, a*, and b* values. The bread crumb structure was determined, using a C-Cell Bread Imaging System (Calibre Control International Ltd., Warrington, U.K.). The central slices from each loaf were used for C-cell testing, and the number of cells, cell wall thickness, and cell diameter were collected.

The texture profiles of the bread crumb were measured, using a TA-XT Plus Texture Analyzer (Stable Micro Systems, Surrey, U.K.) following a previous method [27]. Three slices from each treatment were measured on days 1, 4, and 7, respectively. The first texture profile analysis (TPA) was conducted after 24 h of storage (i.e., day 1).

The moisture content of the bread was determined, following AACC Method 44-15.02. Approximately 2 g crumb was cut from the loaf, accurately weighed, transferred to an aluminum pan, and dried in an air-oven at 135 °C for 3 hr and then weighed again. In order to investigate the effect of different types and amounts of pulse flour on the water retention of the bread, the moisture content was also measured on days 1, 4, and 7, respectively.

2.6. Statistical Analysis

All the experiments and tests were conducted in at least duplicate. Data were analyzed using SAS statistical software, Version 9.4 (SAS Institute, Cary, NC, USA) following Tukey's multiple comparisons test and analysis of variance (ANOVA). The significant level was considered at $p < 0.05$.

3. Results and Discussion

3.1. Flour Properties

The moisture content of the whole wheat flour and pulse flours is summarized in Table 1. The whole wheat flour contained 12.92% moisture, which was much higher than the pulse flours. Among all the pulse flours, green pea flour (8.04%) and chickpea flour (7.95%) had a significantly lower moisture content. The flour moisture content is affected by the post-harvest and environmental condition of the grains and milling process.

Table 1. Proximate composition (as-is) and average particle size of whole wheat and pulse flours.

Flour	Moisture, %	Protein, %	Ash, %	Lipid, %	Total Carb., %	Avg. Particle Size, μm
Whole wheat flour	12.92 ± 0.03 [a]	15.13 ± 0.62 [b]	1.80 ± 0.02 [e]	1.80 ± 0.02 [b]	68.36 ± 0.60 [a]	160.77 ± 1.15 [c]
Commercial yellow pea	10.45 ± 0.10 [b]	24.22 ± 0.31 [a]	2.18 ± 0.01 [c,d]	1.21 ± 0.06 [c]	61.96 ± 0.49 [c]	131.62 ± 12.05 [c]
Yellow pea	9.47 ± 0.04 [c]	19.82 ± 0.45 [a,b]	2.32 ± 0.04 [b,c]	0.84 ± 0.11 [d]	67.56 ± 0.33 [a,b]	242.55 ± 6.26 [b,c]
Green pea	8.04 ± 0.03 [d]	21.78 ± 0.04 [a]	2.38 ± 0.01 [b]	1.18 ± 0.04 [c]	66.62 ± 0.10 [a,b,c]	331.59 ± 15.05 [a,b]
Red lentil	9.80 ± 0.15 [c]	23.72 ± 1.19 [a]	2.10 ± 0.08 [d]	0.96 ± 0.06 [c,d]	63.43 ± 1.20 [b,c]	236.84 ± 0.41 [b,c]
Chickpea	7.95 ± 0.30 [d]	22.19 ± 2.74 [a]	2.66 ± 0.01 [a]	5.10 ± 0.05 [a]	62.10 ± 2.50 [c]	427.01 ± 80.97 [a]

Means with different letters within each column denote significant differences ($p < 0.05$).

The protein, ash, lipid, and total carbohydrate contents are also shown in Table 1. The whole wheat flour had significantly lower protein content (15.13%), compared to the pulse flours. The protein content of pulse flours ranged from 19.82 to 24.22%. The protein contents of the pulse flours are highly variable and determined by their genetic and environmental factors [15]. There was no significant difference in the protein content among the different pulse flours. The ash content of the whole wheat flour and pulse flours ranged from 1.80 to 2.66%. Whole wheat flour possessed the lowest ash content (1.8%), and chickpea flour contained the highest ash content (2.66%) among these six types of flours. The higher ash content in pulse flours could be due to the thicker seed coat of the pulse grains; this can be beneficial by providing more nutrients, such as minerals and fibers [28]. These results were consistent with the previous literature [4], which also showed that pulse flours had a higher protein and fiber content than whole wheat flour. Chickpea presented the highest lipid content (5.10%), similar to the results reported [29]. Chickpea contains highly digestible protein and a high lysine content [18,30]. The carbohydrate content of whole wheat flour was 68.36%, and the pulse flours ranged from 61.96 to 67.56%. Starch and dietary fiber are the two primary components that make up carbohydrates. Most of the polysaccharide is in the form of starch granules in whole wheat flour, and humans have better starch digestibility of cereal products than pulse products [4].

The average particle size of whole wheat and pulse flours is shown in Table 1. Chickpea had a significantly larger particle size (427.0 μm) than the other samples. The average particle size of whole wheat and commercial yellow pea flour was 160.8 and 131.6 μm, respectively, which were much smaller than that of the chickpea flour. There were no significant differences between the average particle size of the yellow pea (242.6 μm), green pea (331.6 μm), and red lentil flour (236.8 μm). The particle size distribution curves are shown in Figure 1. The hull and hardness of the seeds affect the grinding properties of whole pulses. Most of the flours fell in the range of the 53 to 105 μm screen for whole wheat flour, and this is attributed to the endosperm flour fraction [31]. Doblado-Maldonado et al. [31] also indicated that although a smaller particle size of whole wheat flour could release nutrients more effectively, the moderate particle size performed better in the bread structure and quality. Most of the particles of yellow pea (81.95%), green pea (84.23%), and red lentil (76.70%) flour fell between the 149 and 420 μm screen. The chickpea flour had a bimodal distribution, with a smaller peak around 210 μm and a larger peak around 595 μm. The largest average particle size for chickpea flour was due to the higher lipid content, which resulted in the high amount of flour agglomeration on the 595 μm screen, which cannot be sifted during particle size testing. The overall particle size of the lab-ground flour with the hammer mill is larger than the whole wheat flour and commercial yellow pea flour, due to the different milling process and grain properties. The same trend was presented previously [32], and they also found a greater amount of starch damage in the finely ground flours.

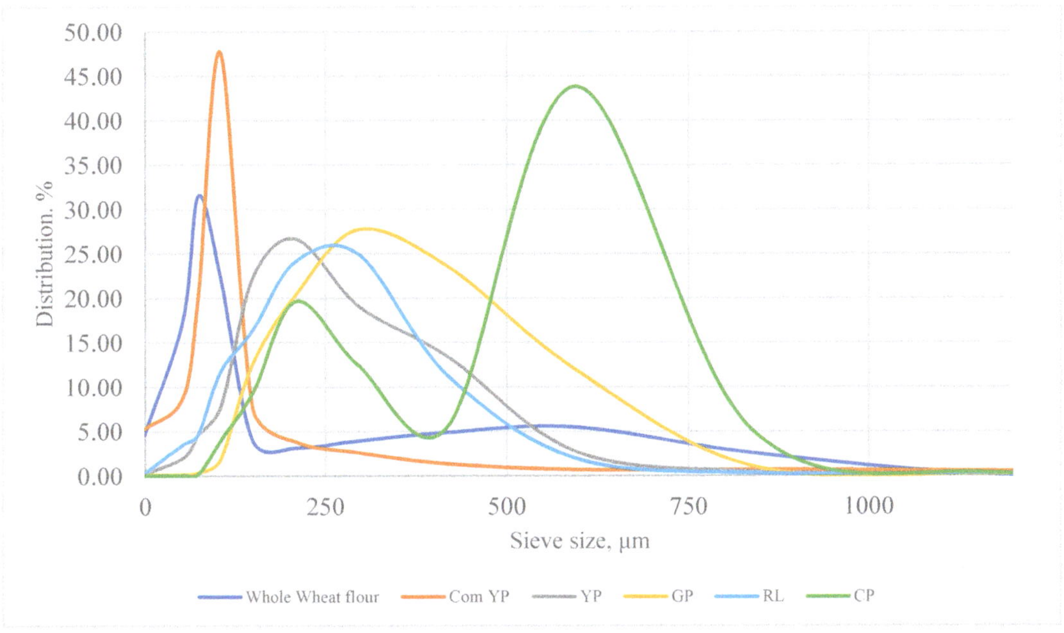

Figure 1. Particle size distribution of whole wheat flour and pulse flour. (CommYP, commercial yellow pea; YP, yellow pea; GP, green pea; RL, red lentil; CP, chickpea.).

3.2. Dough Properties

3.2.1. Mixing Characteristics

The mixing characteristics of the whole wheat dough incorporated with pulse flours are summarized in Table 2. Figure 2 shows an example of the Mixolab graph of composite flours with the addition of 0, 5, 15, and 25% (based on total composite flour weight) yellow pea flour. The mixing properties of dough were determined at 30 °C, and the optimal water absorption was determined by the Mixolab when the mixing torque value reached 1.1 Nm, as specified in the manufacturer's manual. The addition of pulse flours considerably affected the water absorption of the dough. Increasing the amount of pulse flours in the composite presented a higher water absorption value, compared to the whole wheat flour (70.9%), except for 25% commercial yellow pea flour (70.0%). Incorporation of 25% red lentil flour significantly increased the water absorption (74%). This might be due to the larger particle size and non-dehulled coat. There was no significant difference between the yellow pea and chickpea flour on water absorption when the amount of pulse flours increased. The water absorption was significantly increased by adding green pea and red lentil flour, especially with 25% red lentil flour (74%). Adding 25% commercial yellow pea flour presented a lower water absorption value than that with 25% yellow pea flour, which might be due to the smaller particle size. In addition, the proportion of bran in whole wheat flour also influenced the water absorption. Bourré et al. [32] found that the flour and baking properties were affected by the pulse flours' particle size and finer flours presented lower water absorption value, higher viscosities and better crumb texture at 20% substitution of wheat flour. However, this study also indicated that the water absorption is related to the particle size and damaged starch, surface area, protein content, fiber content, and starch content.

Table 2. Mixing characteristics of whole wheat/pulse flour blend from Mixolab.

Sample	Water Absorption, %	Development Time, min	Stability, min	Mechanical Weakening, Nm
Whole wheat	70.90	7.41 ± 0.46 [a,b,c]	9.80 ± 0.42 [a]	0.096 ± 0.00 [a]
5% Comm YP	72.00	7.65 ± 0.57 [a,b,c]	9.10 ± 0.14 [a,b,c]	0.075 ± 0.01 [a]
15% Comm YP	71.80	6.35 ± 0.18 [b,c,d,e]	7.35 ± 0.35 [d,e,f,g]	0.082 ± 0.02 [a]
25% Comm YP	70.00	5.10 ± 0.03 [d,e]	6.10 ± 0.14g [h]	0.076 ± 0.02 [a]
5% YP	71.50	8.09 ± 0.52 [a,b,c]	8.70 ± 0.14 [a,b,c,d,e]	0.081 ± 0.01 [a]
15% YP	71.80	6.77 ± 0.78 [b,c,d]	7.40 ± 0.14 [d,e,f,g]	0.082 ± 0.01 [a]
25% YP	71.80	4.50 ± 0.04 [e]	4.85 ± 0.21 [h]	0.069 ± 0.00 [a]
5% GP	71.80	7.77 ± 0.45 [a,b,c]	8.90 ± 0.28 [a,b,c,d]	0.072 ± 0.02 [a]
15% GP	73.00	7.52 ± 1.06 [a,b,c]	8.15 ± 1.34 [b,c,d,e,f]	0.089 ± 0.01 [a]
25% GP	72.50	4.95 ± 0.32 [d,e]	5.15 ± 0.07 [h]	0.086 ± 0.00 [a]
5% RL	71.80	8.29 ± 0.69 [a,b]	9.65 ± 0.21 [a,b]	0.089 ± 0.01 [a]
15% RL	72.80	6.95 ± 0.21 [a,b,c,d]	7.86 ± 0.07 [d,e,f]	0.106 ± 0.02 [a]
25% RL	74.00	6.18 ± 0.21 [c,d,e]	7.10 ± 0.14 [f,g]	0.110 ± 0.00 [a]
5% CP	71.80	8.88 ± 0.18 [a]	10.05 ± 0.07 [a]	0.091 ± 0.00 [a]
15% CP	72.00	7.21 ± 0.62 [a,b,c]	9.00 ± 0.14 [a,b,c]	0.101 ± 0.01 [a]
25% CP	71.50	6.09 ± 0.54 [c,d,e]	7.20 ± 0.14 [e,f,g]	0.087 ± 0.01 [a]

Means with different letters within each property denote significant differences ($p < 0.05$); CommYP, commercial yellow pea; YP, yellow pea; GP, green pea; RL, red lentil; CP, chickpea.

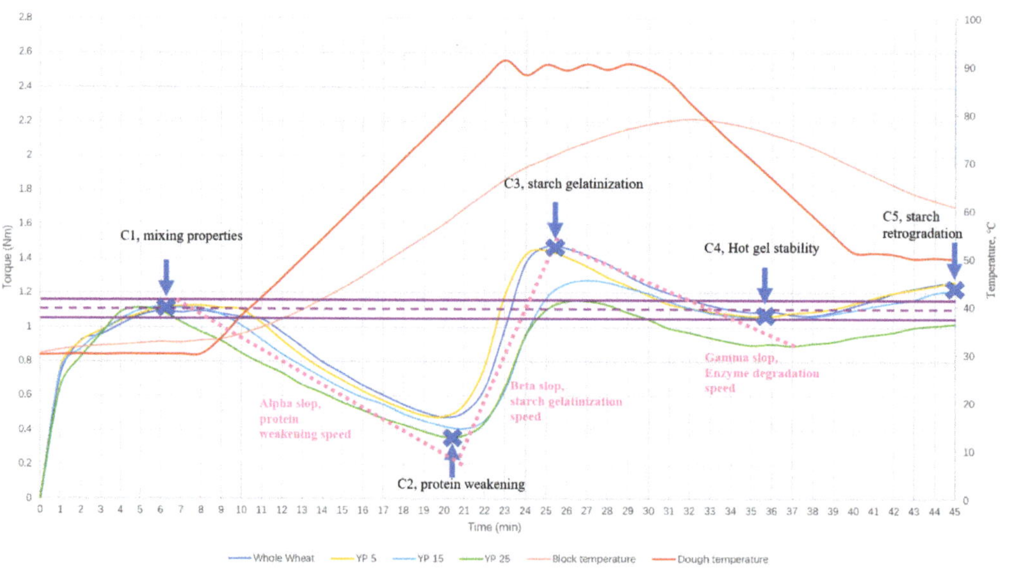

Figure 2. Example Mixolab graph (whole wheat, 5, 15, and 25% yellow pea flour).

The dough development time was affected by the pulse flour addition, especially by adding the yellow pea flour (Table 2). Adding 5% chickpea flour slightly increased the development time of the dough (8.88 min); this result was consistent with a previous study [18]. There was no significant difference between the whole wheat flour (7.41 min) incorporated with 5% of different pulse flours, and no difference was found between adding 15% of pulse flour. Increasing the amount of pulse flour decreased the development time when the dough reached the maximum consistency, especially for the addition of 25% pulse flour; among these, adding 25% of yellow pea flour presented the shortest development time (4.5 min).

The stability of the dough was positively correlated with the development time and affected by the pulse flour addition. The addition of 25% yellow pea flour or green pea flour significantly decreased the dough's stability, and the value was 4.85 and 5.15 min, respectively. These results are also presented in the dough index properties (Figure 3). Yellow pea flour possessed lower protein content than other pulse flours. The addition of 5% chickpea flour presented the highest value of stability (10.05 min), which was similar to the control whole wheat flour (9.80 min). Furthermore, adding 15% of chickpea flour did not significantly affect the stability, and adding chickpea flour showed less influence on dough stability. The stability time of dough indicated the flour strength. Incorporation of pulse flour decreased the dough's strength; this is largely because the dilution of gluten detrimentally affects the development of the gluten network in the whole wheat dough [33]; similar results were also reported by Sadowska et al. [34]. No significant difference was shown in the mechanical weakening value, ranging from 0.069 to 0.110 Nm, which indicated that the protein weakening is similar between different types and amounts of pulse flours in the whole wheat dough. Overall, the addition of 5% pulse flour did not negatively affect the dough mixing properties and demonstrated better potential for the preparation of whole pulse–fortified bakery products.

3.2.2. Pasting Characteristics

The pasting characteristics of the whole wheat dough and whole wheat dough containing pulse flours are shown in Table 3, and some index parameters are presented in Figure 3. The minimum torque values in the protein weakening phase decreased as the amount of pulse flours increased. The highest value of minimum torque was in the whole wheat flour (0.486 Nm). Incorporation with 25% of commercial yellow pea (0.330 Nm), yellow pea (0.351 Nm), green pea (0.376 Nm), red lentil (0.411 Nm), and chickpea flour (0.332 Nm) showed lower minimum torque among different types and amounts of pulse flours. The treatments with commercial yellow pea flour presented lower values in minimum torque than the corresponding incorporation level with lab-ground yellow pea flour. The minimum torque values of lab-ground pulse flours (except for chickpea flour) were positively correlated with particle size. Both flour particle size and substitution level affected the pasting and thermal properties, where increasing the amount of pulse flour decreased the consistency of the dough [14]. Moreover, pulse flour diluted the gluten proteins, which can also result in lower minimum torque value. The thermal weakening parameter was calculated by the torque (C1) in the mixing process minus the minimum torque. There was no significant difference in the value of thermal weakening among treatments containing 5% pulse flour. The extent of thermal weakening was positively correlated with the amount of pulse flour. Tolerance to thermal weakening was important to predict the bread-making quality [35].

During further dough heating in the Mixolab, protein starts unfolding and aggregating; starch starts swelling in the limited available water and ruptures; and amylose, amylopectin, and granule fragments are formed [36]. The addition of pulse flours decreased the paste viscosity of whole wheat doughs as indicated by the minimum torque and peak torque values (Table 3). The peak torque value of whole wheat dough was 1.510 Nm. The peak torque was significantly affected by the substitution level of pulse flours, especially the dough with 25% yellow pea, which had the lowest peak viscosity (1.152 Nm). Increasing the substitution level of pulse flour decreased the peak viscosity (peak torque), due to the higher protein and lower starch content of the pulse flour. The gluten blocks the contact area between separated starch granules and influences the dough gelatinization; gluten absorbs excess water and affects the starch swelling and gelatinization time [37]. The peak viscosity of the pulse flours is in the order of yellow pea flour < green pea flour < chickpea flour < red lentil flour < commercial yellow pea flour. The interaction between starch and protein affects the food matrix through the inter-and intramolecular forces in the macroscopic structure [38].

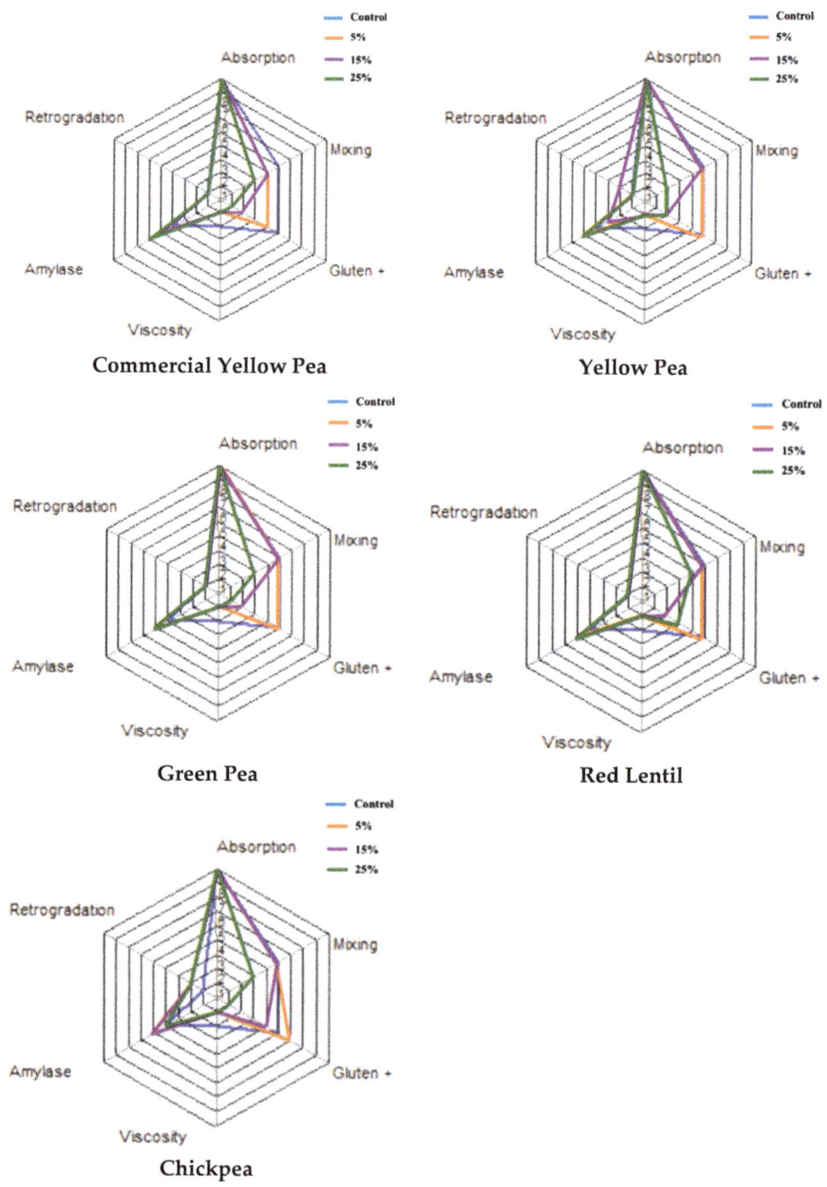

Figure 3. Mixolab index of the whole wheat/pulse flours.

Table 3. Pasting characteristics of whole wheat/pulse flour blend from Mixolab.

Sample	Minimum Torque (Nm)	Thermal Weakening (Nm)	Temperature at Minimum Torque (°C)	Peak Torque (Nm)	Cooking Stability (Nm)	Setback (Nm)
Whole wheat	0.486 ± 0.02 [a]	0.631 ± 0.00 2 [f,g]	58.6 ± 1.6 [a]	1.510 ± 0.03 [a]	1.041 ± 0.04 [b,c,d,e]	0.213 ± 0.02 [a,b,c]
5% Comm YP	0.434 ± 0.01 [b,c,d]	0.641 ± 0.004 [e,f,g]	58.9 ± 1.3 [a]	1.401 ± 0.01 [a,b,c,d,e]	1.063 ± 0.01 [a,b,c,d]	0.171 ± 0.04 [a,b,c]
15% Comm YP	0.379 ± 0.01 [e,f]	0.675 ± 0.006 [c,d,e,f]	58.2 ± 1.2 [a]	1.332 ± 0.00 [c,d,e,f,g]	0.980 ± 0.02 [c,d,e,f]	0.154 ± 0.00 [a,b,c]
25% Comm YP	0.330 ± 0.00 [g]	0.741 ± 0.004 [a,b]	60.5 ± 2.5 [a]	1.322 ± 0.04 [d,e,f,g]	1.018 ± 0.00 [b,c,d,e,f]	0.169 ± 0.0.02 [a,b,c]
5% YP	0.475 ± 0.00 [a]	0.654 ± 0.018 [e,f]	58.6 ± 2.5 [a]	1.425 ± 0.05 [a,b,c,d]	1.073 ± 0.02 [a,b,c,d]	0.186 ± 0.03 [a,b,c]
15% YP	0.423 ± 0.01 [c,d]	0.711 ± 0.002 [b,c]	61.2 ± 0.7 [a]	1.353 ± 0.03 [b,c,d,e,f]	0.993 ± 0.02 [c,d,e,f]	0.199 ± 0.02 [a,b,c]
25% YP	0.351 ± 0.01 [f,g]	0.775 ± 0.013 [a]	58.8 ± 0.4 [a]	1.152 ± 0.02 [h]	0.895 ± 0.00 [f]	0.128 ± 0.00 [c]
5% GP	0.469 ± 0.02 [a,b]	0.650 ± 0.012 [e,f]	59.5 ± 0.1 [a]	1.438 ± 0.06 [a,b,c,d]	1.092 ± 0.00 [a,b,c]	0.195 ± 0.04 [a,b,c]
15% GP	0.430 ± 0.01 [b,c,d]	0.678 ± 0.022 [c,d,e]	59.6 ± 2.3 [a]	1.406 ± 0.05 [a,b,c,d,e]	1.077 ± 0.10 [a,b,c,d]	0.224 ± 0.06 [a,b,c]
25% GP	0.376 ± 0.01 [e,f]	0.750 ± 0.004 [a,b]	60.4 ± 0.6 [a]	1.212 ± 0.00 [g,h]	0.954 ± 0.03 [d,e,f]	0.147 ± 0.05 [b,c]
5% RL	0.477 ± 0.00 [a]	0.651 ± 0.015 [e,f]	61.1 ± 1.6 [a]	1.461 ± 0.03 [a,b]	1.054 ± 0.04 [a,b,c,d]	0.144 ± 0.02 [b,c]
15% RL	0.452 ± 0.01 [a,b,c]	0.671 ± 0.018 [c,d,e,f]	58.7 ± 1.4 [a]	1.399 ± 0.05 [a,b,c,d,e]	1.039 ± 0.01 [b,c,d,e]	0.183 ± 0.00 [a,b,c]
25% RL	0.411 ± 0.01 [d,e]	0.706 ± 0.020 [b,c,d]	60.7 ± 0.7 [a]	1.282 ± 0.03 [e,f,g]	0.922 ± 0.04 [e,f]	0.135 ± 0.02 [b,c]
5% CP	0.467 ± 0.01 [a,b]	0.600 ± 0.005 [g]	59.8 ± 0.4 [a]	1.470 ± 0.01 [a,b]	1.174 ± 0.00 [a]	0.227 ± 0.01 [a,b,c]
15% CP	0.423 ± 0.00 [c,d]	0.664 ± 0.002 [e,f]	58.9 ± 1.3 [a]	1.451 ± 0.01 [a,b,c]	1.138 ± 0.01 [a,b]	0.260 ± 0.01 [a]
25% CP	0.332 ± 0.00 [g]	0.727 ± 0.006 [b]	61.8 ± 0.4 [a]	1.272 ± 0.01 [f,g,h]	0.963 ± 0.02 [d,e,f]	0.242 ± 0.00 [a,b]

Means with different letters within each property denote significant differences ($p < 0.05$); CommYP, commercial yellow pea; YP, yellow pea; GP, green pea; RL, red lentil; CP, chickpea.

The cooking stability was not significantly affected by the different types and substitution levels, except for 25% yellow pea flour (0.895 Nm), and chickpea flour at 5% (1.174 Nm), and 15% (1.138 Nm), compared with the control. The cooking stability, also called hot gel stability, indicates the stability of previously broken starch granules during the heating process [36]. Adding 25% yellow pea flour significantly decreased the hot gel stability, while adding 5 and 15% chickpea flour significantly increased the hot gel stability. The increased cooking stability with chickpea flour might be due to its higher lipid content, which requires further investigation. The setback value was calculated through the torque in starch retrogradation in the cooling phase minus the torque in the hot gel stability phase (C5–C4), and it indicates the potential of straight-chain amylose molecules realigning and forming stable gel structure [39]. The dough incorporated with 25% yellow pea flour had the lowest value (0.128 Nm), while the dough with 15% chickpea flour presented the highest value (0.260 Nm). This might be due to the higher protein content providing better stability for starch granules at high temperatures [40]. The overall tendency of the setback value was slightly increased at the 15% level and then decreased at the 25% level.

3.3. Bread Properties

3.3.1. Bread Volume, C-Cell Structure, and Color Parameters

The specific volume and C-cell properties of whole wheat breads with pulse flours are shown in Table 4, and the pictures of bread products are presented in Figure 4. Increasing the amount of pulse flours slightly decreased the specific volume of the bread, except in the case of 5% yellow pea flour. The bread containing 5% yellow pea flour (3.81 cm^3/g) had a similar specific volume to the control whole wheat bread (3.82 cm^3/g). Among the different pulse flour treatments, the dough containing 5% yellow pea flour presented better pasting properties, good mixing properties and other values similar to the whole wheat control dough, which may explain its better bread volume. The 5% yellow pea bread also had a comparable cell diameter (1.832 mm) and number of cells (2969) as the control bread. Adding 15% of yellow pea flour slightly decreased the specific volume (3.59 cm^3/g) and the number of bread cells (2592), while adding 25% yellow pea flour significantly decreased the number of cells (2220), enlarged the cell volume (2.157 mm), and decreased the bread volume (3.16 cm^3/g). The same trend was also found in other research, which showed that increasing pea flour from 5 to 10% in the formulation decreased the bread volume and cookie spread ratio [41]. The lab-ground yellow pea flour had lower protein content and larger average particle size than the commercial yellow pea flour, which might, in part, contribute to the larger bread specific volume. Adding 25% pulse flours greatly worsened the cell structure (e.g., reduced number of cells, increased cell diameter, thicker cell walls), compared to the control or breads with a lower amount of pulse flours. The number of cells and cell diameter are related to the gluten network quality. Incorporation with higher amount of pulse flours influenced the gluten development and affected the dough stability during the baking process.

Table 4. Baking parameters, specific volume, and C-cell properties of whole wheat/pulse bread.

Sample	Water Abs., %	Mixing Time, min	Specific Volume, cm³/g	Number of Cells	Cell Diameter/mm	Wall Thickness, mm
Whole wheat	75	4.67	3.82 ± 0.07 [a]	2941 ± 54 [a]	1.88 ± 0.04 [b,c,d]	0.420 ± 0.001 [b,c]
5% Comm YP	73	5.00	3.15 ± 0.12 [a,b]	2867 ± 31 [a,b]	1.83 ± 0.01 [c,d]	0.420 ± 0.003 [b,c]
15% Comm YP	70.5	4.50	3.43 ± 0.03 [a,b]	2818 ± 8 [a,b]	1.82 ± 0.01 [c,d]	0.425 ± 0.002 [a,b,c]
25% Comm YP	68	4.67	3.10 ± 0.05 [a,b]	2425 ± 60 [c,d,e]	2.07 ± 0.01 [a,b]	0.443 ± 0.001 [a,b]
5% YP	75	4.67	3.81 ± 0.22 [a]	2969 ± 109 [a]	1.83 ± 0.06 [c,d]	0.423 ± 0.009 [b,c]
15% YP	72.5	3.75	3.59 ± 0.11 [a,b]	2592 ± 34 [b,c,d]	1.99 ± 0.11 [a,b,c]	0.435 ± 0.009 [a,b,c]
25% YP	70	4.17	3.16 ± 0.05 [a,b]	2220 ± 94 [e]	2.16 ± 0.04 [a]	0.450 ± 0.016 [a]
5% GP	75	4.67	3.54 ± 0.06 [a,b]	2825 ± 60 [a,b]	1.75 ± 0.13 [d]	0.415 ± 0.010 [c]
15% GP	72.5	4.50	3.18 ± 0.13 [a,b]	2682 ± 76 [a,b,c]	1.82 ± 0.03 [c,d]	0.421 ± 0.000 [b,c]
25% GP	70	4.00	2.90 ± 0.11 [b]	2285 ± 43 [d,e]	2.01 ± 0.02 [a,b,c]	0.439 ± 0.004 [a,b,c]
5% RL	75	5.00	3.44 ± 0.24 [a,b]	2780 ± 17 [a,b]	1.86 ± 0.07 [b,c,d]	0.428 ± 0.008 [a,b,c]
15% RL	73.5	4.50	3.00 ± 0.11 [a,b]	2732 ± 161 [a,b,c]	1.81 ± 0.07 [c,d]	0.424 ± 0.005 [b,c]
25% RL	72	4.50	2.84 ± 0.05 [b]	2411 ± 132 [c,d,e]	1.89 ± 0.04 [b,c,d]	0.423 ± 0.001 [b,c]
5% CP	75	5.17	3.50 ± 0.14 [a,b]	2919 ± 154 [a,b]	1.85 ± 0.04 [b,c,d]	0.423 ± 0.003 [b,c]
15% CP	72.5	5.67	3.13 ± 0.11 [a,b]	2738 ± 37 [a,b,c]	1.93 ± 0.03 [a,b,c,d]	0.425 ± 0.000 [a,b,c]
25% CP	68	5.33	3.31 ± 0.14 [a,b]	2663 ± 42 [a,b,c]	1.92 ± 0.02 [b,c,d]	0.430 ± 0.000 [a,b,c]

Means with different letters within each property denote significant differences ($p < 0.05$); CommYP, commercial yellow pea; YP, yellow pea; GP, green pea; RL, red lentil; CP, chickpea.

Adding 25% green pea flour (2.90 cm³/g) and 25% red lentil flour (2.84 cm³/g) dramatically decreased the bread volume, compared with the control, which might be because the higher pulse flour content hindered the gluten development. The lower water absorption of pulse flour might also be partially responsible for the lower specific volume of bread. Jekle et al. [38] proposed two interactions between protein and starch in bread baking: the competitive hydration between protein and starch, and formation of the gluten–protein matrix on the starch granule surface. The higher amount of pulse starch resulted in a larger crumb cell wall and a higher amount of open crumb grain [34]. Incorporation of 5 or 15% of green pea flour and red lentil flour into whole wheat flour was acceptable because the volume and structure of bread were not significantly reduced. Unlike the other pulse flours, chickpea flour provided better specific volume and crumb structure at a substitution level of 25% than that of the 15% chickpea level. This might be related to the pasting characteristics of 25% chickpea flour, where the dough had a similar setback value with the whole wheat dough, and the flour had the highest lipid content. Mohammed et al. [42] found that adding 20% of chickpea flour significantly decreased the white bread specific volume. Whole wheat bread dough had a more diluted gluten network than white bread dough, and thus the negative effect might be less obvious in the former system.

Bread crumb color properties in terms of lightness (L*), redness-greenness (a*), and yellowness-blueness (b*) are summarized in Table 5. The addition of pulse flours, especially at 15 or 25%, decreased Hunter L*, which indicates that the breads became darker, such as those containing green pea, red lentil, and chickpea flours. Commercial yellow pea had a smaller particle size and contained the lowest carbohydrate content, showing a lighter color. The color compounds arose from the pigments in the hulls as well as from the Maillard reaction products [42]. The red lentil flour provided a darker red color for the bread crumb, and green pea flour provided a green color for the bread crumb.

3.3.2. Moisture Loss and Texture Properties

The moisture content and moisture loss of bread are shown in Table 6. On days 1, 4 and 7, increasing the substitution level of pulse flours decreased the moisture content. The moisture content of the bread with commercial yellow pea flour significantly decreased with having the storage days increased and was the lowest among all the breads and for all three testing days. The bread supplemented with 5 and 15% yellow pea, green pea, red lentil, and chickpea flour did not differ in moisture content, ranging from 46.02 to 47.06%. Overall, the largest moisture decrease rate was found for the bread with 25% commercial yellow pea flour, and the smallest moisture decrease rate was for 5% chickpea bread. Incorporation with 5% of pulse flour slightly decreased the moisture loss during the one-week storage. This might be due to the slightly higher water absorption requirements of pulse flour. Wang et al. [43] found that protein is more affinitive to water molecules and has stronger water absorption capacity than starch, resulting in the water in the protein being less mobile than in the starch granular. The bread moisture loss was related to several factors, e.g., the gluten content, pulse flours protein content, and pulse flours particle size.

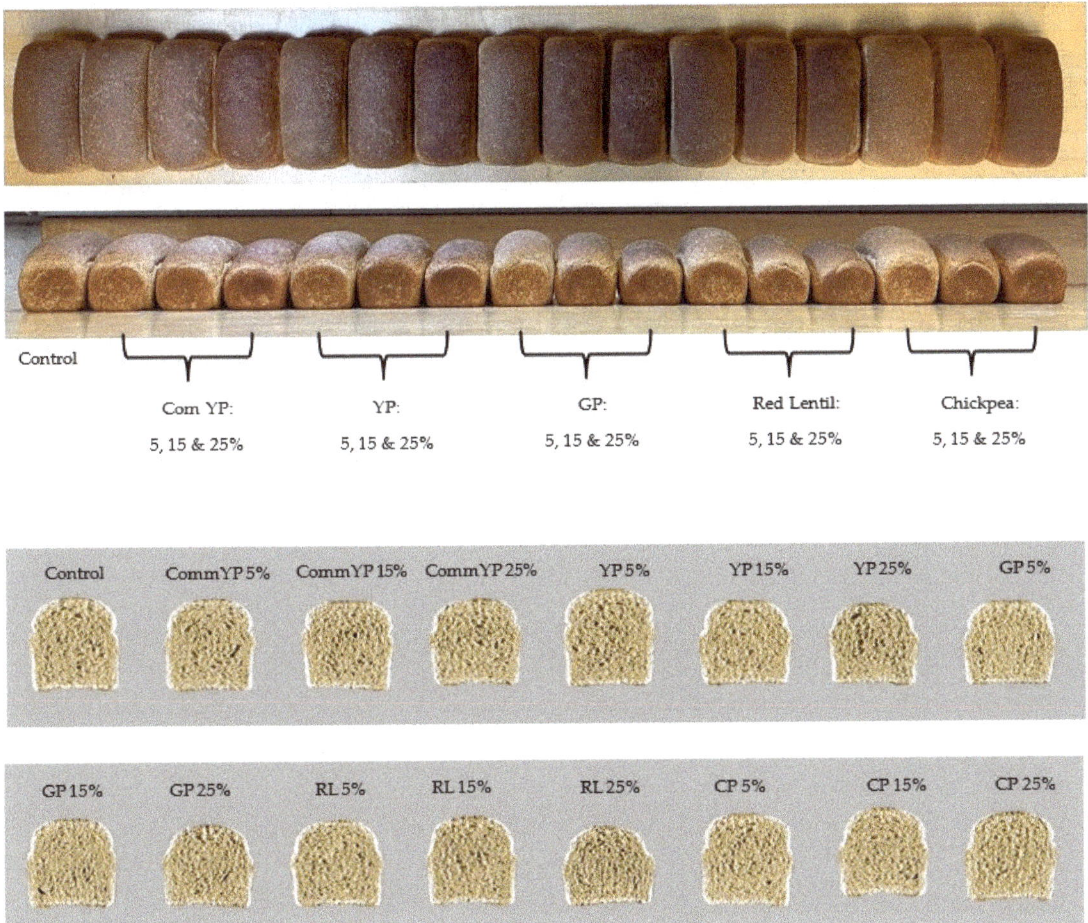

Figure 4. Bread photos and C-cell images. (CommYP, commercial yellow pea; YP, yellow pea; GP, green pea; RL, red lentil; CP, chickpea.).

Table 5. Color parameters of whole wheat/pulse bread crumb.

Bread Sample	L*	a*	b*
Whole wheat	36.25 ± 0.64 [a]	11.46 ± 0.89 [a]	25.90 ± 1.40 [a]
5% Comm YP	36.53 ± 0.69 [a]	12.20 ± 0.12 [a]	26.40 ± 0.54 [a]
15% Comm YP	34.27 ± 2.04 [a,b,c,d]	11.54 ± 0.23 [a]	24.55 ± 0.09 [a]
25% Comm YP	32.21 ± 0.48 [b,c,d]	10.49 ± 1.86 [a]	21.72 ± 1.36 [a]
5% YP	34.81 ± 0.42 [a,b,c]	17.52 ± 8.93 [a]	18.98 ± 10.02 [a]
15% YP	33.96 ± 0.78 [a,b,c,d]	11.45 ± 0.20 [a]	24.05 ± 0.41 [a]
25% YP	31.56 ± 0.00 [c,d]	11.40 ± 0.39 [a]	22.72 ± 0.85 [a]
5% GP	36.21 ± 2.23 [a]	11.65 ± 0.01 [a]	25.20 ± 0.13 [a]
15% GP	33.13 ± 0.06 [a,b,c,d]	11.45 ± 0.39 [a]	23.86 ± 0.88 [a]
25% GP	30.51 ± 0.57 [d]	11.04 ± 0.01 [a]	23.03 ± 0.69 [a]
5% RL	35.57 ± 0.04 [a,b]	11.54 ± 0.37 [a]	24.72 ± 0.04 [a]
15% RL	33.09 ± 0.31 [a,b,c,d]	11.21 ± 0.05 [a]	22.92 ± 0.69 [a]
25% RL	30.59 ± 0.83 [d]	11.29 ± 0.43 [a]	22.45 ± 1.78 [a]
5% CP	36.41 ± 0.09 [a]	12.28 ± 0.76 [a]	25.97 ± 0.23 [a]
15% CP	33.51 ± 1.85 [a,b,c,d]	12.18 ± 0.54 [a]	25.36 ± 1.43 [a]
25% CP	30.74 ± 0.88 [d]	10.94 ± 0.11 [a]	20.68 ± 1.03 [a]

Means with different letters within each property denote significant differences ($p < 0.05$); CommYP, commercial yellow pea; YP, yellow pea; GP, green pea; RL, red lentil; CP chickpea. L*; Lightness; a*; redness-greenness; b*; yellowness-blueness.

Table 6. Moisture content of whole wheat/pulse bread during storage.

Bread Sample	Day 1, %	Day 4, %	Decrease Rate Day 1–Day 4, %	Day 7, %	Decrease Rate Day 4–Day 7, %	Overall Decrease Rate, %
Whole wheat	46.82 ± 0.05 [a,b]	46.37 ± 0.28 [a]	0.96	42.98 ± 0.49 [a,b,c,d]	7.31	8.20
5% Comm YP	46.45 ± 0.47 [a,b,c]	45.92 ± 0.28 [a,b]	1.14	43.16 ± 0.78 [a,b,c,d]	6.01	7.08
15% Comm YP	45.73 ± 0.27 [c,d,e]	44.38 ± 1.03 [a,b]	2.95	41.12 ± 1.01 [d,e]	7.35	10.08
25% Comm YP	44.79 ± 0.14 [f]	43.02 ± 0.94 [b]	3.95	39.67 ± 1.38 [e]	7.79	11.43
5% YP	47.06 ± 0.16 [a]	45.81 ± 1.30 [a,b]	2.66	44.75 ± 1.11 [a]	2.31	4.91
15% YP	46.17 ± 0.03 [a,b,c,d]	44.24 ± 1.28 [a,b]	4.18	41.73 ± 0.67 [b,c,d,e]	5.67	9.62
25% YP	45.58 ± 0.30 [c,d,e,f]	44.64 ± 0.02 [a,b]	2.06	41.53 ± 0.35 [c,d,e]	6.97	8.89
5% GP	46.77 ± 0.08 [a,b]	46.40 ± 0.32 [a]	0.79	44.29 ± 0.77 [a,b,c]	4.55	5.30
15% GP	46.02 ± 0.46 [b,c,d]	45.61 ± 0.20 [a,b]	0.87	41.01 ± 0.90 [d,e]	10.09	10.87
25% GP	45.27 ± 0.07 [d,e,f]	43.50 ± 0.64 [a,b]	3.91	42.06 ± 0.29 [a,b,c,d,e]	3.31	7.09
5% RL	47.00 ± 0.18 [a]	46.12 ± 0.69 [a,b]	1.87	44.54 ± 0.24 [a,b]	3.43	5.23
15% RL	46.71 ± 0.30 [a,b]	45.77 ± 0.80 [a,b]	2.01	44.73 ± 0.03 [a]	2.27	4.24
25% RL	45.94 ± 0.25 [b,c,d]	45.42 ± 0.18 [a,b]	1.13	42.24 ± 1.11 [a,b,c,d,e]	7.00	8.05
5% CP	46.48 ± 0.07 [a,b,c]	45.94 ± 0.54 [a,b]	1.16	45.01 ± 0.50 [a]	2.02	3.16
15% CP	46.17 ± 0.00 [a,b,c,d]	45.18 ± 0.27 [a,b]	2.14	42.47 ± 0.56 [a,b,c,d]	3.78	5.85
25% CP	44.88 ± 0.06 [e,f]	43.92 ± 1.65 [a,b]	2.14	41.46 ± 0.25 [c,d,e]	5.60	7.62

Means with different letters within each property denote significant differences ($p < 0.05$); CommYP, commercial yellow pea; YP, yellow pea; GP, green pea; RL, red lentil; CP, chickpea.

The whole wheat bread hardness was significantly affected by pulse flours and storage time (Table 7). All the breads showed a tendency to increase in hardness as the storage day increased. On the first day, 5% pulse flours and 15% commercial yellow pea flour, yellow pea flour, and green pea flour did not significantly change the hardness of the bread. In particular, adding 5% red lentil flour slightly decreased the hardness of the bread, though not significantly ($p > 0.05$). With an increase in storage time, breads with 5% yellow pea, green pea, red lentil, chickpea flour, and 15% yellow pea did not reveal a significant difference in bread hardness. These results might be related to the starch retrogradation value from the Mixolab analysis since these doughs had a similar starch retrogradation value in the cooling phase. Compared with the bread hardening rate of whole wheat bread, incorporating some pulse flour delayed the hardening change of bread, for example, with 15% yellow pea flour. Increasing the substitution level of pulse flours significantly increased the starch retrogradation value and resulted in a harder texture. Both the polysaccharides and protein in the pulse flour contributed to the bread hardness [44], resulting in the firmer texture of the bread with pulse flour.

Table 7. Hardness of whole wheat/pulse bread during one-week storage.

Bread Sample	Day 1, g	Day 4, g	Increase Rate Day 1–Day 4, %	Day 7, g	Increase Rate Day 4–Day 7, %	Overall Increase Rate, %
Whole wheat	567.06 ± 36.32 [f,g]	853.74 ± 43.06 [e]	50.56	1199.38 ± 260.09 [e]	40.49	111.51
5% Comm YP	712.1 ± 61.83 [e,f,g]	973.47 ± 12.43 [d,e]	36.70	1663.07 ± 90.51 [b,c,d,e]	70.84	133.54
15% Comm YP	725.12 ± 59.52 [d,e,f,g]	1122.67 ± 123.00 [c,d,e]	54.83	1713.21 ± 96.98 [b,c,d,e]	52.60	136.27
25% Comm YP	1056.81 ± 123.45 [a,b,c]	1392.67 ± 144.21 [a,b,c,d]	31.77	2041.58 ± 258.22 [b,c,d]	46.61	93.18
5% YP	580.03 ± 112.71 [f,g]	843.91 ± 28.76 [e]	45.49	1231.22 ± 173.69 [e]	45.89	112.27
15% YP	667.15 ± 112.71 [f,g]	897.78 ± 5.82 [e]	34.57	1138.07 ± 165.77 [e]	26.76	70.59
25% YP	993.23 ± 139.79 [b,c,d]	1410.41 ± 68.88 [a,b,c]	42.00	2191.13 ± 135.79 [b,c]	55.35	120.61
5% GP	722.84 ± 110.58 [e,f,g]	1073.61 ± 112.58 [c,d,e]	48.53	1163.77 ± 109.99 [e]	8.40	61.00
15% GP	822.15 ± 5.88 [c,d,e,f]	1219.51 ± 141.75 [b,c,d,e]	48.33	1907.13 ± 76.45 [b,c,d]	56.38	131.97
25% GP	1219.74 ± 132.70 [a,b]	1552.77 ± 283.37 [a,b]	27.30	2276.93 ± 220.38 [a,b]	46.64	86.67
5% RL	550.99 ± 2.28 [g]	1051.97 ± 40.25 [c,d,e]	90.92	1460.46 ± 169.68 [d,e]	38.83	165.06
15% RL	992.63 ± 94.40 [b,c,d]	1362.45 ± 184.44 [b,c,d]	37.26	1588.22 ± 66.76 [c,d,e]	16.57	60.00
25% RL	1296.06 ± 105.56 [a]	1793.29 ± 118.36 [a]	38.36	2367.41 ± 81.26 [b,c,d]	32.01	82.66
5% CP	734.98 ± 62.23 [d,e,f,g]	1131.76 ± 175.65 [b,c,d,e]	53.99	1409.70 ± 38.27 [d,e]	24.56	91.80
15% CP	963.08 ± 56.84 [b,c,d,e]	1460.00 ± 71.59 [a,b,c]	51.60	2873.91 ± 249.33 [a]	96.84	198.41
25% CP	1029.22 ± 102.67 [a,b,c]	1224.73 ± 272.05 [b,c,d,e]	19.00	1971.11 ± 72.52 [b,c,d]	60.94	91.51

Values are expressed as the mean and SD of three measurements. Means with different letters within each property denote significant differences ($p < 0.05$); CommYP, commercial yellow pea; YP, yellow pea; GP, green pea; RL, red lentil; CP, chickpea.

4. Conclusions

Pulse flour particle size and content of protein and carbohydrate greatly affected the dough mixing and pasting properties. The addition of pulse flours increased the water absorption of the dough compared to whole wheat flour alone, except for 25% commercial yellow pea flour. Increasing substitution level of pulse flours decreased dough stability; however, the dough stability of composite flours containing chickpea was better than the other flours at the same substitution level. Dough thermal weakening increased, and minimum torque values decreased during the protein weakening phase as the amount of pulse flours increased. The smaller particle size of commercial yellow pea flour had a more negative effect on the dough and bread properties, compared to the lab-ground yellow pea flour with a larger particle size. Adding 5% of lab yellow pea flour did not obviously affect the bread volume, structure, or texture. Incorporating 25% of green pea flour or red lentil flour significantly decreased the bread specific volume. Increasing the substitution level of pulse flours decreased the moisture content and increased the hardness of the bread. Overall, adding up to 5% of pulse flours was acceptable, having minimal effect on the dough properties and bread quality and improving the nutritional value of whole grain bread. Compared with commercial yellow pea flour, lab-ground yellow pea flour is considered more suitable for bread preparation; up to 15% of yellow pea flour is acceptable to substitute whole wheat flour. Future studies are recommended to understand the effect of pulse flour particle size on bread-making performance and the sensory properties of composite flour as well as to further elucidate some unique properties of chickpea flour as compared to other pulse flours.

Author Contributions: Conceptualization, Y.Z. and Y.L.; methodology, Y.Z. and Y.L.; software, Y.Z. and Y.L.; validation, Y.Z., R.H., and Y.L.; formal analysis, Y.Z. and Y.L.; investigation, Y.Z., R.H., M.T., K.S., and Y.L.; resources, M.T., K.S., and Y.L.; data curation, Y.Z. and R.H.; writing—original draft preparation, Y.Z.; writing—review and editing, Y.Z., R.H., M.T., K.S., and Y.L.; visualization, Y.Z.; supervision, Y.L.; project administration, Y.L.; funding acquisition, Y.L. All authors have read and agreed to the published version of the manuscript.

Funding: This research was funded by the USDA Pulse Crop Health Initiative (PCHI), grant number 58-3060-0-051" and "The APC was funded by THE USDA PCHI".

Institutional Review Board Statement: Not applicable.

Informed Consent Statement: Not applicable.

Data Availability Statement: The data presented in this study are available on request from the corresponding author.

Acknowledgments: This is contribution No. 21-268-J from the Kansas Agricultural Experimental Station. Mention of trade names or commercial products in this publication is solely for the purpose of providing specific information and does not imply recommendation or endorsement by the U.S. Department of Agriculture. The USDA is an equal opportunity provider and employer.

Conflicts of Interest: The authors declare that there are no known conflicts of interest.

References

1. Zucco, F.; Borsuk, Y.; Arntfield, S.D. Physical and nutritional evaluation of wheat cookies supplemented with pulse flours of different particle sizes. *LWT Food Sci. Technol.* **2011**, *44*, 2070–2076. [CrossRef]
2. Hall, C.; Hillen, C.; Robinson, J.G. Composition, nutritional value, and health benefits of pulses. *Cereal Chem.* **2017**, *94*, 11–31. [CrossRef]
3. Millar, K.A.; Gallagher, E.; Burke, R.; McCarthy, S.; Barry-Ryan, C. Proximate composition and anti-nutritional factors of fava-bean (*Vicia faba*), green-pea and yellow-pea (*Pisum sativum*) flour. *J. Food Compos. Anal.* **2019**, *82*, 103233. [CrossRef]
4. Rebello, C.J.; Greenway, F.L.; Finley, J.W. Whole grains and pulses: A comparison of the nutritional and health benefits. *J. Agric. Food Chem.* **2014**, *62*, 7029–7049. [CrossRef]
5. Smith, C.E.; Mollard, R.C.; Luhovyy, B.L.; Harvey Anderson, G. The effect of yellow pea protein and fibre on short-term food intake, subjective appetite and glycaemic response in healthy young men. *Br. J. Nutr.* **2012**, *108*, S74–S80. [CrossRef] [PubMed]
6. Bazzano, L.A.; Thompson, A.M.; Tees, M.T.; Nguyen, C.H.; Winham, D.M. Non-soy legume consumption lowers cholesterol levels: A meta-analysis of randomized controlled trials. *Nutr. Metab. Cardiovasc. Dis.* **2011**, *21*, 94–103. [CrossRef] [PubMed]
7. Padhi, E.M.T.; Ramdath, D.D. A review of the relationship between pulse consumption and reduction of cardiovascular disease risk factors. *J. Funct. Foods* **2017**, *38*, 635–643. [CrossRef]
8. Roy, F.; Boye, J.I.; Simpson, B.K. Bioactive proteins and peptides in pulse crops: Pea, chickpea and lentil. *Food Res. Int.* **2010**, *43*, 432–442. [CrossRef]
9. Curran, J. The nutritional value and health benefits of pulses in relation to obesity, diabetes, heart disease and cancer. *Br. J. Nutr.* **2012**, *108*, S1–S2. [CrossRef]
10. Tyler, R.; Wang, N.; Han, J. Introduction to the focus issue on pulses. *Cereal Chem.* **2017**, *94*, 1. [CrossRef]
11. Tebben, L.; Shen, Y.; Li, Y. Improvers and functional ingredients in whole wheat bread: A review of their effects on dough properties and bread quality. *Trends Food Sci. Technol.* **2018**, *81*, 10–24. [CrossRef]
12. Jiang, D.; Peterson, D.G. Identification of bitter compounds in whole wheat bread. *Food Chem.* **2013**, *141*, 1345–1353. [CrossRef]
13. Meybodi, N.M.; Mirmoghtadaie, L.; Sheidaei, Z.; Mortazavian, A.M. Wheat Bread: Potential Approach to Fortify its Lysine Content. *Curr. Nutr. Food Sci.* **2019**, *15*, 630–637. [CrossRef]
14. Marchini, M.; Carini, E.; Cataldi, N.; Boukid, F.; Blandino, M.; Ganino, T.; Vittadini, E.; Pellegrini, N. The use of red lentil flour in bakery products: How do particle size and substitution level affect rheological properties of wheat bread dough? *LWT* **2021**, *136*, 110299. [CrossRef]
15. Man, S.; Păucean, A.; Muste, S.; Pop, A. Effect of the Chickpea (*Cicer arietinum* L.) Flour Addition on Physicochemical Properties of Wheat Bread. *Bull. Univ. Agric. Sci. Vet. Med. Cluj Napoca Food Sci. Technol.* **2015**, *72*, 41–49. [CrossRef]
16. Millar, K.A.; Barry-Ryan, C.; Burke, R.; McCarthy, S.; Gallagher, E. Dough properties and baking characteristics of white bread, as affected by addition of raw, germinated and toasted pea flour. *Innov. Food Sci. Emerg. Technol.* **2019**, *56*, 102189. [CrossRef]
17. Shrestha, A.K.; Noomhorm, A. Comparison of physico-chemical properties of biscuits supplemented with soy and kinema flours. *Int. J. Food Sci. Technol.* **2002**, *37*, 361–368. [CrossRef]
18. Mohammed, I.; Ahmed, A.R.; Senge, B. Dough rheology and bread quality of wheat-chickpea flour blends. *Ind. Crops Prod.* **2012**, *36*, 196–202. [CrossRef]
19. Setia, R.; Dai, Z.; Nickerson, M.T.; Sopiwnyk, E.; Malcolmson, L.; Ai, Y. Impacts of short-term germination on the chemical compositions, technological characteristics and nutritional quality of yellow pea and faba bean flours. *Food Res. Int.* **2019**, *122*, 263–272. [CrossRef]
20. Cereals & Grains Association. *AACC Approved Methods of Analysis*, 11th ed.; Method 46-30.01: Crude Protein—Combustion Method; Cereals & Grains Association: St. Paul, MN, USA, 1999.
21. Cereals & Grains Association. *AACC Approved Methods of Analysis*, 11th ed.; Method 08-01.01: Ash—Basic Method; Cereals & Grains Association: St. Paul, MN, USA, 2009. [CrossRef]
22. Cereals & Grains Association. *AACC Approved Methods of Analysis*, 11th ed.; Method 44-19.01: Moisture—Air-Oven Method, Drying at 135°; Cereals & Grains Association: St. Paul, MN, USA, 1999.
23. Cereals & Grains Association. *AACC Approved Methods of Analysis*, 11th ed.; Method 30-10.01: Crude Fat in Flour, Bread, and Baked Cereal Products Not Containing Fruit; Cereals & Grains Association: St. Paul, MN, USA, 1999.
24. Cereals & Grains Association. *AACC Approved Methods of Analysis*, 11th ed.; Method 54-60.01: Determination of Rheological Behavior as a Function of Mixing and Temperature Increase in Wheat Flour and Whole Wheat Meal by Mixolab; Cereals & Grains Association: St. Paul, MN, USA, 2010. [CrossRef]
25. Cereals & Grains Association. *AACC Approved Methods of Analysis*, 11th ed.; Method 10-10.03: Optimized Straight-Dough Bread-Baking Method; Cereals & Grains Association: St. Paul, MN, USA, 1999.

26. Cereals & Grains Association. *AACC Approved Methods of Analysis*, 11th ed.; Method 10-05.01: Guidelines for Measurement of Volume by Rapeseed Displacement; Cereals & Grains Association: St. Paul, MN, USA, 1999.

27. Tebben, L.; Li, Y. Effect of xanthan gum on dough properties and bread qualities made from whole wheat flour. *Cereal Chem.* **2019**, *96*, 263–272. [CrossRef]

28. Summo, C.; De Angelis, D.; Ricciardi, L.; Caponio, F.; Lotti, C.; Pavan, S.; Pasqualone, A. Nutritional, physico-chemical and functional characterization of a global chickpea collection. *J. Food Compos. Anal.* **2019**, *84*, 103306. [CrossRef]

29. Dalgetty, D.D.; Baik, B.K. Isolation and characterization of cotyledon fibers from peas, lentils, and chickpeas. *Cereal Chem.* **2003**, *80*, 310–315. [CrossRef]

30. Wani, S.A.; Kumar, P. Comparative study of chickpea and green pea flour based on chemical composition, functional and pasting properties Comparative Study of Chickpea and Green Pea Flour Based on Chemical. *J. Food Res. Technol.* **2014**, *2*, 124–129.

31. Doblado-Maldonado, A.F.; Pike, O.A.; Sweley, J.C.; Rose, D.J. Key issues and challenges in whole wheat flour milling and storage. *J. Cereal Sci.* **2012**, *56*, 119–126. [CrossRef]

32. Bourré, L.; Frohlich, P.; Young, G.; Borsuk, Y.; Sopiwnyk, E.; Sarkar, A.; Nickerson, M.T.; Ai, Y.; Dyck, A.; Malcolmson, L. Influence of particle size on flour and baking properties of yellow pea, navy bean, and red lentil flours. *Cereal Chem.* **2019**, *96*, 655–667. [CrossRef]

33. Roccia, P.; Ribotta, P.D.; Pérez, G.T.; León, A.E. Influence of soy protein on rheological properties and water retention capacity of wheat gluten. *LWT* **2009**, *42*, 358–362. [CrossRef]

34. Sadowska, J.; Błaszczak, W.; Fornal, J.; Vidai-Valverde, C.; Frias, J. Changes of wheat dough and bread quality and structure as a result of germinated pea flour addition. *Eur. Food Res. Technol.* **2003**, *216*, 46–50. [CrossRef]

35. Simsek, S.; Ohm, J.B.; Cariou, V.; Mergoum, M. Effect of flour polymeric proteins on dough thermal properties and breadmaking characteristics for hard red spring wheat genotypes. *J. Cereal Sci.* **2016**, *68*, 164–171. [CrossRef]

36. Rosell, C.M.; Collar, C.; Haros, M. Assessment of hydrocolloid effects on the thermo-mechanical properties of wheat using the Mixolab. *Food Hydrocoll.* **2007**, *21*, 452–462. [CrossRef]

37. Xu, F.; Liu, W.; Liu, Q.; Zhang, C.; Hu, H.; Zhang, H. Pasting, thermo, and Mixolab thermomechanical properties of potato starch–wheat gluten composite systems. *Food Sci. Nutr.* **2020**, *8*, 2279–2287. [CrossRef]

38. Jekle, M.; Mühlberger, K.; Becker, T. Starch-gluten interactions during gelatinization and its functionality in dough like model systems. *Food Hydrocoll.* **2016**, *54*, 196–201. [CrossRef]

39. Gadallah, M.G.E. Rheological, organoleptical and quality characteristics of gluten-free rice cakes formulated with sorghum and germinated chickpea flours. *Food Nutr. Sci.* **2017**, *08*, 535–550. [CrossRef]

40. Shevkani, K.; Kaur, A.; Kumar, S.; Singh, N. Cowpea protein isolates: Functional properties and application in gluten-free rice muffins. *LWT* **2015**, *63*, 927–933. [CrossRef]

41. Kamaljit, K.; Baljeet, S.; Amarjeet, K. Preparation of bakery products by incorporating pea flour as a functional ingredient. *Am. J. Food Technol.* **2010**, *5*, 130–135. [CrossRef]

42. Mohammed, I.; Ahmed, A.R.; Senge, B. Effects of chickpea flour on wheat pasting properties and bread making quality. *J. Food Sci. Technol.* **2014**, *51*, 1902–1910. [CrossRef]

43. Wang, X.; Choi, S.G.; Kerr, W.L. Water dynamics in white bread and starch gels as affected by water and gluten content. *LWT* **2004**, *37*, 377–384. [CrossRef]

44. Aider, M.; Sirois-Gosselin, M.; Boye, J.I. Pea, Lentil and Chickpea Protein Application in Bread Making. *J. Food Res.* **2012**, *1*, 160. [CrossRef]

 *processes*

Article

Physicochemical and Nutritional Evaluation of Bread Incorporated with Ayocote Bean (*Phaseolus coccineus*) and Black Bean (*Phaseolus vulgaris*)

Rosa María Mariscal-Moreno [1,*], **Cristina Chuck-Hernández** [2], **Juan de Dios Figueroa-Cárdenas** [3] **and Sergio O. Serna-Saldivar** [2]

[1] Universidad Iberoamericana Ciudad de México, Health Department, Prolongación Paseo de la Reforma 880, Col. Lomas de Santa Fe, Álvaro Obregón, Mexico City 01219, Mexico

[2] Tecnologico de Monterrey, School of Engineering and Sciences, Av. Eugenio Garza Sada 2501, Monterrey 64849, Mexico; cristina.chuck@tec.mx (C.C.-H.); sserna@tec.mx (S.O.S.-S.)

[3] CINVESTAV-Unidad Querétaro, Libramiento Norponiente No. 2000, Fraccionamiento Real de Juriquilla, Querétaro 76230, Mexico; jfigueroa@cinvestav.mx

* Correspondence: rosa.mariscal@ibero.mx; Tel.: +52-1-442-1057986

Abstract: The objective of this study was to examine the physicochemical composition, thermal properties, quality, and sensorial characteristics of bread with substitution of wheat flour with ayocote bean (*Phaseolus coccineus*) or black bean (*Phaseolus vulgaris*) flours at 10, 20, and 30%. Ayocote and black bean contain 21.06 and 23.94% of protein, 3.06 and 5.21% of crude fiber, and 3.1 and 5.21% of ash, respectively, directly influencing bread composition. Bread with ayocote and black bean presented higher values in those components in contrast with control bread. The protein content increased in a range of 14–34%; ash increased by 10% to double, and crude fiber also increased. In vitro protein digestibility was similar for bread with 10% of substitution and control, and it decreased in samples with 30% of wheat substitution. Thermal analysis by DSC denoted that the addition of those legumes reduces retrogradation, as seen in 45.33–65.65 °C endotherm, producing higher endotherms of amylose-lipid complexes and protein denaturalization. Finally, the addition of black bean and ayocote bean decreased specific volume when the replacement percentage was 30% black bean and 20 and 30% for ayocote. An increase in nutrient content without sensorial properties affectation could be observed in substitution around 10 and 20%.

Keywords: ayocote bean; black bean; bread; protein digestibility; sensorial properties

Citation: Mariscal-Moreno, R.M.; Chuck-Hernández, C.; Figueroa-Cárdenas, J.d.D.; Serna-Saldivar, S.O. Physicochemical and Nutritional Evaluation of Bread Incorporated with Ayocote Bean (*Phaseolus coccineus*) and Black Bean (*Phaseolus vulgaris*). *Processes* **2021**, *9*, 1782. https://doi.org/10.3390/pr9101782

Academic Editors: Yonghui Li and Shawn (Xiaorong) Wu

Received: 31 August 2021
Accepted: 24 September 2021
Published: 6 October 2021

Publisher's Note: MDPI stays neutral with regard to jurisdictional claims in published maps and institutional affiliations.

1. Introduction

Bread is a dietary staple in human nutrition. Research on bread is globally conducted to improve its nutritional value (macronutrients: carbohydrates, proteins, fat, and dietary fibers; micronutrients: minerals and vitamins), health-supporting bioactive compounds, sensory acceptability, shelf life, and affordability [1], additionally, some studies have focused on provide options to celiatics individuals. Several studies have shown that vegetable-based products contain a significant amount of nutrients (proteins, vitamins and minerals) as well as functional and bioactive compounds and that in this context bread enrichment with non-cereal flours could have a positive effect on their nutritional value [2,3]. The possibility of bread enrichment with legumes facilitates increased consumption in the home and adaptation to social, economic, and cultural changes. Research concerning new alternative plant protein sources is mostly focused on legumes [4]. Legumes are of interest to food technologists and the food industry due to their functionality, nutritional characteristics, and protein content. Some underutilized legumes are locally consumed, despite having high potential as a source of bioactive compounds and health-promoting food. Ayocote bean (*Phaseolus coccineus* L.) is a native legume from the Mexican highlands. It is a poorly studied species, despite its importance for human feeding and its potential for breeding because of its disease resistance and against some abiotic factors as cold tolerance [5]. Ayocote has been found to be a promising source of proteins (185.3 mg/kg), carbohydrates (677.6 mg/kg), fiber (67.4 mg/kg) and minerals. Glutamatic acid was the most abundant amino acid (32.2 to 35.8 g/kg). The purple variety contains the highest amount of total phenolic compounds (2075.9 mg GAE/kg), total flavonoids (1612.9 mg QE/kg) and total anthocyanins (1193.2 mg CGE/kg) [6].

Additionally, the common bean (*Phaseolus vulgaris* L.) is the most important food legume used for direct consumption worldwide. Compared with other food crops, it has one of the widest ranges of variation in growth habits, seed characteristics (size, shape, and color), maturation times, and adaptation [7]. Bean proteins are rich in lysine, being an excellent complement to cereal proteins such as rice or corn, which are deficient in this amino acid [8]. And those legumes are adequate sources of complex carbohydrates, usually the predominant fraction is starch (50–60 g/100 g) and dietary fiber [9]. More importantly, the starch has slow digestion properties mainly attributed to the amylose content of beans have been cataloged as low-medium GI foods [10]. Recently, the bioactive compounds associated to beans, mainly constituted by saponins and flavonoids, have shown potential to prevent chronic diseases and cancer [11].

The food industry is interested in the development on cereal products with higher nutritional value. This work investigates the effects of the substitution of wheat flour by ayocote and black bean (AB and BB, respectively) in bread production on the thermal properties of composite flours, the bread's proximal composition, in vitro protein digestibility (IPD) as well as color and sensorial parameters of the final products. This use coul figure as an alternative application for underutilized legumes.

2. Materials and Methods

2.1. Ayocote Beans and Black Beans

Commercial grade wheat bread flour was used in this study (brand: San Antonio®, Tres estrellas, Toluca, México). The wheat flour had a proximal composition of 12.78% of protein and 13.87% of moisture. A commercial dry yeast (*Saccharomyces cerevisiae*) was used (Tradi-Pan®, Lesaffre, Toluca, México).

One variety of purple ayocote beans (AB) and one black bean (BB) were assayed in the present study. Both were purchased from a local market in Actopan, Hidalgo, México. Physical dimension of beans, namely length, width and thickness were measured for 50 seeds using a digital Vernier caliper with an accuracy of 0.01 mm. BB seeds measured 1.06 ± 0.05 cm (length), 0.6 ± 0.06 cm (width) and 0.43 ± 0.05 cm (thickness). AB measured 1.76 ± 0.06 cm (length), 1.16 ± 0.06 cm (width) and 0.80 ± 0.10 cm (thickness). Seeds were ground using a 900 W Nutribullet® blender (Nutribullet, Los Ángeles, CA, USA) to obtain flour. It was then sieved through a 40 US mesh until all portions passed and was stored at room temperature in polyethylene bags until analysis.

2.2. Flour and Bread Preparation

Four wheat-legume flour mixtures were prepared with 0, 10, 20, and 30% (total flour weight basis) for each bean. Flours were mixed for 3 min, and bread were elaborated accordingly with Yeh et al. [12] with slight modifications: double of yeast was added, and butter was replaced for vegetable oil (8 g/100 g of flour). Bread formulation is presented in Table 1 for each sample. The ingredients were mixed for 5 min, in a blender (Kitchen Aid, Mod K5SSWH, Wirlpool, Apodaca, NL, México) equipped with a spiral hook blade. The same method was applied for all samples. Fermentation was done in a fermentation cabinet at 25 °C with 60% RH, 10 min. Next, the dough (100 g) was again kneaded for 10 min and put in a greased mold (23.11 cm × 9.9 cm × 6.1 cm) for bread, increasing its volume for 20 min. Afterward, doughs were simultaneously baked in a pre-heated oven (Mod. HSPU, Sanson, Naucalpan, Edo de Mex, México) at 200 °C, for 15 min. After baking, the bread was cooled down to room temperature 1 h and then they were stored in polyethylene bags for 24 h and subsequently cut into slices (1.27 cm thick) using an electric knife (type EK08, Hamilton Beach, Mexico City, CDMX, México). Slices were used for sensorial analysis one day after baking and other slices were dried at room temperature until moisture around 12% and were milled for further analysis.

Table 1. Formula (in baker's percent) of breads added with black bean and ayocote bean.

Ingredients	Control Bread	Bread with Black Bean			Bread with Ayocote Bean		
		10	20	30	10	20	30
Wheat flour	100	90	80	70	90	80	70
Ayocote flour	-	10	20	30	-	-	-
Black bean flour	-	-	-	-	10	20	30
Yeast	1	1	1	1	1	1	1
Milk power	4	4	4	4	4	4	4
Sugar	8	8	8	8	8	8	8
Salt	1.5	1.5	1.5	1.5	1.5	1.5	1.5
Water	63	63	63	63	63	63	63
Butter	4	4	4	4	4	4	4

2.3. Proximal Analysis

Chemical characterization was made in flours and bread samples, following official methods: moisture, ash, protein, crude fat, and fiber according to AACC 44–15.02, 08–01.01, 46–13.01, 30–25.01, and 32–10.01, respectively (AACC 2000). Carbohydrates were determined by difference [100 − (crude protein + crude fat + ash+ crude fiber)].

The nutritional value of protein bread was calculated using conversion factors according to EU Regulation No 1169/2011 [11] on the provision of food information to consumers: (1) Carbohydrates (except polyols): 4 kcal·g^{-1}; (2) Protein, 4 kcal·g^{-1}; (3) Fat, 9 kcal·g^{-1}; (4) Fiber, 2 kcal·g^{-1}

2.4. Thermal Properties of Bread Incorporated with Black Bean and Ayocote Bean

Before carrying the thermal analysis, breads were dried at 40 °C for 72 h until the moisture ranged from 12.5–12.0%, three hours post-baking. Then they were milled and sieved through a 60 US mesh until all samples passed the mesh. This process was done in triplicate for each sample. Thermal analyses were conducted as reported by Santiago-Ramos et al. [13] using a differential scanning calorimeter (DSC1 model 821, Mettler Toledo, Greifensee, Switzerland) previously calibrated with indium. Five milligrams of the sample were weighed in an aluminum pan, and deionized water was added until the moisture content reached 60%, using an empty aluminum pan as a reference. The sample was heated from 30 to 130 °C at a rate of 10 °C min^{-1}. Transition endotherms were characterized by their onset, peak, and final transition temperatures and enthalpy change. Each treatment was analyzed in triplicate.

2.5. In Vitro Protein Digestibility (IPD)

This determination was assessed with the procedure reported by Hsu et al. 1977 [14], briefly described: a 50 mL quantity of protein suspension of 6.25 mg/mL of protein was prepared and adjusted to pH 8. The protein suspension was prepared with dry bread adjusting weights accordingly with protein bread content. As was described previously, bread samples were dried, grinding and subsequently sieved with a 40 US mesh until the complete sample passed. Separately, a multi-enzymatic solution was prepared by mixing trypsin (1.6 mg/mL), chymotrypsin (3.1 mg/mL), and *S. griseus* protease (1.3 mg/mL). The protein solution was maintained in agitation, and 5 mL of the multi-enzymatic solution was added. The pH drop was recorded over a 10 min period using a potentiometer previously calibrated with standard pH solutions. Enzymes used were porcine pancreatic trypsin type IX-S (T4799, Sigma Aldrich, Burlington, MA, USA), bovine pancreatic chymotrypsin type

II (C4129, Sigma Aldrich), and *S. griseus* protease type XIV (P5147, Sigma Aldrich). The percent of protein digestibility was calculated using the following equation [14]:

$$IPD = 210.464 - 18.103 \times (pH \text{ at } 10 \text{ min digestion})$$

2.6. Bread Characterization

2.6.1. Color

For each bread, crumb color was analyzed accordingly with Figuera-Cárdenas et al. [15] using a MiniScan XE Plus colorimeter (Hunter Lab, Virginia, United States, bidirectional illumintation 45°/0°, Illuminants D 65 and observer 2°).

Values reordered were: L* (lightness), a* (redness: green to red), and b* (yellowness: blue to yellow). Calibration was carried out using a standard white tile manufactured by the provider. Results were the average of 10 measures of bread per formula measured one day post-baking.

2.6.2. Specific Volume

Regarding specific volume, bread were weighed after cooling for 1 h, and their volume (cm^3) was determined by the millet displacement method. The specific volume (cm^3/g) was calculated as loaf volume/bread weight [16].

2.7. Sensorial Evaluation

Sensory evaluation was undertaken on 1 st-day post-baking with 40 semi-trained panelists. Panelists were familiar with white bread. Written consent was obtained from each subject after explaining the procedure of them. Sensory evaluation was conducted in a laboratory with adequate light and space based on the methodology described by Millar et al. [17] with some modifications.

Briefly, samples were placed on white polystyrene plates labeled with random codes and presented to consumers in a randomized order. A 60 s time laps were employed between each sensory palate to reduce sensory fatigue, and potable water was used as a palate cleanser between tastings. Each panelist assessed all three samples of each bread with legume and was asked to indicate his or her opinion about odor, texture, color, and flavor of the products, using a 5-point hedonic scale (from 5: like extremely to 1: dislike extremely). The samples could be re-tasted as often as desired.

2.8. Statistical Analysis

Statistical differences were assessed using one-way ANOVA followed by Tukey's multiple comparison test at a significance level of 0.05. The statistical analysis was performed using the JMP 5.0.1 software (SAS, Cary, NC, USA).

3. Results and Discussion

3.1. Beans and Bread Chemical Composition

Table 2 presents the proximal composition of BB flour, AB flour, and bread elaborated with flour mixes (wheat flour and BB or AB). Ayocote bean and black bean reported different protein content, with ayocote depicting the higher percentage (23.94%). Other authors have reported a range of 19.5–31.0% of crude protein in beans [18]. Alvarado-López et al. [6] considerate ayocote as a promising source of protein. Protein content in bread increased with the addition of the legumes in a range of 15–34%. In general, legume proteins are rich in lysine and threonine [19]; both essential amino acids are limited in durum wheat [20]. Ash content for bread added with both legumes was higher, indicating an increment in mineral content, as Hoehnel et al. [4] presented previously, legumes flours presented higher mineral content. The same pattern was reported for crude fiber. Both legumes improved the chemical composition of bread, demonstrating the possibility of legumes utilization to increase the nutritional quality of bread in terms of protein, fiber, and ash content, especially in bread added with more AB and BB flour levels, but this impact in sensorial and quality attributes as presented below. Finally, energy value was evaluated, with an increment in this parameter because beans have high energy values. This variable in bread with AB and BB increased mainly by protein coming from AB and BB. Percentage of calories provided from protein presented a range of 16.33–18.94 in breads added with at AB and BB flours, in contrast with the control where kilocalories provided from protein were 14.50%. A similar pattern was presented for fiber, where the percentage of kilocalories coming from this nutrient increased from 0.32 to 0.39–1.22%. In contrast, kilocalories coming from carbohydrates and fat decreased in bread when wheat flour was replaced with AB and BB. The contribution of kilocalories from carbohydrates decreased from 60.22 in control bread to a range of 53.68 to 58.38% in breads with AB and BB%. Regarding energy, similar values for breads were identified previously by Ahmad et al. [21]. They concluded that protein and fiber kilocalories were higher in cumin and caraway seeds. Additionally, the difference in fat is because seed presented higher fat content than beans studied in this project.

Table 2. Proximate composition of black bean (BB), ayocote bean (AB), control bread and breads with 10, 20 and 30% of wheat flour substitution with BB and AB in dry weight basis *.

Samples	Crude Protein (%)	Ash (%)	Crude Fiber (%)	Crude Fat (%)	Carbohydrates (%)	Energy Value (Kcal/100 g)
Black bean (*Phaseolus vulgaris* L.) flour	21.06 ± 0.13	4.20 ± 0.04	3.06 ± 0.07	1.68 ± 0.26	59.43 ± 1.22	342.20 ± 0.12
Ayocote bean (*Phaseolus coccineus* L.) flour	23.94 ± 0.36	3.70 ± 0.13	5.21 ± 0.53	1.31 ± 0.02	56.36 ± 1.28	343.41 ± 0.34
Control bread	12.60 ± 0.06 d	0.19 ± 0.01 c	0.57 ± 0.09 c	9.63 ± 0.31 b	52.31 ± 1.43 a	347.45 ± 0.05 f
Bread with black bean (*Phaseolus vulgaris* L.)						
10%	14.67 ± 0.12 c	0.23 ± 0.03 c	0.71 ± 0.12 c	11.53 ± 0.35 a	48.86 ± 1.55 ab	359.31 ± 0.10 c
20%	15.23 ± 0.09 bc	0.27 ± 0.02 b	0.98 ± 0.17 c	11.67 ± 0.37 a	50.70 ± 1.31 ab	370.71 ± 0.07 b
30%	15.72 ± 0.06 ab	0.34 ± 0.07 a	1.24 ± 0.13 b	11.90 ± 0.18 a	49.98 ± 0.35 ab	372.38 ± 0.09 a
Bread with ayocote bean (*Phaseolus coccineus* L.)						
10%	14.84 ± 0.10 bc	0.21 ± 0.04 c	1.61 ± 0.22 b	8.62 ± 0.45 c	50.32 ± 1.18 ab	341.44 ± 0.35 g
20%	16.33 ± 0.10 a	0.27 ± 0.01 b	2.03 ± 0.45 a	10.32 ± 0.18 b	48.20 ± 1.79 b	355.06 ± 0.08 e
30%	16.88 ± 0.09 a	0.39 ± 0.04 a	2.18 ± 0.28 a	9.63 ± 0.29 b	49.47 ± 1.63 ab	356.43 ± 0.09 d

* The values represent the mean ± standard deviation. Values with different letters in the same column for each sample are significantly different ($p \leq 0.05$).

3.2. Thermal Properties of Bread Incorporated with Black Bean and Ayocote Bean

Results for thermal properties: retrogradation, gelatinization, and amylose lipid complexes + protein digestibility of black bean, ayocote bean, and bread can be observed in Tables 3 and 4. Differences in thermal profiles were associated with variations in chemical composition and quality protein fractions. Retrogradation transition temperatures To_R, Tp_R, and Tf_R and enthalpy for control bread and bread with 10 and 20% of black bean were 46.27 °C, 54.23 °C, 63.62 °C, and 0.25 J/g (in average) Xie et al. [22] declared that amylopectin melting peak temperatures were 60–70 °C for bread stored at high temperature. Some samples did not present retrogradation endotherm because those bread had low staling time. Only in control and samples with10 and 20% of BB was this endotherm observed; those differences are governed because in those samples, amylopectin presented a high availability to reassociate in an ordered structure because amylopectin formed a smaller number of links with protein and lipids. Addition of 30% black bean and ayocote bean to bread limited the bread retrogradation. Abdel-Aal and Hucl. [22] indicated that protein slowed down the bread staling process. Additionally, an increase in fat content in bread increases amylose lipid complexes (Table 4), associated with reducing the retrogradation process [23]. Gelatinization temperatures (To_g, Tp_g, and Tf_g) for black bean and ayocote bean on average were 67.76 °C, 79.23 °C, and 88.63 °C; similar values were identified previously by Ahmad et al. [21] in black bean. Also, gelatinization endotherm presented differences among control bread and samples with beans. Control presented a higher endotherm than the other samples. Gelatinization was higher in control bread, followed by bread with lower ayocote and black bean concentration; endotherm for control was undetectable because starch was completely gelatinized. Lower gelatinization temperatures are correlated with lower starch digestibility. In the third thermal transition (Table 4), the temperature range observed for raw beans were at 94–104.6 °C (average), this is similar to data reported by Santiago-Ramos et al. [23]. Some authors have reported that this endotherm corresponds to the melting of amylose lipid complexes, and denaturalization of heat resistant protein fractions. For breads added with ayocote and black bean, this endotherm presented higher enthalpy values, this could be attributed to the addition of ayocote and black bean, indicating that those breads have more protein, protein content and ΔH of amylose lipid and protein denaturalization presented a correlation (r) of 0.45, also these breads also presented more amylose lipid complexes. Protein and fat were increased by BB and AB addition, and the beans impacted chemical bread composition, also when more fat is added if amylose is not saturated, it can form amylose lipid complexes, which are classified as part of the resistant starch (Type 5).

Table 3. Retrogradation and gelatinization properties of black bean (BB), ayocote bean (AB), and breads by DSC analysis *.

	Melting of Retrograded Starch				Starch Gelatinization			
Sample	ToR (°C)	TpR (°C)	TfR (°C)	ΔH (J/g)	To g (°C)	Tpg (°C)	Tf g (°C)	ΔH (J/g)
Black bean flour					73.44	85.80	93.48	3.15
Ayocote bean flour					62.08	72.66	83.78	4.49
Control bread	45.83 ± 1.87 a	55.32 ± 0.35 a	65.65 ± 0.47 a	0.26 ± 0.08 a				
Bread with black bean (*Phaseolus vulgaris* L.)								
10%	47.67 ± 2.13 a	53.55 ± 0.10 b	62.25 ± 0.21 b	0.21 ± 0.05 a	70.93 ± 0.28 bc	80.03 ± 0.17 b	86.13 ± 0.22 c	0.14 ± 0.01 c
20%	45.33 ± 0.48 a	53.83 ± 0.35 b	62.96 ± 0.48 b	0.28 ± 0.05 a	67.40 ± 0.39 e	81.63 ± 0.49 a	86.46 ± 0.57 c	0.27 ± 0.05 bc
30%					73.10 ± 0.54 a	79.83 ± 0.12 b	84.51 ± 0.34 d	0.56 ± 0.08 a
Bread with ayocote bean (*Phaseolus coccineus* L.)								
10%					66.72 ± 0.48 d	74.50 ± 0.12 c	89.12 ± 0.47 b	0.18 ± 0.04 c
20%					72.44 ± 0.44 ab	73.18 ± 0.23 d	90.12 ± 0.34 ab	0.22 ± 0.10
30%					70.35 ± 1.01 c	74.25 ± 0.18 c	90.17 ± 0.12 a	0.44 ± 0.05 ab

* The values represent the mean ± standard deviation. Values with different letters in the same column for each sample are significantly different ($p \leq 0.05$). ToR: onset retrogradation temperature; TpR: peak retrogradation temperature; TfR: Final retrogradation temperature; AH: enthalpy; Tog: onset gelatinization temperature; Tpg: peak gelatinization temperature; Tfg: final gelatinization temperature.

Table 4. Amylose lipid and protein denaturalization of black bean, ayocote bean, and breads by DSC analysis *.

	Melting of Amylose-Lipid Complexes + Protein Denaturalization			
Sample	$To_{AMLC+PD}$ (°C)	$Tp_{AMLC+PD}$ (°C)	$Tf_{AMLC+PD}$ (°C)	ΔH (J/g)
Black bean flour	97.70 ± 0.34	101.95 ± 0.54	106.17 ± 0.32	1.65 ± 0.09
Ayocote bean flour	90.21 ± 0.26	96.47 ± 0.41	103.02 ± 0.17	1.71 ± 0.06
Control bean	100.05 ± 0.14 b	106.48 ± 0.47 a	113.20 ± 0.20 a	0.78 ± 0.04 d
Bread with black bean (*Phaseolus vulgaris* L.)				
10%	89.28 ± 0.27 e	105.57 ± 0.45 a	109.55 ± 0.55 bc	0.68 ± 0.07 d
20%	88.25 ± 0.25 f	102.43 ± 0.60 b	109.83 ± 0.63 bc	0.97 ± 0.06 d
30%	101.97 ± 0.50 a	106.47 ± 0.27 a	109.90 ± 0.49 b	1.38 ± 0.08 c
Bread with ayocote bean (*Phaseolus coccineus* L.)				
10%	94.34 ± 0.56 c	102.46 ± 0.26 b	108.12 ± 0.12 cd	1.48 ± 0.19 c
20%	95.14 ± 0.34 c	103.13 ± 0.33 b	108.63 ± 0.63 bcd	2.08 ± 0.08 b
30%	93.17 ± 0.27 d	103.59 ± 0.97 b	106.97 ± 1.14 d	2.50 ± 0.18 a

* The values represent the mean ± standard deviation. Values with different letters in the same column for each sample are significantly different ($p \leq 0.05$). $To_{AMLC+PD}$: onset temperature of amylose lipid complexes + protein denaturalization; $Tp_{AMLC+PD}$: peak temperature of amylose lipid complexes + protein denaturalization; $Tf_{AMLC+PD}$: final temperature of of amylose lipid complexes + protein denaturalization.

3.3. In Vitro Protein Digestibility

As previously described, in vitro protein digestibility was determined with a multienzymatic method. Protein digestibility is defined as the ability of an organism to hydrolyze, absorb, and use amino acids linked by peptide bonds [24]. Figure 1 shows the results of in vitro protein digestibility of control bread, bread added with ayocote bean and black bean, and raw beans.

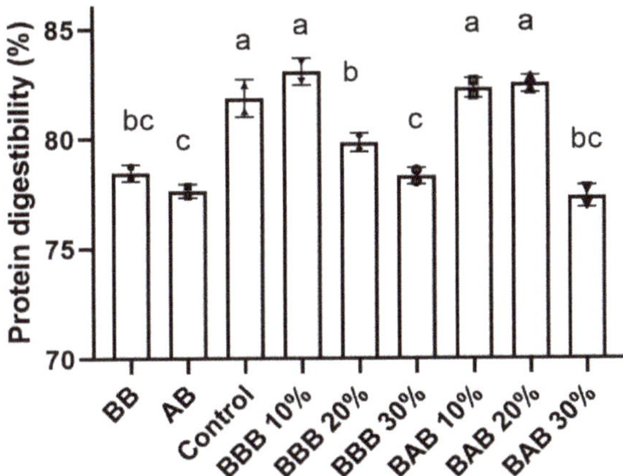

Figure 1. Protein digestibility of black bean, ayocote bean and breads. BB: Black bean flour; AB: Ayacote bean flour; BBB Bread with black bean; BAB: Bread with ayacote beans.

Raw beans and bread with 30% of wheat flour substitution presented lower protein digestibility than bread with 10 and 20% of substitution. Breadmaking is a complex process that involves various steps, [25] investigated the protein digestibility changes during breadmaking using gluten-containing and gluten-free flours. Protein digestibility is shown to increase during fermentation/proofing. Additionally, [26] argue that food processing techniques have been demonstrating to improve the nutritional quality of plant proteins by eliminating or inactivating the compounds that can be diminished in vitro protein digestibility. For this reason, it seems that protein digestibility was incremented by inactivation of antinutritional compounds and fermentation, in samples with 10 and 20% of AB and 10% of BB have similar IPD than control. However, when concentrations are greater than those percentages, there is a downgrade in IPD, as [25] identified, when there is an increase in fiber content, there is a reduction in IPD, IPD and fiber content presented a negative correlation (r = −0.6; $p \leq 0.05$). Samples with AB presented a reduction in higher concentrations (>20%) because AB has higher fiber content. This behavior has been reported previously by other authors, more enzyme accessibility to the protein could be improved by more AB and BB processing to gain more protein enzyme access and increase IPD; however, it plays a key role in the use of non-thermal processes to get a better nutritional profile without degrading nutraceutical and vitamins.

3.4. Bread Characterization

3.4.1. Color

Color is a key parameter that is related to consumer acceptance and is a quality parameter for bread. Additionally, beans color is associated with the existence of phenolic compounds, such as flavonol glycosides, anthocyanins and condensed tannins, those components are considered health-promoting compounds, phenolic compounds contribute to antioxidative activities, removing free radicals, chelating metal catalysis, activating antioxidant enzymes and inhibited oxidases [6]. In this study, bread with different formulations were evaluated for color. Bread denoted color differences, with values were summarized in Table 5. The parameter L* corresponds to the lightness of the bread, a trait that decreased gradually with the addition of ayocote bean and black bean, mainly by coat color of both; however, as can be observed in Table 5, black bean influences more in this parameter compared with ayocote bean flour with values ranging from 54.56 to 58.57 and 62.26 to 63.55, respectively. As other authors have reported, bread L* values between 51–100 indicate light products [27], as depicted in Table 5. However, the influence of BB and AB in L is clear as the L value decreased when legume flour increases.

Table 5. Color and sensory evaluation of breads *.

Samples	%	Color			Sensorial Analysis [b]			
		L*	A*	B*	Odor	Flavor	Texture	Color
Control		65.25 ± 0.18 a	5.07 ± 0.07 a	15.32 ± 0.07 a	4.25 ± 0.48 a	4.15 ± 0.48 a	4.15 ± 0.63 a	3.85 ± 0.81 a
Black bean (*Phaseolus vulgaris* L.) breads	10	58.57 ± 0.10 d	2.80 ± 0.05 d	10.70 ± 0.12 e	3.80 ± 0.69 ab	3.80 ± 0.69 a	3.75 ± 0.71 a	3.85 ± 0.81 ab
	20	58.51 ± 0.41 d	1.95 ± 0.08 e	8.83 ± 0.14 f	2.95 ± 0.94 bc	2.90 ± 0.85 b	2.50 ± 0.88 b	2.90 ± 0.78 cd
	30	54.56 ± 0.51 e	1.63 ± 0.08 f	6.98 ± 0.16 g	1.85 ± 1.08 d	1.95 ± 0.88 c	2.30 ± 0.86 b	2.05 ± 0.75 e
Ayocote (*Phaseolus coccineus* L.) breads	10	63.55 ± 0.30 b	4.72 ± 0.15 b	14.55 ± 0.24 b	3.95 ± 0.68 a	3.79 ± 1.05 a	3.55 ± 0.71 a	3.80 ± 0.73 a
	20	62.51 ± 0.28 c	4.12 ± 0.01 c	13.10 ± 0.09 c	2.20 ± 1.00 cd	2.45 ± 0.75 bc	2.35 ± 0.67 b	3.00 ± 0.65 cd
	30	62.26 ± 0.46 c	3.90 ± 0.05 c	12.43 ± 0.10 d	1.80 ± 1.00 d	2.10 ± 0.78 c	2.20 ± 0.85 b	2.55 ± 0.51 de

* The values represent the mean ± standard deviation. Values with different letters in the same column for each sample are significantly different ($p \leq 0.05$). [b] Evaluated on a scale of 1 to 5, where 5 was the maximum value.

Additionally, positive values for a* and b* indicate redness and yellowness, respectively [27]. Control bread presented a b* value of 15.32; in contrast, bread added with AB presented a range from 12.43 to 14.55, and bread with BB showed a range from 6.98 to 10.7. Lower b* values were observed in bread with major wheat flour substitution because yellowness is given by wheat flour. Values obtained for a* parameter indicated that bread added with AB are more reds than other bread. Ayocote had a coat with more red pigments as anthocyanins. Hence, it can be concluded that BB and AB contribute to changing the color values of the final product.

3.4.2. Specific Volume

Specific volume is a quality parameter in bread. Bread with 10 and 20% of added BB presented no differences in specific volume compared with control; the same was true for 10% AB bread (Figure 2). However, when BB was used at 30% and AB above 20%, it resulted in a significantly lower volume. The addition of black and ayocote bean decreased the gluten concentration in breads. This protein plays a crucial role in guaranteeing wheat's bakery quality and influences water absorption, cohesion, viscosity, extensibility, elasticity, resistance deformation, tolerance to kneading, ability to gas retention and dough strengthening properties [28,29]. Besides, those breads contain more fiber, as was observed in the chemical composition (Table 2), and according to [30], this addition contributes to gluten dilution in the food matrix. As explained before, the gluten protein forms a three-dimensional network that entraps carbon dioxide in bread and in the case of breads with high bean flour addition, the network was reduced, and less CO_2 could be maintained decreasing volume. This can also influence the texture and softness of the bread.

3.5. Sensorial Analysis

Results of sensory evaluation of breads samples containing different beans levels are shown in Table 5. Four parameters (odor, texture, color, and flavor) were evaluated to determine the breads' acceptance. The first parameter evaluated was the odor, which is an essential sensorial parameter in organoleptic characteristics. According to data presented in Table 5 the panelist preferred control bread, it obtained 4.25 on a scale where 5 was the maximum, followed in acceptance for breads with 10% with a 3.80 and 3.95 for black bean and ayocote addition, in contrast, an addition of 30% of both legumes caused the least acceptance (1.80 to 1.85), which can be related to the characteristic odor of BB and AB, more detectable at this concentration. The taste had a similar pattern that smell; it seems that legumes used are detectable for panelists at a concentration above 20%. For texture, panelists preferred control and breads at 10%, followed by 20 and 30%. Texture by sensorial analysis presented a negative correlation with fiber (r = −0.92 and r = −0.98) and protein content (r = −0.93 and, they presented content and r= −0.98) for breads with black bean and ayocote bean respectively. As defined before, the texture was influenced negatively by gluten dilution by other macromolecules added from beans, avoiding air retention and affecting softness. A similar pattern has been reported before by Ahmad et al. [21] and Hoehnel et al. [4]. It has been reported that although fiber has a positive nutritional effect, it can reduce the expansion of the gas cells leading to a lower volume of loaves and smaller porosity of the crumbs [2].

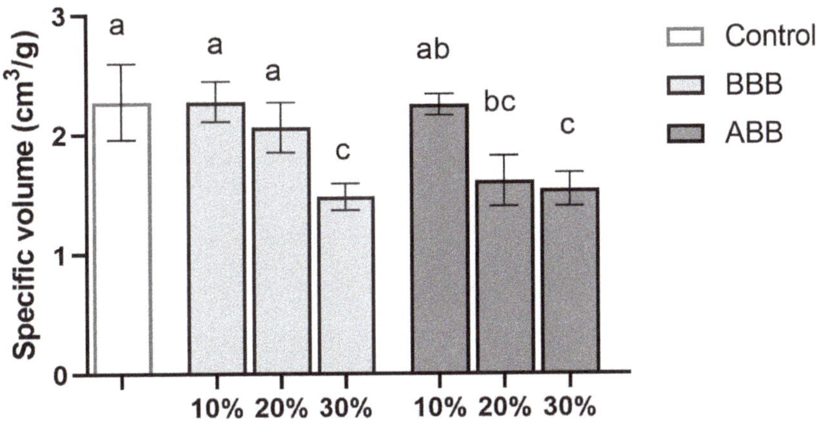

Figure 2. The volume of control bread and breads added with black bean and ayocote bean. BBB: Bread with black beans and ABB: Bread with ayocote beans.

Finally, panelists could detect color differences, like changes measured by a colorimeter. Moreover, as explained above, those color modifications are associated with pigments present in the bean coatings. Therefore, the sensorial analysis showed that it could substitute 10% black bean and ayocote bean without perceived changes in a sensorial panel. Other studies indicated that chestnut and coconut can be used at percentages up to 15% similar to the consumer acceptance level seen in this study [2].

4. Conclusions

Breads fortified with BB and AB offer people the option to improve the nutritional quality of their diet, specifically the protein content, fiber, and fat content, and is a good option for low-income people. Additionally, an ash increment also denoted a higher mineral content, however, mineral-specific analysis is needed to further evaluate this aspect. The energy content was higher for breads with bean flour addition, and most of this calorie increment came from protein, fiber, and fat. Moreover, those breads added with beans showed similar protein digestibility, except when used at levels higher than 20%. Thermal analysis suggests that gelatinization temperature is reduced, and consequently, starch digestibility too. Finally, the addition of black bean and ayocote impacts some quality characteristics, as color and volume. Volume was affected when the substitution level was above 20%. Regarding color, it changed in all breads with BB and AB; however, it does not influence panelist acceptance by sensory evaluation when the addition is below 20%.

Author Contributions: Conceptualization, R.M.M.-M.; methodology, R.M.M.-M., C.C.-H., J.d.D.F.-C. and S.O.S.-S.; formal analysis, R.M.M.-M., C.C.-H., J.d.D.F.-C. and S.O.S.-S.; writing—original draft preparation, R.M.M.-M. and C.C.-H.; writing—review and editing, C.C.-H., J.d.D.F.-C. and S.O.S.-S. All authors have read and agreed to the published version of the manuscript.

Funding: R.M.M.-M. received financial support from the for the publishing fees from Universidad Iberoamericana.

Institutional Review Board Statement: Sensorial evaluation was performed according with IFST (Institute of Food Science and technology) Guidelines for Ethical and Professional Practices for the Sensory Analysis of Foods.

Informed Consent Statement: Informed consent was obtained from all subjects involved in the study.

Data Availability Statement: The data presented in this study are available on request from the corresponding author.

Acknowledgments: Authors are grateful with Universidad Iberoamericana Ciudad de Mexico. And to Alondra Ramírez Hernández, Jose Juan Veles Medina and Rosa Selene Espiricueta Candelaria to their support for this work. Additional thanks are due the Centro de Agua para América Latina y el Caribe del Tecnológico de Monterrey.

Conflicts of Interest: The authors declare no conflict of interest.

References

1. Dewettinck, K.; van Bockstaele, F.; Kuhen, F.; Van de Walle, D.; Courtens, T.M.; Gellynck, X. Nutritional value of bread: Influence of processing, food interaction and consumer perception. *J. Cereal Sci.* **2008**, *48*, 243–257. [CrossRef]
2. Raczyk, M.; Kruszewski, M.; Michalowska, D. Effect of Coconut and Chestnut Flour Supplementations on Texture, Nutritional and Sensory Properties of Baked Wheat Based. *Molecules* **2021**, *26*, 4641. [CrossRef]
3. Krupa-Kozak, U.; Drabińska, N.; Baczek, N.; Šimková, K.; Starowicz, M.; Jeliński, T. Application of broccoli leaf powder in gluten-free bread: An innovative approach to improve its bioactive potential and technological quality. *Foods* **2021**, *10*, 819. [CrossRef]
4. Hoehnel, A.; Axel, C.; Bez, J.; Arendt, E.K.; Zannini, E. Comparative analysis of plant-based high-protein ingredients and their impact on quality of high-protein bread. *J.Cereal Sci.* **2019**, *89*, 102816. [CrossRef]
5. Ruíz-Salazar, R.; Mayek-Pérez, N.; Vargas-Vázquez, M.L.; Hernández-Delgado, S.; Muruaga-Martínez, J. Análisis de la estructura poblacional del frijol Ayocote (*Phaseolus coccineus* L.) mediane AFLP. *Polibotánica* **2019**, *46*, 13–24. [CrossRef]
6. Alvarado-López, A.N.; Gómez-Oliván, L.M.; Heredia, J.B.; Baeza-Jiménez, R.; Garcia-Galindo, H.S.; Lopez-Martinez, L.X. Nutritional and bioactive characteristics of Ayocote bean (*Phaseolus coccienus* L.): An underutilized legume harvested in Mexico. *CYTA-J. Food* **2019**, *17*, 199–206. [CrossRef]
7. Available online: http://www.fao.org/documents/card/en/c/f961bcd6-85db-405e-af70-3ed044f1b1d7/ (accessed on 29 August 2021).
8. Gepts, P. *Phaseolus vulgaris* (Beans). In *Encyclopedia of Genetics*; Academic Press: Cambridge, MA, USA, 2001; pp. 1444–1445.
9. Brummer, Y.; Kaviani, M.; Tosh, S.M. Structural and functional characteristics of dietary fibre in beans, lentils, peas and chickpeas. *Food Res. Int.* **2015**, *67*, 117–125. [CrossRef]
10. Rizkalla, S.W.; Bellisle, F.; Slama, G. Health benefits of low glycaemic index foods, such as pulses, in diabetic patients and healthy individuals. *Br. J. Nutr.* **2002**, *88*, 255–262. [CrossRef]
11. De la Rosa-Millán, J.; Heredia-Olea, E.; Perez-Carrillo, E.; Guajardo-Flores, D.; Serna-Saldívar, S.R.O. Effect of decortication, germination and extrusion on physicochemical and in vitro protein and starch digestion characteristics of black beans (*Phaseolus vulgaris* L.). *LWT* **2019**, *102*, 330–337. [CrossRef]
12. Yeh, L.T.; Charles, A.L.; Ho, C.T.; Huang, T.C. A novel bread making process using salt-stressed Baker's yeast. *J. Food Sci.* **2009**, *74*, S399–S4002. [CrossRef]
13. Santiago-Ramos, D.; de Figueroa-Cárdenas, J.D.; Véles-Medina, J.J.; Mariscal-Moreno, R.M. Changes in the thermal and structural properties of maize starch during nixtamalization and tortilla-making processes as affected by grain hardness. *J. Cereal Sci.* **2017**, *74*, 72–78. [CrossRef]
14. Hsu, H.W.; Vavak, D.L.; Satterlee, L.D.; Miller, G.A. A Multienzyme Technique for Estimating Protein Digestibility. *J. Food Sci.* **1977**, *42*, 1269–1273. [CrossRef]
15. Figueroa Cárdenas, J.D.; Godinez, M.G.A.; Méndez, N.L.V.; Guzmán, A.L.; Acosta, L.M.F.; González-Hernández, J. Fortificacion y evaluacion de tortillas de nixtamal. *Arch. Latinoam. Nutr.* **2001**, *51*, 293–302.
16. Tomic, J.; Torbica, A.; Popovic, L.; Rakita, S.; Zivancev, D. Breadmaking potential and proteolytic activity of wheat varieties from two production years with different climate conditions. *Food Feed Res.* **2015**, *42*, 83–90. [CrossRef]
17. Millar, K.A.; Barry-Ryan, C.; Burke, R.; Hussey, K.; McCarthy, S.; Gallagher, E. Effect of pulse flours on the physiochemical characteristics and sensory acceptance of baked crackers. *Int. J. Food Sci. Technol.* **2017**, *52*, 1155–1163. [CrossRef]
18. Carbas, B.; Machado, N.; Oppolzer, D.; Ferreira, L.; Brites, C.; Rosa, E.A.S.; Barros, A.I.R.N.A. Comparison of near-infrared (NIR) and mid-infrared (MIR) spectroscopy for the determination of nutritional and antinutritional parameters in common beans. *Food Chem.* **2020**, *306*, 125509. [CrossRef] [PubMed]

19. Alonso, R.; Orúe, E.; Zabalza, M.J.; Grant, G.; Marzo, F. Effect of extrusion cooking on structure and functional properties of pea and kidney bean proteins. *J. Sci. Food Agric.* **2000**, *80*, 397–403. [CrossRef]
20. Abdel-Aal, E.S.M.; Hucl, P. Amino acid composition and in vitro protein digestibility of selected ancient wheats and their end products. *J. Food Compos. Anal.* **2002**, *15*, 737–747. [CrossRef]
21. Ahmad, B.S.; Talou, T.; Straumite, E.; Sabovics, M.; Kruma, Z.; Saad, Z.; Hijazi, A.; Mera, O. Protein bread fortification with cumin and caraway seeds and by-product flour. *Foods* **2018**, *7*, 28. [CrossRef]
22. Xie, F.; Dowell, E.; Sun, X.S. Using Visible and Near-Infrared Reflectance Spectroscopy and Differential Scanning Calorimetry to Study Starch, Protein, and Temperature Effects on Bread Staling. *Cereal Chem.* **2004**, *81*, 249–254. [CrossRef]
23. Santiago-Ramos, D.; Figueroa-Cárdenas, J.D.; Véles-Medina, J.J.; Salazar, R. Physicochemical properties of nixtamalized black bean (*Phaseolus vulgaris* L.) flours. *Food Chem.* **2017**, *240*, 456–462. [CrossRef]
24. De Jongh, H.H.J.; Broerse, K. Application Potential of Food Protein Modification. In *Advances in Chemical Engineering*, 1st ed.; Nawaz, Naveed, S., Eds.; In TEch: London, UK, 2012; Volume 1, p. 584.
25. Wu, T.; Taylor, C.; Nebl, T.; Ng, K.; Bennett, L.E. Effects of chemicpp. al composition and baking on in vitro digestibility of proteins in breads made from selected gluten-containing and gluten-free flours. *Food Chem.* **2017**, *233*, 514–524. [CrossRef] [PubMed]
26. Gomes Almeida Sá, C.B.; Franco, M.Y.M.; Mattar, B.A.C. Food processing for the improvement of plant proteins digestibility. *Crit. Rev. Food Sci. Nutr.* **2020**, *60*, 3367–3386. [CrossRef]
27. Lafarga, T.; Álvarez, C.; Bobo, G.; Aguiló-Aguayo, I. Characterization of functional properties of proteins from Ganxet beans (*Phaseolus vulgaris* L. var. Ganxet) isolated using an ultrasound-assisted methodology. *LWT* **2018**, *98*, 106–112. [CrossRef]
28. Lazaridou, A.; Duta, D.; Papageorgiou, M.; Belc, N.; Biliaderis, C.G. Effects of hydrocolloids on dough rheology and bread quality parameters in gluten-free formulations. *J. Food Eng.* **2007**, *79*, 1033–1047. [CrossRef]
29. Wieser, H. Chemistry of gluten proteins. *Food Microbiol.* **2007**, *24*, 115–119. [CrossRef] [PubMed]
30. Martínez, M.M.; Díaz, Á.; Gómez, M. Effect of different microstructural features of soluble and insoluble fibres on gluten-free dough rheology and bread-making. *J. Food Eng.* **2014**, *142*, 49–56. [CrossRef]

Article

Effect of Gaseous Chlorine Dioxide Treatment on the Quality Characteristics of Buckwheat-Based Composite Flour and Storage Stability of Fresh Noodles

Zhiyuan Cheng [1], Xiaoping Li [1,*], Jingwei Hu [1], Xin Fan [1], Xinzhong Hu [1], Guiling Wu [2] and Yanan Xing [2]

[1] College of Food Engineering and Nutritional Science, Shaanxi Normal University, Xi'an 710062, China; zhiyuanhi@126.com (Z.C.); 18392535049@163.com (J.H.); 15091478392@163.com (X.F.); hxinzhong@126.com (X.H.)
[2] Cereal Industrial Technology Academy, Hebei Jinshahe Flour and Noodle Group/Hebei Cereal Food Processing Technology Innovation Centre, Xingtai 054100, China; vmm1021@126.com (G.W.); nanyaxing@126.com (Y.X.)
* Correspondence: xiaopingli@snnu.edu.cn; Tel.: +86-158-0929-8810

Abstract: In this study, the effects of gaseous chlorine dioxide treatment on the physicochemical properties of buckwheat-based composited flour (buckwheat-wheat-gluten) and shelf-life of fresh buckwheat noodles (FBNs), as well as the textural qualities and sensory properties of noodles were investigated. Chlorine dioxide treatment significantly reduced the total plate count (TPC) and the total flavonoids content in the mixed flour ($p < 0.05$), but the whiteness, development time and stability time were all increased. During storage, the microbial growth and darkening rate of FBNs made from chlorine dioxide treated buckwheat-based composite flour (CDBF) were delayed significantly, slowing the deterioration and improving storage stability of buckwheat noodles. In addition, chlorine dioxide treatment had no apparent adverse effect on the cooking loss and sensory characteristics during noodle storage. This finding would provide a new concept for the production of "low bacterial buckwheat-based flour" and have important consequences for the application of gaseous chlorine dioxide in food industry.

Keywords: gaseous chlorine dioxide; buckwheat-based composited flour; fresh buckwheat noodle; shelf-life and quality characteristics

Citation: Cheng, Z.; Li, X.; Hu, J.; Fan, X.; Hu, X.; Wu, G.; Xing, Y. Effect of Gaseous Chlorine Dioxide Treatment on the Quality Characteristics of Buckwheat-Based Composite Flour and Storage Stability of Fresh Noodles. *Processes* **2021**, *9*, 1522. https://doi.org/10.3390/pr9091522

Academic Editor: Yonghui Li

Received: 1 August 2021
Accepted: 23 August 2021
Published: 27 August 2021

Publisher's Note: MDPI stays neutral with regard to jurisdictional claims in published maps and institutional affiliations.

1. Introduction

Buckwheat is a kind of small coarse pseudocereal that belongs to the family Polygonaceae, which contains common buckwheat (*Fagopyrum esculentum*) and tartary buckwheat (*Fagopyrum tataricum* (L.) *Gaertn*). It is known as both medicine and food. Buckwheat is rich in resistant starch, dietary fiber, vitamins and minerals, which can prevent colon cancer, diabetes and obesity [1]. Due to its excellent nutritional value and unique protein composition, buckwheat is suitable for making gluten-free products, especially for celiac disease people [2]. Specifically, the Tartary buckwheat has more Vitamin B, crude fiber, minerals and flavonoids than those in common buckwheat [3], so it has received considerable attention as a functional food material in the food industry.

Noodle, the staple food in China, Korea and Japan, is widely enjoyed throughout the world because of its convenience and palatability [4]. In the past, because dried noodles and instant noodles were easy to preservation, many people chose to buy them. However, more and more consumers now prefer to the fresh noodles because of their good taste and nutritional quality [5]. To improve the nutrients and function of traditional wheat flour noodles, the production of buckwheat noodles is attracting a great deal of attention and buckwheat noodles have been consumed widely in most Asia countries and Italy. However, because of the lack of gluten in Tartary buckwheat, pure buckwheat noodles have poor processing and texture properties. To solve the problem, buckwheat noodles usually contain at least 20% wheat flour or other food additives to substitute gluten. In addition, because fresh buckwheat noodles have high water content and initial microorganisms, it has a shorter shelf-life when compared to the wheat noodles. Therefore, to better realize commercial production, some measures need to be taken to extend the shelf-life of fresh buckwheat noodles.

At present, many studies have been conducted to extend the shelf life of fresh noodles, through the application of chemical preservatives and physical preservation method (microwave, irradiation and modified packaging condition), that inhibited the microbial growth in noodle [6]. However, there methods have more or less some defects, such as consumers' doubts about the security of chemical preservatives, the unsatisfactory effect of one physical treatment, unsolved source problems with high microbial content in flour [7,8]. Therefore, alternative methods that can reduce or inhibit microbial growth in the flour and extend the shelf-life of noodles are still needed.

Chlorine dioxide, a strong and highly efficient oxidant, can be used in both aqueous and gaseous formulations to disinfect food and for food preservation [9,10]. It has been reported that chlorine dioxide can effectively inactivate insect, bacteria, fungi, viruses, spore and toxigenic molds. Stabilized ClO_2 is legally allowed to be applied with surface treatment of fresh fruit and vegetables and aquatic product in China [11]. Early literature reports pointed out that chlorine dioxide has replaced agene as the more commonly used improver in wheat flour in the United States after extensive feeding experiments with animal and human being that had shown no clinical evidence of toxicity. [12,13]. In a recent literature, Liu et al. indicated that the aqueous chlorine dioxide treatment prolonged the shelf-life and improved the quality characteristics of fresh noodles [14]. When compared with aqueous form, the gaseous chlorine dioxide is attracting increasing attention due to its many advantages, including ease of mixing with air, powerful antimicrobial activity, low corrosivity to using equipment and low environmental impact [9,10]. Therefore, in recent years, chlorine dioxide gas has become one of the most effective disinfectants in the food industry [15]. In addition, it has been widely studied in the preservation of fresh fruits and vegetables, control of fungal contamination in stored grains and green coffee beans [9,16]. However, there are only a few studies on the effect of gaseous chlorine dioxide on the quality of buckwheat flour and the storage stability of fresh buckwheat noodles (FBNs). Therefore, the objective of this work was to obtain helpful information for the possible application of gaseous chlorine dioxide on tartary buckwheat flour and prolong the shelf-life of FBNs.

2. Materials and Methods

2.1. Materials

Tartary buckwheat flour was manufactured by Saixue Grain and Oil Industry and Trade Co., Ltd. (Dingbian, Shaanxi, China). The content of moisture and protein in which were 11.64% and 12.62% (on a dry basis). The commercial wheat flour (12.10% moisture and 13.67% protein) was purchased from CR Vanguard supermarket. Salt was purchased from the local market. ClO_2 powder was purchased from Zhangda Technology Development Co., Ltd., Tianjin, China. All chemical reagents used in this research was of analytical grade.

2.2. Chlorine Dioxide Processing

Before being treated, buckwheat flour (306 g), wheat flour (294 g) and wheat gluten (60 g) were uniformly mixed to form a mixed flour. The treatment of mixed flour with chlorine dioxide was carried out in a test chamber (3 m^3) specially designed to produce chlorine dioxide. The gas is released through the reaction between water and solid ClO_2 powder in the test chamber. A fan is placed in the chamber to spread the chlorine dioxide more quickly and evenly. Different treatments included the following combinations: 20 g of ClO_2 powder and 1000 mL of sterile distilled water into a polypropylene container to generate concentrations of 76 ppm after 30 min (low treatment); 40 g of ClO_2 powder and 1000 mL of sterile distilled water to generate concentrations of 152 ppm after 30 min (medium treatment); and 60 g of ClO$_2$ powder and 1000 mL of sterile distilled water to generate concentrations of 232 ppm after 30 min (high treatment). The mixed flour samples were exposed to the corresponding chlorine dioxide gas for 30 min to ensure a better treatment effect. After being treated, the flour was put into sterile plastic packaging bags and stored at room temperature until use.

2.3. Determination of Total Flavonoid Content

Total flavonoid contents in CDBF were determined using method described by Bai et al. [17] with a slight modification. The total flavonoid content was calculated based on a calibration curve of rutin standards (y = 17.246x + 0.028, R^2 = 0.9944) and expressed as mg/g of dry weight.

2.4. Color of Mixed Flour and Noodle Sheet

The color values (L*, a* and b*) of the flour samples and fresh noodle sheets were measured using a chromameter (NS800, 3nh Technology Co., Ltd., Shenzhen, China). The noodle sheets were cut into pieces of approximately 5 cm in diameter and measured at 0, 24, 48 and 72 h.

2.5. Viscosity Analysis

Pasting properties of CDBF were measured using a Rapid Viscosity Analyzer (RVA, model 4500, Newport Scientific, Warriewood, Australia) by the AACC 76-21 method. Suspensions were made by blending pure deionized water (25 g) and the flour samples (3 g), and the mixtures were uniformly mixed using the plastic paddle right before the RVA test, and then the tests were conducted in a programmed heating and cooling cycle.

2.6. Swelling Power and Solubility

The swelling power and solubility of flour were determined according to the method described by [18] with a slight modification. Flour sample (0.5 g) was transferred into centrifuge tube and add 15 mL of distilled water, then the mixture was placed in a water bath at 95 °C for 30 min, with constant stirring. Then, the mixture was centrifugated at 3500 rpm for 20 min after being cooled at room temperature. Eventually, the supernatant was decanted, and the tubes were weighted, the sediment weight was determined and the solubility (S, g/100 g) and swelling power (SP, g/100 g) were calculated as follows:

$$S(g/100) = \text{Dry supernatant weight (g)/dry sample weight (g)} \tag{1}$$

$$SP(g/100g) = \frac{\text{Sediment (g)} \times 100}{\text{Dry sample weight (g)} \times (100 - S)} \tag{2}$$

2.7. Dough Rheological Characteristics

Rheological characteristics of mixed flour samples were determined by Branbender farinograph (Model: JFZD; Dongfujiuheng Instrument Technology Co., Ltd., Beijing, China). Water absorption, dough development time, dough stability and degree of softening were automatic recorded by computer.

2.8. Scanning Electron Microscopy (SEM) Observation of CDBF

Tabletop scanning electron microscopy (Hitachi Consumer Marketing (China) Ltd., Shanghai, China) was used to obtain the micrographs of mixed flour. Before scanning mixed flour were coated with gold under vacuum.

2.9. Preparation of Fresh Buckwheat Noodles

Before the experiment, the lab and packing materials were sterilized using ultraviolet radiation. All instruments were disinfected with 75% alcohol. The buckwheat noodle formula consisted of 110 g of flour and 35 mL of distilled water, and 3 g of NaCl. The noodle dough was formed with a Kitchen Aid Mixer (KM336, KENWOOD, Dongguan,

China) with the following mixing parameters: medium speed mixing (the second gear) for 210 s, high speed mixing (the third gear) for 60 s and low speed mixing (the first gear) for 150 s, then the obtained dough crumbles were transferred into a sterilized plastic bag to rest for 20 min. Subsequently, the rested crumbles passed through an experimental noodle machine (Ningbo Hantenite Machinery Co., Ltd., Ningbo, China) for 4 times folds, and the initial roller gap was set at 2.4 mm and then reduced from 2.0 mm to 1.5 mm to obtain dough sheets. The strands of the noodles were 2 mm in width and 1.5 mm in thickness. The fresh noodles were stored for 0, 24, 48 and 72 h at 25 °C for further research.

2.10. Determination of TPC

Total plate count (TPC) was examined according to the food microbiological examination standard of China (GB 4789.2-2016) with some modifications. The mixed flour and FBNs samples (1.0 g) were placed into 9 mL of 0.85% aseptic physiological saline (PA), and the mixture was shaken for 1 min. Gradient dilutions were prepared using PA and 100 µL of the appropriate dilutions were poured onto sterile plate count agar plates to determine the TPC. Then, the plates were incubated at 37 °C for 48 h, and the bacterial colonies were counted.

2.11. Cooking Properties

To determine the cooking loss, 25 g noodles were placed into 500 mL of boiling distilled water for 3 min until the hardcore of noodle strands disappeared. Next, the cooked noodles were rinsed in cold water for 1 min. Then, the cooking water was collected into the volumetric flask (500 mL). An aliquot of 50 mL was added to a pre-weighted beaker and dried in an oven at 105 °C till constant weight. The cooking loss was calculated as the percentage of the raw noodle.

$$\text{cooking loss } (\%) = \frac{5 \times \text{Constant weight}}{\text{Weight of raw noodle} \times (1 - \text{moisture content of rawnoodle})} \times 100\% \qquad (3)$$

2.12. Texture Properties of Cooked Noodles

Textural properties of cooked noodles were measured using a TA-XT Plus Texture Analyzer (Stable Micro System, Surrey, UK) according to method described by Guo et al. [19] with some modifications. Briefly, fresh buckwheat noodles (30 g) were cooked in 500 mL of boiling distilled water until the optimal cooking time. The optimal test conditions with the P 36/R probe for tensile test as follows: strain, 75%; pretest, test and post-test speed, 2 mm/s, 1 mm/s, 1 mm/s, respectively.

2.13. Determination of pH

The noodle pH value was measured by a pH meter according to the method proposed by Guo et al. [8].

2.14. Sensory Evaluation

Sensory evaluations were performed by 10 well-trained team members. The 9-point hedonic rating scale was used to assess overall acceptability, with "9" means "very much", "1" means "very little" and "5" means "acceptable". These analyses were performed in a standard sensory room. The final result is obtained by averaging the values obtained by each people.

2.15. Statistical Analysis

All experimental data represent the mean of at least three replicates and standard deviation (SD). Significant differences were verified by one-way analysis of variance (ANOVA) using SPSS statistical software (25.0, SPSS Inc., Chicago, IL, USA). $p < 0.05$ was considered statistically significant using Duncan's test.

3. Results and Discussion

3.1. Microbial and Physicochemical Changes in Mixed Flour

3.1.1. Microbial Count

The most critical factor for the perishability of fresh noodles is the high level of initial microorganisms in the flour [4]. As a strong oxidant, chlorine dioxide is recognized by the Food and Agriculture Organization of the United Nations (FAO) as a highly effective and safe chemical disinfectant used for food preservation. Table 1 shows the total plate count (TPC) of chlorine dioxide treated and untreated buckwheat-based composite flour. There was a significant ($p < 0.05$) decrease in TPC of mixed flour with the increase of treating concentration. The initial TPC of flour without chlorine dioxide treatment was 4.35 $\log_{10}$ CFU/g. It decreased to 3.51 $\log_{10}$ CFU/g when chlorine dioxide concentration increased to 232 ppm. This result shows that chlorine dioxide treatment can effectively reduce the initial bacterial quantity in buckwheat-based composite flour. Similar decreasing patterns were observed in carrots and tomatoes [20]. The decrease in TPC may be attributed to the damage effect of chlorine dioxide on microbial cells, thus inhibiting the microbial growth [10]. Furthermore, Guo et al. also pointed out that ClO_2 could disrupt the protein synthesis in microorganisms [21]. Therefore, gaseous chlorine dioxide can be used to produce low-bacteria flour in the food industry.

Table 1. TPC and physiochemical of buckwheat-based composite flour as affected by chlorine dioxide treatment.

Quality Parameters		Exposure Maximum Concentration			
		0 ppm	76 ppm	152 ppm	232 ppm
TPC (Log CFU g-1)		4.35 ± 0.07 [a]	4.01 ± 0.07 [b]	3.63 ± 0.07 [c]	3.51 ± 0.02 [c]
Color	L*	83.18 ± 0.01 [c]	83.20 ± 0.01 [c]	83.52 ± 0.01 [b]	84.08 ± 0.01 [a]
	a*	−0.04 ± 0.01 [c]	−0.05 ± 0.01 [c]	0.41 ± 0.01 [a]	0.36 ± 0.00 [b]
	b*	12.85 ± 0.01 [a]	11.32 ± 0.01 [d]	11.69 ± 0.01 [c]	11.74 ± 0.01 [b]
Farinograph Parameters	Water Absorption (%)	60.2 ± 0.2 [a]	60.3 ± 0.02 [a]	58.3 ± 0.01 [b]	58.3 ± 0.0 [b]
	Development Time (min)	6.95 ± 0.07 [d]	7.4 ± 0.15 [c]	9.0 ± 0.0 [b]	9.5 ± 0.12 [a]
	Stability Time (min)	8.20 ± 0.00 [d]	8.45 ± 0.07 [c]	10.2 ± 0.02 [b]	10.8 ± 0.04 [a]
	Degree of Softening (BU)	97.50 ± 2.12 [a]	87.5 ± 0.71 [b]	77.0 ± 0.42 [c]	75.0 ± 0.32 [c]
	Farinograph Quality Number (FQN)	102.0 ± 1.4 [d]	114.5 ± 0.71 [c]	135.0 ± 0.46 [b]	139.0 ± 1.4 [a]
RVA	Peak Time (min)	5.97 ± 0.05 [a]	5.80 ± 0.10 [ab]	5.67 ± 0.00 [b]	5.64 ± 0.05 [b]
	Peak Visc. (cP)	1169.00 ± 2.83 [a]	1138.00 ± 1.41 [b]	1098.00 ± 16.87 [c]	1105.50 ± 2.12 [c]
	Trough Visc. (cP)	1069.50 ± 2.12 [a]	1039.50 ± 3.54 [b]	1022.50 ± 17.68 [b]	1015.50 ± 3.54 [b]
	Setback (cP)	809.50 ± 12.02 [a]	761.00 ± 12.73 [b]	718.00 ± 24.04 [b]	725.00 ± 15.56 [b]
	Final Visc. (cP)	1879.00 ± 9.90 [a]	1800.50 ± 16.26 [b]	1740.50 ± 41.72 [b]	1740.50 ± 19.09 [b]
Swelling Power (g/100 g)		7.95 ± 0.02 [b]	8.14 ± 0.06 [ab]	8.49 ± 0.58 [ab]	8.77 ± 0.03 [a]
Solubility (g/100 g)		13.39 ± 0.70 [b]	13.36 ± 0.17 [b]	14.75 ± 0.88 [b]	16.61 ± 0.13 [a]
TFC (mg/g)		7.43 ± 0.09 [a]	6.94 ± 0.00 [b]	6.74 ± 0.23 [bc]	6.42 ± 0.09 [c]

Note: The data in the table are the mean ± SD. Means with different small letter superscripts within the same rows are significantly different at $p < 0.05$.

3.1.2. The Change of Total Flavonoids Content

The total flavonoid content (TFC) in control and chlorine dioxide treated mixed flours are shown in Table 1. The control flour had the highest total flavonoids content (7.43 mg/g). The TFC in mixed flour decreased to 6.94, 6.74 and 6.42 mg/g, respectively, when the exposure concentration was 76, 152 and 232 ppm. The decrease in total flavonoid content in buckwheat flour may be due to the loss of biological compounds caused by the strong oxidation of chlorine dioxide [17]. In this study, 86% total flavonoids remained after 232 ppm chlorine dioxide treatment.

3.1.3. Swelling Power and Solubility

The swelling power (SP) and solubility (S) data of the buckwheat flour are presented in Table 1. Higher swelling power is usually required for the smooth surface and elastic texture of cooking noodles [4]. The SP and S of mixed flour with chlorine dioxide treatment showed higher values than the control. The increase of the swelling power of chlorine dioxide oxidized starch was due to the introduction of hydrophilic carboxyl groups [22]. Guo et al. suggested that carboxyl is the leading functional group produced in the oxidation process of starch [22]. The negative charges of carboxyl groups repel each other, causing starch granules swelling when heated in water. The solubility increases significantly with the increase of treating concentration. This may be due to the depolymerization and structural weakening of starch particles, and the solubility increases after oxidation [18].

3.1.4. Color Analysis

The color of flour is related to consumers' acceptance of the flour itself and its noodle products. Table 1 presents the color parameters in chlorine dioxide treated and untreated mixed flour. In this study, after chlorine dioxide treatment, the L* value of mixed flour increased. In contrast, the b* value decreased significantly, indicating the treated flour appeared brighter and less yellow than the control. This result could be explained by the fact that chlorine dioxide can break down the natural yellow compounds (carotenoid and flavone) found in flour. This is consistent with the decrease in total flavonoids content observed in flour. In general, the color of CDBF is more visually acceptable and commercially valuable than untreated flour.

3.1.5. Pasting Properties

Pasting parameters of chlorine dioxide treated buckwheat-based composite flours are shown in Table 1, and RVA pasting curves are presented in Figure 1. According to Figure 1, it can be seen that with the concentration increasing of chlorine dioxide, the pasting temperature of the mixed flour decreased slightly, indicating that the concentration of the treatment used in this study did not inhibit the water absorption of the flour [23]. The trough viscosity (1069.50 ± 2.12 to 1015.50 ± 3.54) and final viscosity (1879.00 ± 9.90 to 1740.50 ± 19.09) decreased significantly with the increase of the treatment concentration. The peak viscosity (from 1169.00 ± 2.83 to 1098.00 ± 16.87) decreased when the treating concentration increased from 0 to 152 ppm, however, in which a small increase of peak viscosity (from 1098.00 ± 16.87 to 1105.50 ± 2.12) was observed when the treatment concentration increased from 152 to 232 ppm. The peak viscosity of the 0-152 ppm samples may be due to the oxidation of chlorine dioxide, which caused partial glycosidic bond cleavage, and in turn the decrease of starch molecular weight. Finally, the partially degraded starch could not maintain its particle integrity and was not resistant to shear, resulting in noodle viscosity decrease [24]. However, when treating concentration increase to 232 ppm, the increase of peak viscosity of samples may be caused by excessive oxidation which promoted a large amount of protein polymerization [25]. In addition, after chlorine dioxide treatment, the setback value of flour showed a decreased trend first (152 ppm) and then increased (232 ppm), indicating that excessive oxidation treatment would lead to easy retrogradation of starch. These results showed that treatment of flour with appropriate concentration of chlorine dioxide will be helpful to improve the gelatinization property of mixed flour.

3.1.6. Farinograph Properties Analysis

The farinograph indices of the mixed flour treated with different chlorine dioxide concentrations are presented in Table 1. After chlorine dioxide treatment, the water absorption of mixed flour decreased slightly with increasing treating concentration. However, the development time, stability time and farinograph quality number (FQN) of dough were significantly increased by chlorine dioxide treatment, indicating that the treated flours would form a stronger dough structure than un-treated flour. This may be due to the aggregation of proteins and the combination of proteins and starches that contributed to forming a stronger gluten network. The oxidation of chlorine dioxide can lead to cross-linking between proteins, thus increasing the strength of the dough [16]. In addition, Table 1 also showed that chlorine dioxide treatment decreased the degree of softening of the flour. The degree of softening suggests the rate of structural damage of the dough during the stirring process and it is inversely proportional to the strength of gluten. The lower the degree of softening, the easier the dough is to process. Present results indicated that chlorine dioxide treatment could enhance the processing property of mixed flour, which agrees with an early research that shows the effects of chlorine dioxide on the baking qualities of flour [26].

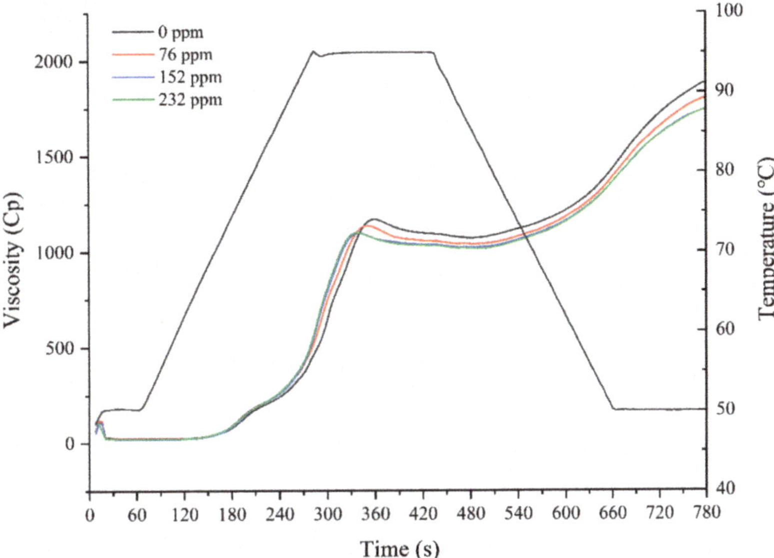

Figure 1. RVA pasting curves of buckwheat-based composite flour treated by gas chlorine dioxide at different concentrations.

3.1.7. Microstructure of Mixed Buckwheat Flour

Microstructures of the mixed buckwheat flour were examined by tabletop SEM and presented in Figure 2. At least three types of particle morphology including big, smooth and elliptical particle, aggregation of small particle and disperse tiny particles with different shapes and sizes could be observed in the control sample, which might be starch granules (big and smooth), buckwheat flour (aggregation of the small particle) and the protein bodies or fragments of protein matrix (disperse small particle). In addition, the disperse small particles might also contain fiber components and minerals [27]. After chlorine dioxide treatment, the disperse small particles in the mixed flour gradually decreased and the aggregation of small particle increased. In addition, it can be seen that surface of some starch granules emerged with spots and cracks. The effect of chlorine dioxide treatment on the morphology of flour particles may be caused by the oxidation of sulfhydryl and hydroxyl groups to form disulfide bonds and carboxyl groups [28]. In the early study, chlorine dioxide was used as the improver of wheat flour in the United States, indicating that it can increase the disulfide bonds of wheat flour. In the previous literature, Guo et al. found that the basic morphology of oxidized corn starch granules prepared by using chlorine dioxide as oxidant was not destroyed completely [22], but chlorine dioxide can penetrate into starch granules and cause damage on the surface and inside of starch granules.

3.2. Microbial and Physicochemical Changes in FBNs

3.2.1. Microbial Growth

Changes of TPC in FBNs during storage were presented in Figure 3A. The TPC value of 10^6 CFU/g in fresh noodles was considered as the cut-off point whether spoiled or unspoiled. Therefore, when TPC exceeds 1×10^6 CFU/g, the detection of microbial quantity is terminated. FBNs made from CDBF exhibited lower initial TPC at 0 h, which might be due to the low TPC in raw materials. During storage, the TPC of the control group increased rapidly and TPC exceeded 10^6 CFU/g after 48 h. With the increase of chlorine dioxide concentration, microbial growth of noodles made from CDBF was inhibited gradually. Moreover, the TPC of noodles made from flours treated with high concentration (232 ppm) of chlorine dioxide was still less than 10^6 CFU/g after 96 h because of the lower initial microbial content. However, the total number of molds in buckwheat noodles exceeded the limit ($\geq$150). Therefore, 232 ppm of chlorine dioxide can extend the shelf life of FBNs to 72 h. The reduction of microorganisms in mixed flour is helpful for the prolong of the shelf life of FBNs. This phenomenon may be due to the damaging effect of chlorine dioxide on microbial cells, thereby delaying the growth of microbes. In the previous literature, Han et al. found gaseous chlorine dioxide can effectively control pests and fungal contamination in stored grain without reducing the viability of rice seed [16]. In recent years, gaseous chlorine dioxide has been used widely in the food industry as a fungicide [15]. Thus, we could conclude that gaseous chlorine dioxide treatment can be seen as an effective method to diminish the initial TPC level in FBNs by suppressing TPC count in flour and extending the shelf-life of buckwheat noodles.

3.2.2. Color Changes in Noodle Sample

Color is considered the primary determinant of the sales of noodle products, and it also reflects the changes in the quality of FBNs during storage. Changes in the L* value of noodles made from CDBF were showed in Figure 3B. Date was recorded at 24 h intervals during a period of 72 h storage. The initial L* value of noodles made from CDBF increased significantly compared to that of the control. The initial L* value of control (53.90) was lower than that of treated groups (54.92 at 76 ppm, 55.66 at 152 ppm and 55.85 at 232 ppm). This was mainly because chlorine dioxide treatment inhibits polyphenol oxidase activity (PPO) [29], thereby reducing the degree of enzymatic darkening. Chen et al. also reported that the treatment of fresh-cut asparagus lettuce with ClO_2 aqueous solution could reduce its PPO activity [30]. With the extension of storage time, the L* value of the control noodles decreased faster than that in the chlorine dioxide treated group and reduced to 48.05 (control group) at the end of 48 h. In contrast, the L* value of chlorine dioxide treated samples were 49.25 at 76 ppm, 50.80 at 152 ppm and 51.61 at 232 ppm. In addition, during the storage period, the high-concentration (232 ppm) treatment delayed the decrease in L* value more significantly. This result indicated that FBNs darkening could be inhibited by treating flour with chlorine dioxide.

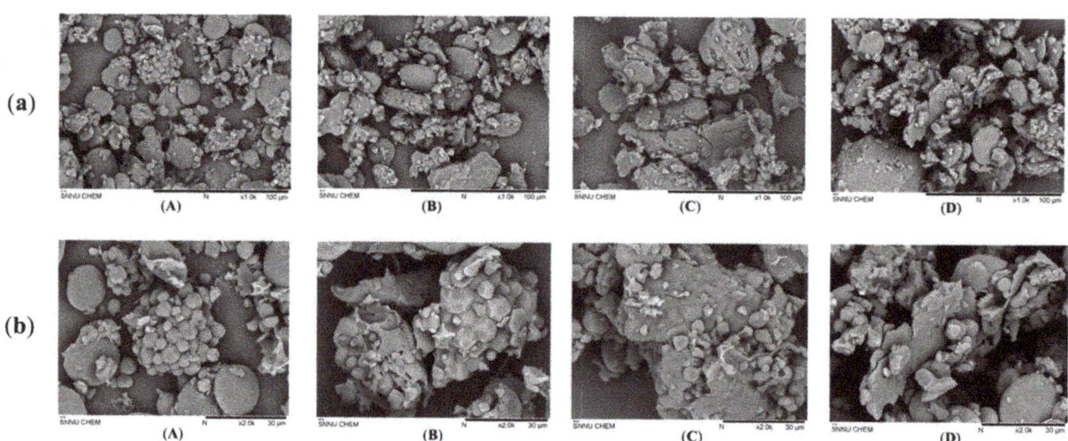

Figure 2. Microstructure ((**a**) 1000× magnification; (**b**) 2000× magnification) of buckwheat-based composite flour treated with chlorine dioxide at different treatment concentration: (**A**) 0 ppm; (**B**) 76 ppm; (**C**) 152 ppm; (**D**) 232 ppm.

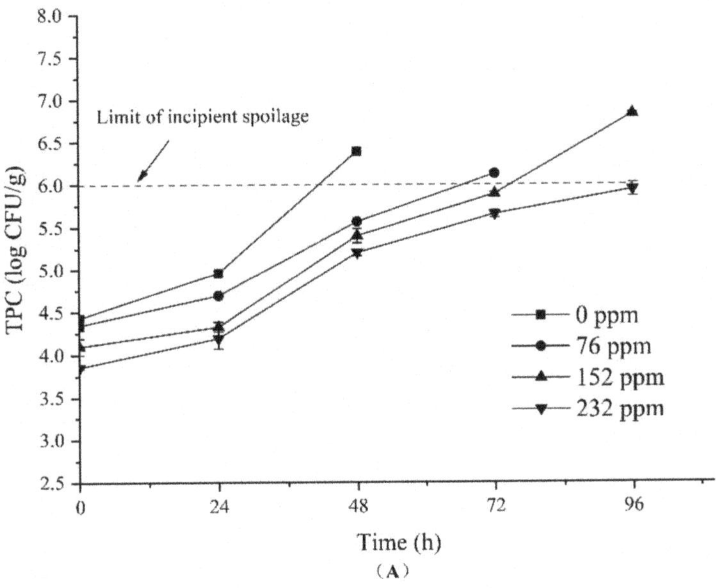

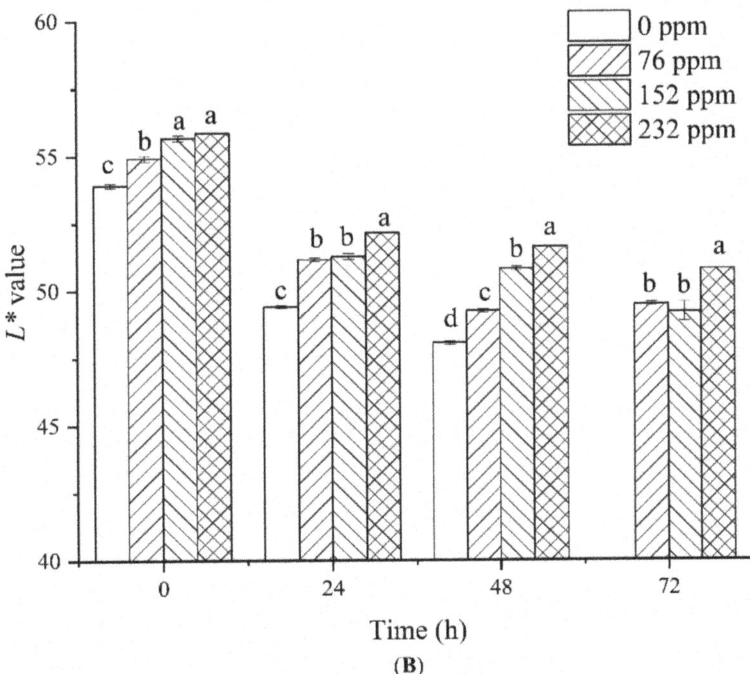

Figure 3. *Cont.*

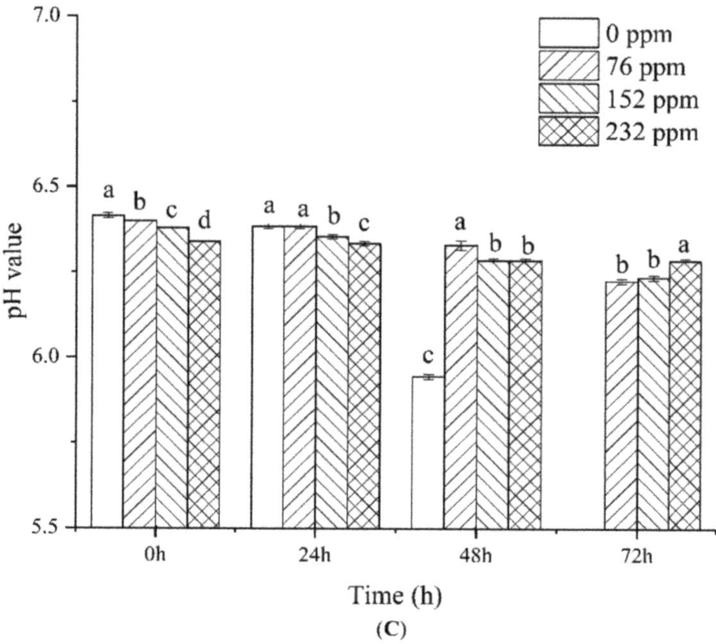

(C)

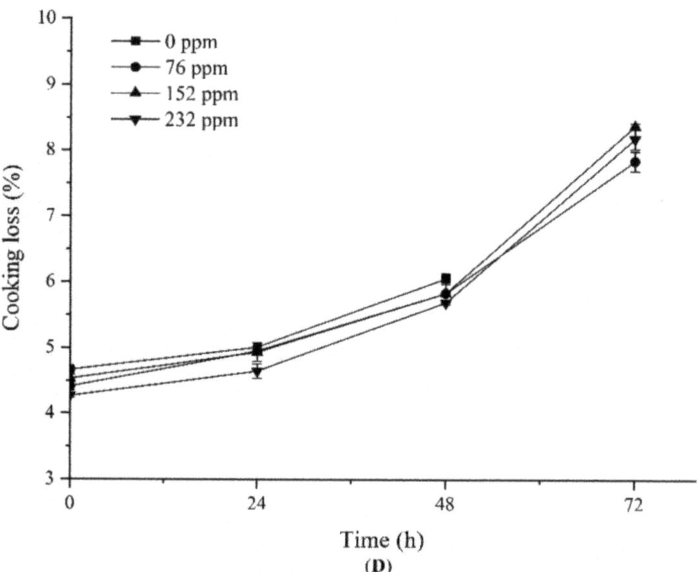

(D)

Figure 3. Effect of chlorine dioxide gas treatment on the microbial growth (**A**), L* value (**B**), pH value (**C**) and cooking loss (**D**) in FBNs during storage. Different lowercase letters indicate significant differences at $p < 0.05$.

3.2.3. pH Changes in FBNs during Storage

Figure 3C showed the pH changes in FBNs during storage. The initial pH values of chlorine dioxide treated samples were slightly lower compared to the control group.

During storage, the pH value of all FBNs showed a continuous downward trend, and the pH value of the control group decreased faster than other treatment groups, especially from 24 h to 48 h This phenomenon was attributed to the growth of noodle bacteria at the suitable pH [31]. In carbohydrate-rich foods, microorganisms can use carbohydrates to produce acid, thereby reducing the pH value of the food. Therefore, it is common to observe a decrease in the pH of noodles during storage [32]. For FBNs made from CDBF, the lower initial pH values than the control group could be attributed to the oxidation effect of chlorine dioxide on the lipid and starch. Lyu et al. regarded the progress of lipid oxidation could be accelerated by chlorine dioxide [33], and as described in the microstructure of mixed flour, oxidization of carboxymethyl might lead to decrease of initial pH value of noodles. The reduction of pH in FBNs made from CDBF were slower than that in the control group due to the inhibition of microorganism growth. At the end of storage, the pH values of noodle made from 232 ppm group were higher when compared to the other groups.

3.2.4. The Cooking Loss of FBNs

Cooking loss of FBNs during the storage was presented in Figure 3D. Cooking loss is defined as the amount of soluble matter in the cooking process of noodles. It indicates the extent of damage of noodles and their ability of structural maintenance during the cooking procedure [8]. As shown in Figure 3D, the cooking loss of FBNs was significantly reduced ($p < 0.05$) with the increase of chlorine dioxide concentration when the storage at 0 d, which were 4.66 (control group), 4.41 (76 ppm), 4.53 (152 ppm) and 4.27 (232 ppm), respectively. In addition, the cooking loss of all noodles increased significantly with the extension of storage time. With the same storage time, the cooking loss of the untreated noodles was always higher than that of chlorine dioxide treated group. It is speculated that the oxidation effect of chlorine dioxide could strengthen the gluten network in FBNs (that can be confirm by faringograph properties data) through disulfide bond induced protein polymerization, which finally led to the cooking loss decrease of FBNs during storage. The starch granules were tightly packed by protein polymers during the cooking process, therefore, further reduced the cooking loss of the noodles. However, the growth and reproduction of microorganisms in noodles during storage may cause protein depolymerization and damage to the gluten network, resulting the increase the cooking loss of noodles [8].

3.2.5. Texture Properties Analysis

Changes in textural properties of FBNs during storage are shown in Table 2. The results showed that FBNs made from CDBF had higher hardness and springiness, and lower adhesiveness than the control. With storage time increasing, hardness and springiness of all the noodles decreased, and adhesiveness increased. However, compared with control noodles, the textural parameters of FBNs made from CDBF changed more slowly during the same storage period. Obviously, the texture deterioration of noodle was correlated with the growth rate of microorganisms, which further indicated that the proliferation and metabolism of microorganisms were the main cause of FBNs deterioration. The rapid proliferation of microorganisms in the control sample led to the quick deterioration of texture quality. This result is consistent with a previous study, that found that microbial activities caused the deterioration of texture characteristics of noodle products during storage [5].

Table 2. Effect of chlorine dioxide gas treatment on the changes of texture properties of FBNs during storage.

Texture Profile	Storage Time/h	Exposure Maximum Concentration			
		0 ppm	76 ppm	152 ppm	232 ppm
Hardness(g)	0	3709.59 ± 11.24 [b]	3758.72 ± 51.08 [b]	3866.66 ± 10.25 [a]	3888.29 ± 48.16 [a]
	24	3493.44 ± 20.77 [b]	3706.77 ± 70.88 [a]	3648.60 ± 40.50 [a]	3668.75 ± 33.83 [a]
	48	3318.94 ± 3.92 [b]	3418.52 ± 48.08 [b]	3600.70 ± 71.58 [a]	3433.07 ± 4.67 [b]
	72	-	3325.06 ± 54.88 [b]	3553.15 ± 0.08 [a]	3546.55 ± 9.42 [a]
Adhesiveness(g/s)	0	59.10 ± 1.12 [a]	56.19 ± 0.06 [b]	55.61 ± 0.61 [b]	57.16 ± 0.20 [b]
	24	77.00 ± 0.63 [a]	59.07 ± 0.69 [d]	63.80 ± 1.06 [c]	67.14 ± 0.41 [b]
	48	77.29 ± 0.50 [a]	71.06 ± 0.59 [b]	63.99 ± 0.65 [c]	70.04 ± 0.72 [b]
	72	-	71.47 ± 0.41 [a]	69.71 ± 0.15 [b]	72.01 ± 0.67 [a]
Springiness	0	0.91 ± 0.00 [b]	0.92 ± 0.00 [a]	0.92 ± 0.00 [a]	0.92 ± 0.00 [a]
	24	0.90 ± 0.01 [a]	0.91 ± 0.00 [a]	0.90 ± 0.00 [a]	0.91 ± 0.01 [a]
	48	0.85 ± 0.01 [c]	0.88 ± 0.01 [b]	0.90 ± 0.01 [a]	0.90 ± 0.00 [a]
	72	-	0.84 ± 0.01 [b]	0.89 ± 0.01 [a]	0.89 ± 0.01 [a]

Note: The data in the table are the mean ± SD. Means with different small letter superscripts within the same rows are significantly different at $p < 0.05$. '-' Not detected.

3.2.6. Sensory Evaluation

Figure 4 shows the change in sensory properties of uncooked buckwheat noodles in terms of overall acceptability during storage. No significant chlorine dioxide odor was detected in the noodle samples during the whole storage period. The overall acceptability score of the control group on the first day was 8.6, slightly higher than that of the treatment group. The growth of noodle microbes caused unacceptable changes in taste, smell, appearance and any combination of these factors, resulting in spoilage [5]. With the extension of storage time, the overall acceptability scores of all samples showed a downward trend. Specifically, the sensory score (overall acceptability) of the control noodles decreased rapidly during the storage, which is related to the results of TPC. The rapid growth of microorganisms in noodles would have an adverse effect on the flavor of noodles, thus affecting the sensory evaluation. When the overall acceptability score is below 5, the product is defined as unacceptable. At the end of storage, the overall acceptability score of 76 ppm group was less than 5; however, the overall acceptability score of 232 was higher than 5. The reason may be that the protection of brightness and the inhibition of microbial growth are more evident in the chlorine dioxide treatment at 232 ppm. The results showed that high concentration of chlorine dioxide could prolong the shelf-life of FBNs to 72 h.

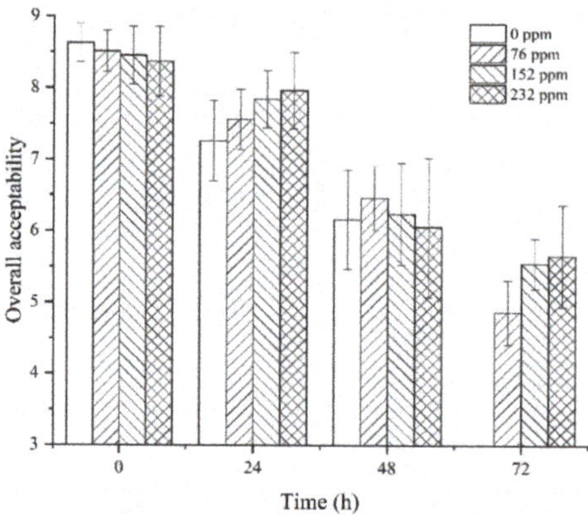

Figure 4. Overall acceptability of uncooked FBNs made from CDBF at different treatment concentration during storage.

4. Conclusions

The effects of gaseous chlorine dioxide treatment on microorganism, physicochemical properties of CDBF and the quality of CDBN were evaluated in this study. It was found that chlorine dioxide treatment reduced the microorganisms and the total flavonoids content but caused increases of lightness and farinograph properties of the mixed buckwheat flour. In addition, fresh noodles made from CDBF have a longer shelf life and better quality characteristics during storage. The application of gaseous chlorine dioxide provides a new concept for the preservation of flour and fresh noodles. This study provides an effective raw material processing method for making fresh buckwheat noodles. To further improve the quality stability of buckwheat noodles during storage, more research is needed.

Author Contributions: Conceptualization, X.L.; methodology, X.L. and X.H.; software, J.H. and X.F.; investigation, Z.C. and J.H.; data curation, Z.C.; writing—original draft preparation, Z.C.; visualization, Z.C.; funding acquisition, G.W. and Y.X. All authors have read and agreed to the published version of the manuscript.

Funding: This research was supported by the National Natural Science Foundation of China (Grant No. 31470093), the Science and Technology Project of Xi'an City of China (Grant No. 20193046YF034NS034) and the Key Research and Development Programs of Shaanxi Province of China (Grant No. 2018TSCXL-NY-03-02).

Institutional Review Board Statement: Not applicable.

Informed Consent Statement: Not applicable.

Data Availability Statement: Data generated or analyzed during this study are included in this published article.

Conflicts of Interest: This study does not have any conflict of interest.

References

1. Ahmed, A.; Khalid, N.; Ahmad, A.; Abbasi, N.A.; Latif, M.S.Z.; Randhawa, M.A. Phytochemicals and biofunctional properties of buckwheat: A review. *J. Agric. Sci.* **2014**, *152*, 349–369. [CrossRef]
2. Giménez-Bastida, J.A.; Piskuła, M.; Zieliński, H. Recent advances in development of gluten-free buckwheat products. *Trends Food Sci. Technol.* **2015**, *44*, 58–65. [CrossRef]
3. Qin, P.Y.; Wang, Q.; Shan, F.; Hou, Z.H.; Ren, G.X. Nutritional composition and flavonoids content of flour from different buckwheat cultivars. *Int. J. Food Sci. Technol.* **2010**, *45*, 951–958. [CrossRef]
4. Li, M.; Zhu, K.X.; Wang, B.W.; Guo, X.N.; Peng, W.; Zhou, H.M. Evaluation the quality characteristics of wheat flour and shelf-life of fresh noodles as affected by ozone treatment. *Food Chem.* **2012**, *135*, 2163–2169. [CrossRef] [PubMed]
5. Li, M.; Ma, M.; Zhu, K.X.; Guo, X.N.; Zhou, H.M. Delineating the physico-chemical, structural, and water characteristic changes during the deterioration of fresh noodles: Understanding the deterioration mechanisms of fresh noodles. *Food Chem.* **2017**, *216*, 374–381. [CrossRef] [PubMed]
6. Li, M.; Zhu, K.X.; Guo, X.; Peng, W.; Zhou, H.M. Effect of water activity (a(w)) and irradiation on the shelf-life of fresh noodles. *Innov. Food Sci. Emerg.* **2011**, *12*, 526–530. [CrossRef]
7. Li, M.; Peng, J.; Zhu, K.X.; Guo, X.N.; Zhang, M.; Peng, W.; Zhou, H.M. Delineating the microbial and physical–chemical changes during storage of ozone treated wheat flour. *Innov. Food Sci. Emerg.* **2013**, *20*, 223–229. [CrossRef]
8. Guo, X.N.; Jiang, Y.; Xing, J.J.; Zhu, K.X. Effect of ozonated water on physicochemical, microbiological, and textural properties of semi-dried noodles. *J. Food Process. Preserv.* **2020**, *44*, e14404. [CrossRef]
9. Lee, H.; Ryu, J.H.; Kim, H. Antimicrobial activity of gaseous chlorine dioxide against Aspergillus flavus on green coffee beans. *Food Microbiol.* **2020**, *86*, 103308. [CrossRef]
10. Sun, X.; Baldwin, E.; Bai, J. Applications of gaseous chlorine dioxide on postharvest handling and storage of fruits and vegetables—A review. *Food Control* **2019**, *95*, 18–26. [CrossRef]
11. Sun, C.; Zhu, P.; Ji, J.; Sun, J.; Tang, L.; Pi, F.; Sun, X. Role of aqueous chlorine dioxide in controlling the growth of Fusarium graminearum and its application on contaminated wheat. *LWT* **2017**, *84*, 555–561. [CrossRef]
12. Meredith, P.; Sammons, H.G.; Frazer, A.C. Studies on the effects of treatment with chlorine dioxide on the properties of wheat flour. I.—The chemical composition of protein of treated flours. *J. Sci. Food Agric.* **1956**, *7*, 361–370. [CrossRef]
13. Moran, T.; Pace, J.; Mcdermott, E.E. Interaction of Chlorine Dioxide with Flour: Certain Chemical Aspects. *Nature* **1953**, *171*, 103–106. [CrossRef] [PubMed]
14. Liu, Z.G.; Xu, X.M.; Jin, Z.Y. Study on effect of chlorine dioxide and hydrogen peroxide on the preservation of wet raw noodles. *Sci. Technol. Food Ind.* **2008**. [CrossRef]

15. Park, S.H.; Kang, D.H. Influence of surface properties of produce and food contact surfaces on the efficacy of chlorine dioxide gas for the inactivation of foodborne pathogens. *Food Control.* **2017**, *81*, 88–95. [CrossRef]
16. Han, G.D.; Kwon, H.; Kim, B.H.; Kum, H.J.; Kwon, K.; Kim, W. Effect of gaseous chlorine dioxide treatment on the quality of rice and wheat grain. *J. Stored Prod. Res.* **2018**, *76*, 66–70. [CrossRef]
17. Bai, Y.P.; Guo, X.N.; Zhu, K.X.; Zhou, H.M. Shelf-life extension of semi-dried buckwheat noodles by the combination of aqueous ozone treatment and modified atmosphere packaging. *Food Chem.* **2017**, *237*, 553–560. [CrossRef]
18. Obadi, M.; Zhu, K.X.; Peng, W.; Sulieman, A.A.; Mohammed, K.; Zhou, H.M. Effects of ozone treatment on the physicochemical and functional properties of whole grain flour. *J. Cereal Sci.* **2018**, *81*, 127–132. [CrossRef]
19. Guo, X.N.; Wu, S.H.; Zhu, K.X. Effect of superheated steam treatment on quality characteristics of whole wheat flour and storage stability of semi-dried whole wheat noodle. *Food Chem.* **2020**, *322*, 126738. [CrossRef]
20. Bridges, D.F.; Rane, B.; Wu, V.C.H. The effectiveness of closed-circulation gaseous chlorine dioxide or ozone treatment against bacterial pathogens on produce. *Food Control* **2018**, *91*, 261–267. [CrossRef]
21. Guo, Q.; Wu, B.; Peng, X.Y.; Wang, J.D.; Li, Q.P.; Jin, J.; Ha, Y.M. Effects of chlorine dioxide treatment on respiration rate and ethylene synthesis of postharvest tomato fruit. *Postharvest Biol. Technol.* **2014**, *93*, 9–14. [CrossRef]
22. Guo, J.T.; Huang, Y.C.; Zhang, J.; Yin, J.W. Preparation of Oxidized Starch Using Environment Friendly Chlorine Dioxide as Oxidant. *Int. J. Food Eng.* **2014**, *10*, 243–249. [CrossRef]
23. Li, M.; Sun, Q.J.; Zhu, K.X. Delineating the quality and component changes of whole-wheat flour and storage stability of fresh noodles induced by microwave treatment. *LWT* **2017**, *84*, 378–384. [CrossRef]
24. Sandhu, K.S.; Kaur, M.; Singh, N.; Lim, S.T. A comparison of native and oxidized normal and waxy corn starches: Physicochemical, thermal, morphological and pasting properties. *LWT Food Sci Technol.* **2008**, *41*, 1000–1010. [CrossRef]
25. Hu, Y.M.; Wang, L.J.; Zhu, H.; Li, Z.G. Superheated steam treatment improved flour qualities of wheat in suitable conditions. *J. Food Process. Preserv.* **2017**, *41*, 10. [CrossRef]
26. Ferrari, C.; Hutchinson, W.; Croze, B.; Mecham, D. Flour bleaching with chlorine dioxide. *Cereal Chem.* **1941**, *18*, 699.
27. Aguilera, Y.; Esteban, R.M.; Benitez, V.; Molla, E.; Martin-Cabrejas, M.A. Starch, Functional Properties, and Microstructural Characteristics in Chickpea and Lentil as Affected by Thermal Processing. *J. Agric. Food Chem.* **2009**, *57*, 10682–10688. [CrossRef]
28. Tharanathan, R.N. Starch-value addition by modification. *Crit. Rev. Food Sci. Nutr.* **2005**, *45*, 371–384. [CrossRef] [PubMed]
29. Chen, S.J.; Wang, H.O.; Wang, R.R.; Fu, Q.Q.; Zhang, W. Effect of gaseous chlorine dioxide (ClO_2) with different concentrations and numbers of treatments on controlling berry decay and rachis browning of table grape. *J. Food Process. Preserv.* **2018**, *42*, e13662. [CrossRef]
30. Chen, Z.; Zhu, C.H.; Zhang, Y.; Niu, D.B.; Du, J.H. Effects of aqueous chlorine dioxide treatment on enzymatic browning and shelf-life of fresh-cut asparagus lettuce (*Lactuca sativa* L.). *Postharvest Biol. Technol.* **2010**, *58*, 232–238. [CrossRef]
31. Bonafaccia, G.; Marocchini, M.; Kreft, I. Composition and technological properties of the, flour and bran from common and tartary buckwheat. *Food Chem.* **2003**, *80*, 9–15. [CrossRef]
32. Ghaffar, S.; As, A.; Abu Bakar, F.; Karim, R.; Saari, N. Microbial Growth, Sensory Characteristic and pH as Potential Spoilage Indicators of Chinese Yellow Wet Noodles from Commercial Processing Plants. *Am. J. Appl. Sci.* **2009**, *6*, 1059–1066. [CrossRef]
33. Lyu, F.; Zhao, Y.L.; Shen, K.J.; Zhou, X.X.; Zhang, J.Y.; Ding, Y.T. Using Pretreatment of Carbon Monoxide Combined with Chlorine Dioxide and Lactic Acid to Maintain Quality of Vacuum-Packaged Fresh Beef. *J. Food Qual.* **2018**, 3158086. [CrossRef]

Article

Quality Characteristics and Antioxidant Activity of Fresh Noodles Formulated with Flour-Bran Blends Varied by Particle Size and Blend Ratio of Purple-Colored Wheat Bran

Gyuna Park [1], Hyejin Cho [1], Kyeonghoon Kim [2] and Meera Kweon [1,3,*]

[1] Department of Food Science and Nutrition, Pusan National University, Busan 46241, Korea; honeychuchu@pusan.ac.kr (G.P.); tyu117@pusan.ac.kr (H.C.)
[2] Wheat Team, National Institute of Crop Science, Rural Development Administration, Jeonju 55365, Korea; k2h0331@korea.kr
[3] Kimchi Institute, Pusan National University, Busan 46241, Korea
[*] Correspondence: meera.kweon@pusan.ac.kr; Tel.: +82-51-510-2716

Abstract: This study explored the noodle-making performance of flour blends with different particle sizes and blending ratios of purple-colored wheat bran and their antioxidant properties. The bran particle size was reduced using an ultra-centrifugal mill equipped with 1, 0.5, and 0.2 mm sieves. The damaged starch and swelling capacity of the bran were analyzed. Quality of the flour-bran blends at different blending ratios was analyzed by solvent retention capacity (SRC). Noodles made from the blends and their corresponding antioxidant activities were examined. The damaged starch and swelling capacity of bran were higher for smaller particles than for larger particles. Water and sodium carbonate SRC values of blends increased as the bran particle size decreased. The smaller the bran particles incorporated in the cooked noodles, the greater firmness and springiness measured. The antioxidant activity of noodles made with blends reflected better embedding of the small particles of bran than the large particles into noodle sheets. Small bran particles significantly enhanced noodles' quality and antioxidant activity at higher blending ratios than large bran particles. Particle size reduction of bran enhanced the noodle-making performance of flour blended with purple-colored wheat bran; this could increase the utilization of bran to produce noodles with health benefits.

Keywords: purple-colored wheat; wheat bran; particle size; blending ratio; fresh noodles; antioxidant properties

Citation: Park, G.; Cho, H.; Kim, K.; Kweon, M. Quality Characteristics and Antioxidant Activity of Fresh Noodles Formulated with Flour-Bran Blends Varied by Particle Size and Blend Ratio of Purple-Colored Wheat Bran. *Processes* **2022**, *10*, 584. https://doi.org/10.3390/pr10030584

Academic Editors: Yonghui Li and Shawn/Xiaorong Wu

Received: 23 February 2022
Accepted: 15 March 2022
Published: 17 March 2022

Publisher's Note: MDPI stays neutral with regard to jurisdictional claims in published maps and institutional affiliations.

1. Introduction

Wheat (*Triticum aestivum*) is a typical crop used in various food products worldwide. Wheat bran, a by-product generated during wheat milling, usually accounts for 14–19% of the grain weight and consists of 37–52% total dietary fiber [1,2]. Bran has nutritional properties and provides many functional benefits, including antioxidant activity [3,4]. The main antioxidative components in wheat bran are polyphenols (mainly phenolic acids) [5]. In particular, purple- or black-colored wheat contains more anthocyanin and polyphenol compounds in the wheat bran and aleurone layers than does common wheat, providing a more significant nutritional advantage [6,7].

Despite these benefits, bran in whole wheat flour (WWF) presents challenges to producing high-quality bakery or noodle products. Wheat bran negatively influences the qualities of bread dough by physically interfering with the gluten matrix of the dough [8,9]. It also destroys the starch–gluten matrix of the dough and leads to lower firmness of cooked noodles and higher loss of solids in cooking water with increased bran content [10]. Aravind et al. [11] reported that wheat bran in pasta negatively affects the cooking and sensory quality of pasta due to increased cooking loss and swelling index, leading to unfavorable sensory acceptance and appearance. Additionally, whole wheat products are of less attractive quality and have lower sensory acceptance than do refined wheat products [12,13]. Accordingly, the current intake of whole wheat grain products is still much lower than the recommended level worldwide.

Noodles are a wheat-based food widely used as a meal substitute [14]. The development of noodles made from WWF can be an effective way to increase the demand for low-calorie healthy foods and promote high-fiber foods. Noodles made with WWF provide enhanced nutritional benefits compared to those with regular flour, but they are not appealing to many consumers because of their negative sensory properties such as texture and taste. When bran is blended with refined wheat flour for noodle production, it is challenging to increase the blending ratio of bran because of its large particle size and high water absorption compared to wheat flour itself; this hinders dough development. Numerous methods such as fermentation [15], enzyme treatment [16,17], germination [18], particle size control by milling [19,20], and physical treatments including hydration, autoclaving, and freezing [5,21,22] have been investigated to solve these issues.

Among these methods, particle size reduction of bran by milling can alter its physical and functional properties [23–25], causing structural rather than chemical modification [20]. Particle size reduction of wheat bran can improve the quality of wheat-based products. Fine bran is less destructive to dough-mixing and allows better gluten network formation than does coarse bran [26]. Particle size reduction of wheat bran has a minimal effect on dough characteristics, leading to high sensory acceptability of the resulting noodles [10,27]. However, the effect of the bran particle size on the volume of breads is somewhat contradictory due to additional effects of bran composition and type [9,19].

The majority of the research on bran particle size was performed for brans with different particle sizes prepared with different pulverizers [24] or by sifting with various sieves after pulverization [3,28]. Memon et al. [28] reported the direct influence of WWF particle size on the distribution of phenolic acid, carbohydrate, protein, crude fiber, ash, crude fat, and moisture in the three commercial wheat varieties when the WWF was fractionated by shifting with a series of five sieves. In comparison, it is expected that bran milled with the same pulverizer equipped with different sieves separately might differ in only particle size, not in the distribution of nutritional components. However, the research on the effect of bran particle size segregated from the effect of the distribution of nutritional components in bran is rare and worth investigating. The National Institute of Crop Science in Korea recently released a purple-colored wheat cultivar, "Ariheuk", as value-added wheat for enhancing health benefits. Avarzed et al. [29] reported that the purple-colored wheat bran has significantly higher total phenolic and anthocyanin content and antioxidant activity than common wheat bran. In terms of this new cultivar, it is worth investigating the fresh noodle-making performance of flour-bran blends by increasing the blending ratio of bran and reducing its particle size using the same pulverizer equipped with different sieves.

The present study explored the effect of particle size reduction of purple-colored wheat bran on the hydration properties, solvent retention capacity (SRC), and dough-mixing properties of flour-bran blends with different bran blending ratios. In addition, the fresh noodle-making performance of the blends and the antioxidant activity of the noodles were assessed.

2. Materials and Methods

2.1. Materials

Korean domestic wheat flour milled from the cultivar "Keumkang" and purple-colored wheat bran obtained from the cultivar "Ariheuk" used in this study were supplied by the National Institute of Crop Science in Korea. To produce noodles, salt (Samyang, Seoul, Korea) was purchased from a local market. To analyze SRC, sodium carbonate (Duksan, Seoul, Korea) was used as extra-pure-grade reagent.

2.2. Size Reduction of Purple-Colored Wheat Bran and Measurement of Bran Particle Size

To measure the particle size of the bran before milling, the bran (300 g) was separated with three sieves (DAIHAN Scientific, Wonju, Korea) with 2.0, 1.0, and 0.5 mm openings, respectively. The bran passed through each sieve was weighed, and the percentage of each fraction was: 5% $\leq$ 0.5 mm, 0.5 mm < 24% < 1.0 mm, 1.0 mm < 68% < 2.0 mm, 2.0 mm $\leq$ 3%. To reduce the bran particle size, bran was milled using an ultra-centrifugal mill (POWTEQ FM200, Beijing, China), which was equipped with a 12-tooth rotor and three-ring individually inserted sieves (1.0, 0.5, and 0.2 mm), at a centrifugal speed of 15,000 rpm. Brans of different sizes were obtained and labeled as L (1.0 mm), M (0.5 mm), and S (0.2 mm). The original bran sample was labeled VL because it was much larger than L, M, and S bran samples.

The particle size of the milled bran was measured using a laser-scattering particle size analyzer (Beckman Coulter LS 13 320, Fullerton, CA, USA) equipped with a vacuum delivery system for dry samples and the maximum particle size limit of 2000 μm.

2.3. Analysis of Physicochemical Properties of Purple-Colored Wheat Bran

The moisture content of the bran samples was measured according to Method 44-15.02 [30]. The total starch content of only VL bran sample and the damaged starch content of all bran samples were measured using the Megazyme Total Starch Assay Kit (K-TSTA-100A, Megazyme International, Wicklow, Ireland) and the Starch Damage Assay Kit (K-SDAM) by Method 76-31.01 [30], respectively.

The swelling capacity of the bran samples was determined according to the method reported by Jacobs et al. [20] with slight modifications. Bran (750 mg) was soaked in deionized water (7.5 mL) in a 10 mL graduated cylinder for 60 min to absorb water and swell. The volume of the swollen bran is referred to as the swelling capacity.

2.4. Preparation of Flour-Bran Blends with Different Blending Ratios

"Keumkang" flour and "Ariheuk" bran were blended at ratios of 9:1, 7:3, and 5:5 and denoted by F90-B10, F70-B30, and F50-B50, respectively. In addition, four flour-bran blends with different blending ratios were prepared with four bran sizes (S, M, L, and VL). Flour alone was designated as F100-B0.

2.5. Solvent Retention Capacity of Flour-Bran Blends

SRC analysis of the flour-bran blends was performed according to Method 56-1.02 [30]. SRC in only water and sodium carbonate solutions was tested because the SRC values of wheat bran in lactic acid and sucrose solutions tend to be under or overestimated [31]. Flour-bran blend (5 g) was added to pre-weighed 50 mL conical tubes. Two solvents, 25 g of distilled water, and 5% (w/w) sodium carbonate solution were prepared separately in each tube. Each solution was poured into a tube containing the flour-bran blend, and each tube was shaken every 5 min for 20 min to sufficiently disperse the flour-bran blend. The suspension of each flour-bran blend was centrifuged at $1000 \times g$ for 15 min using a centrifuge (LaboGene1248, Gyrozen Inc., Daejeon, Korea), and the supernatant was discarded. The tube containing the pellet was weighed, and the SRC value (%) was calculated according to Method 56-1.02.

2.6. Preparation of Fresh White-Salted Noodles Formulated with the Flour-Bran Blends

Fresh noodles were prepared using the method described by Moon et al. [32]. The flour-bran blend (100 g) was placed in the bowl of a micro mixer (National Manufacturing Inc., Lincoln, NE, USA), and 2 g of salt and 30–43.3 g of distilled water, based on the water absorption capacity measured by water SRC value of each blend, was added (Table 1). The flour-bran blend was mixed with salt and water for 15 min, and the prepared dough was placed in a plastic bag and rested for 30 min in a resting chamber (Phantom M301 Combi, Phantom Korea, Hanam, Korea) at 35 °C and 85% relative humidity. The rested dough was sheeted into thicknesses of 3.0, 2.0, and 1.5 mm continuously with a noodle-maker

(SN-88, Samwoo Industrial Co., Daegu, Korea) and then cut (4.0 mm width) to produce fresh noodles. One strand of each noodle from the fresh noodles prepared with flour-bran blends at different blending ratios and bran sizes was aligned in parallel together and photos were taken to observe the appearance of fresh noodles.

Table 1. Ingredients and formula used for preparing fresh noodles made with flour-bran blends at different ratios.

Ingredients	Formula (g)			
	F100-B0 [1]	F90-B10	F70-B30	F50-B50
Flour	100.0	90.0	70.0	50.0
Wheat bran	0.0	10.0	30.0	50.0
Water	30.0	32.0	37.4	43.3
Salt	2.0	2.0	2.0	2.0

[1] F100-B0, Keumkang flour; F90-B10, F70-B30, and F50-B50: blends of Keumkang flour and Ariheuk bran at ratios of 9:1, 7:3, and 5:5, respectively.

2.7. Color Measurements of Fresh Noodle Sheets and Cooked Noodles

Before cutting a noodle sheet of 1.5 mm thickness into noodles, the color of the noodle sheet was measured using a colorimeter (CR-20, Konica Minolta, Tokyo, Japan). The color parameters of brightness (L^*), redness (a^*), and yellowness (b^*) were repeatedly measured five times and calculated as average values.

To measure the color of cooked noodles, 15 g of fresh noodles were placed in 500 mL of boiling water and boiled for 15 min according to the methods reported by Moon et al. [32] and Wang and Kweon [33]. The cooking water was drained separately into a beaker to measure turbidity. The color of the cooked noodles was measured for five noodle strands arranged side by side. The remaining cooked noodles were used to measure texture.

2.8. Measurement of Turbidity of Cooking Water

The turbidity of the cooking water after boiling noodles was measured by absorbance using a spectrophotometer (X-ma 6100PC, Human Corporation, Seoul, Korea) at 675 nm.

2.9. Analysis of Textural Property of Fresh and Cooked Noodles

As a representative method to measure the texture of fresh noodles, the extensibility of the noodles was measured using a texture analyzer (CT3, Brookfield, Middleboro, MA, USA), and the measurement conditions were as follows: test mode, tension; pretest speed, 2 mm/s; test speed, 3.3 mm/s; probe, Kieffer rig (TA-KF); target value, 20 mm. The average value was calculated from ten measurements.

The texture of the cooked noodles was also analyzed using a texture analyzer on five strands of noodles. The measurement conditions were as follows: test mode, TPA; pretest speed, 2 mm/s; test speed, 1 mm/s; probe, Asian noodle rig (TA 7); deformation%, 70. Firmness, cohesiveness, springiness, and chewiness were measured, and the average value was calculated from five measurements.

2.10. Measurement of Total Polyphenol Content of Fresh Noodles

The total polyphenol content (TPC) was determined by a slight modification of the method described by Yu and Beta [34]. Fresh noodles were freeze-dried and ground using a grinder (WSG-9100, Joong San Co., Seoul, Korea). Ground noodles (2 g) were extracted twice with 20 mL of 80% methanol in a 50 mL tube by shaking for 90 min with a rotator (CN/VM-80, Miulab, Hangzhou, China). The noodle mixture was sonicated in an ice-filled sonicator (LK-U105, LK Lab Korea, Namyangju, Korea) at 40 kHz for 30 min in the dark. After centrifuging the noodle mixture at $12,000 \times g$ and 4 °C for 15 min, the supernatant was filtered with 90 mm filter paper (Qualitative Filter Paper No.2, ADVATEC, Tokyo, Japan) and stored at −20 °C. The TPC of the noodles was determined using Folin–Ciocalteu reagent (Sigma, St. Louis, MO, USA). The noodle extract (0.2 mL) was oxidized with

10-times-diluted Folin–Ciocalteu reagent (1.5 mL) for 5 min and neutralized with 1.5 mL sodium carbonate solution (60 g/L). After 90 min, the absorbance of the reacted noodle extract was measured at 725 nm against a blank of distilled water using a spectrophotometer (X-ma 6100PC, Human Corporation, Seoul, Korea). Gallic acid (Sigma, St. Louis, MO, USA) was used as the standard. The TPC of fresh noodles was expressed as mg GAE (gallic acid equivalent)/100 g.

2.11. Measurement of Total Anthocyanin Content of Fresh Noodles

The total anthocyanin content (TAC) of fresh noodles was determined using the method described by Yu and Beta [34]. Acidified methanol (methanol: 1.0 N HCl = 85:15 (v/v), Ph = 1) was used to extract anthocyanins from the noodles. The procedure to react the noodle mixture with Folin–Ciocalteu reagent was the same as that used for TPC. Absorbance was measured at 535 nm. Cyanidin-3-glucoside was used as a standard in an 80% methanol blank. The results were expressed as mg C3GE (cyaniding-3-glucoside equivalent)/100 g sample.

2.12. Antioxidant Activity of Fresh Noodles—ABTS Radical Scavenging Capacity

Antioxidant activity as indicated by the ABTS radical scavenging capacity of fresh noodles was determined by a slight modification of the method of Yu and Beta [34]. First, 10 mL of ABTS reagent was diluted with approximately 990 mL of distilled water. Then, antioxidant activity of phenolic and anthocyanin extracts (50 µL) from the noodles was evaluated by adding 1.85 mL of the diluted ABTS reagent. Absorbance was measured at t = 30 min. The absorbance of the reaction solution was measured at 734 nm using a spectrophotometer (X-ma 6100PC, Human Corporation, Seoul, Korea) after 30 min on an 80% methanol blank. For absorbance at t = 0 min, 1.85 mL of the diluted ABTS reagent was added to 100 µL of 80% methanol. A standard curve was generated based on different Trolox concentrations vs. %ABTS decolorization, and the ABTS value was expressed as µL TE (Trolox equivalents)/100 g sample.

2.13. Antioxidant Activity of Fresh Noodles—DPPH Radical-Scavenging Capacity

Antioxidant activity as indicated by DPPH radical-scavenging capacity of fresh noodles was determined by a slight modification of the method of Yu and Beta [34]. A 55 µL/L DPPH reagent was prepared by dilution in 100% methanol. After adding 3.9 mL of DPPH solution to 0.1 mL of noodle extract, the absorbance was measured at t = 30 min at 515 nm using a spectrophotometer (X-ma 6100PC, Human Corporation, Seoul, Korea) on an 80% methanol blank. After adding 3.9 mL of DPPH solution to 0.1 mL of 80% methanol, the absorbance was measured at t = 0 min. A standard curve was generated based on different Trolox concentrations vs. % DPPH scavenging activity. DPPH value was expressed as µL TE/100 g sample.

2.14. Statistical Analysis

All data were obtained from measurements in at least triplicate. Differences between samples (significance threshold, $p < 0.05$) were analyzed by ANOVA and Tukey's HSD test using SPSS 22.0 (SPSS Inc., Armonk, New York, NY, USA).

3. Results and Discussion

3.1. Particle Size of Purple-Colored Wheat Bran Milled with an Ultra-Centrifugal Mill

The particle size distributions of the reduced sizes of the brans are shown in Figure 1. All bran samples (L, M, and S) with reduced particle size showed a unimodal shape with small humps. These are notably distinct in particle size, demonstrating an appropriate selection for exploring the effect of particle size reduction of bran in this study. The bran samples exhibited a significantly increased portion of smaller particles due to particle size reduction. The calculated mean particle sizes of the L, M, and S bran samples were 584.0,

255.2, and 123.1 µm, respectively. The particle sizes of the brans were significantly different, potentially resulting in different noodle-making performance [10,27,35].

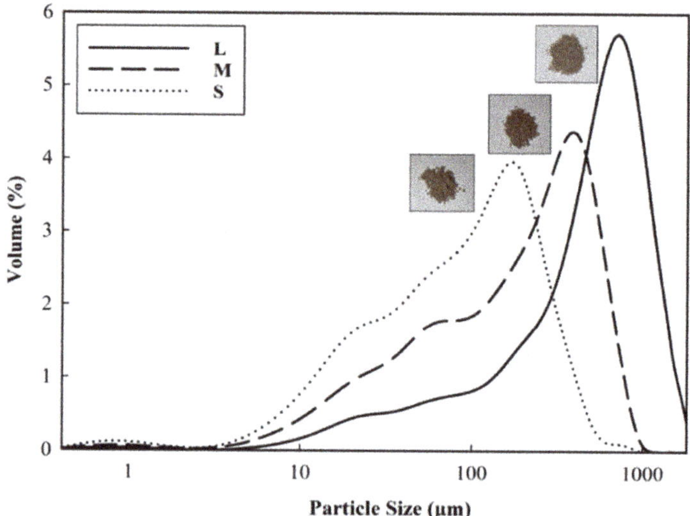

Figure 1. The particle size distribution of wheat bran with size reduction determined by laser diffraction. L, M, and S indicate the brans with large, medium, and small particle sizes.

3.2. Physicochemical Properties of Purple-Colored Wheat Bran

Physicochemical properties of bran samples are presented in Table 2. The moisture content of the samples was 9.0–12.2% and decreased as the bran size decreased. This decrease was due to the moisture loss by frictional heat produced in milling wheat bran because more frictional heat is produced when making bran of smaller sizes [36].

Table 2. Moisture and damaged starch contents and swelling capacities of purple-colored wheat bran with different sizes.

Bran Size	Moisture Content (%)	Damaged Starch Content (%)	Swelling Capacity (%)
VL [1]	12.2 ± 0.0 [d(2)]	1.9 ± 0.0 [a]	58.7 ± 0.0 [a]
L	11.9 ± 0.2 [c]	2.7 ± 0.0 [b]	74.2 ± 2.0 [b]
M	10.2 ± 0.1 [b]	3.0 ± 0.1 [c]	80.4 ± 0.8 [bc]
S	9.0 ± 0.1 [a]	4.0 ± 0.1 [d]	85.3 ± 3.5 [d]

[1] VL, L, M, and S indicate the brans with very large, large, medium and small particle sizes. [2] Results are expressed as mean ± SD. Values with the different letters within the same column are significantly different ($p < 0.05$) according to Tukey's HSD test.

The VL bran contained approximately 23.9% of total starch. The damaged starch content of the bran samples is shown in Table 2; it increased with a decrease in bran size. The damaged starch content of VL bran was 1.9%, and that of S bran was 4.0%, which could affect water absorption, dough property, and noodle quality of the bran samples. Fu [14] reported that the increased damaged starch in wheat flour increased water absorption and decreased dough-mixing stability. The noodles formulated with the flour-bran blend with S bran required increased water for dough development. With increasing damaged starches in flour, noodles may increase cooking loss and become gummier and less firm [37].

The swelling capacity of the bran significantly increased with decreasing bran size (Table 2). The swelling capacity of VL bran was 58.7%, and that of S bran was 85.3%, reflecting an increase in hydration properties as the particle size decreased. Even L bran showed a dramatic increase in swelling capacity, confirming the significant effect of bran size reduction on swelling power. Sanz Penella et al. [8] reported that decreased wheat bran size could increase the farinograph water absorption value through more water interaction via hydrogen bonding in the fiber structure. However, other studies have shown the opposite impact: larger particles of wheat bran have greater swelling capacity than smaller particles in terms of hydration properties [20,38,39]. In addition, the hydration properties of bran vary by differences in the composition of bran depending on the wheat cultivar and growing environment [19,40] and in the distribution of chemical components between the bran layers [41,42].

3.3. Solvent Retention Capacity Values of Flour-Bran Blends

The SRC values of the flour-bran blends are listed in Table 3. Flour alone without bran exhibited 60.9% water SRC and 78.7% sodium carbonate SRC values. For the flour-bran blends with VL bran, the water SRC values were 65.1 for F90-B10, 76.0 for F70-B30, and 88.6% for F50-B50. With an increase in the bran blending ratio, the water SRC values of the blends increased significantly. For the same blending ratio of bran, the water SRC values of the blends increased (e.g., from 76.6 to 78.8% for F70-B30) as the bran size decreased from L to S. Cai et al. [19], De Bondt et al. [24], and Habuš et al. [39] showed an inconsistent result on water-holding/retention capacity of bran due to differences in milling methods used to reduce the bran particle size and in wheat classes tested. De Bondt et al. [24] explained an increase in water SRC upon particle size reduction by an increase in specific surface area.

Table 3. Moisture content and solvent retention capacity values of flour-bran blends with different bran sizes and blend ratios.

Flour-Bran Blend	Bran Size	SRC (%)	
		Water	Sodium Carbonate
F100-B0 [1]	-	60.9 ± 0.2 [a(2)]	78.7 ± 0.1 [a]
F90-B10	VL [3]	65.0 ± 0.3 [b]	82.7 ± 0.0 [b]
	L	65.4 ± 0.1 [bc]	83.2 ± 0.4 [bc]
	M	65.8 ± 0.1 [bc]	84.1 ± 0.3 [bc]
	S	66.5 ± 0.1 [c]	84.7 ± 0.0 [c]
F70-B30	VL	76.0 ± 0.4 [d]	94.6 ± 0.4 [d]
	L	76.6 ± 0.1 [d]	95.8 ± 0.1 [d]
	M	76.9 ± 0.2 [d]	97.5 ± 0.4 [e]
	S	78.8 ± 0.3 [e]	99.3 ± 0.1 [f]
F50-B50	VL	88.6 ± 0.1 [f]	106.0 ± 0.2 [g]
	L	89.5 ± 0.1 [fg]	110.8 ± 0.3 [h]
	M	89.8 ± 0.0 [fg]	113.2 ± 0.4 [i]
	S	90.3 ± 0.4 [g]	113.8 ± 0.3 [i]

[1] F100-B0, Keumkang flour; F90-B10, F70-B30, and F50-B50: blend of Keumkang flour and Ariheuk bran at ratios of 9:1, 7:3, and 5:5, respectively. [2] Results are expressed as mean $\pm$ SD. Values with the different letters within the same column are significantly different ($p < 0.05$) according to Tukey's HSD test. [3] VL, L, M, and S indicate the brans with very large, large, medium, and small particle sizes.

The sodium carbonate SRC values were 82.7 for F90-B10, 94.6 for F70-B30, and 106.0% for F50-B50. As the blending ratio increased, the sodium carbonate SRC values of the blends also increased, showing a similar trend to the water SRC values. Within the same blending ratio, the sodium carbonate SRC values of the blends increased as the bran size decreased. As a representative example, the sodium carbonate SRC values of F70-B30 increased from 95.8% for L bran to 99.3% for S bran. Water SRC and sodium carbonate SRC are indicators of water absorption and the contribution of damaged starch, respectively. The sodium carbonate SRC solution (5% w/w) has a pH $\approx$ 12.0, which is above the pK of starch hydroxyl groups. Under this condition, damaged starch can be easily solvated by sodium carbonate solution and shows exaggerated swelling [31]. Sodium carbonate SRC values were affected significantly by reducing the bran size due to the increased contribution of damaged starch [43,44]. Overall, the SRC results suggested an increased amount of water required for making noodle dough with the flour-bran blends upon increasing the bran blending ratio.

3.4. Appearance and Texture of Fresh Noodles

The appearance of fresh noodles prepared from flour-bran blends with different blending ratios and bran sizes is shown in Figure 2. The noodles became dark as the bran blending ratio increased and had a smooth surface as the bran size reduced. In addition, the noodles with the smallest size of bran appeared the bran to be embedded uniformly by adhering firmly in the noodles.

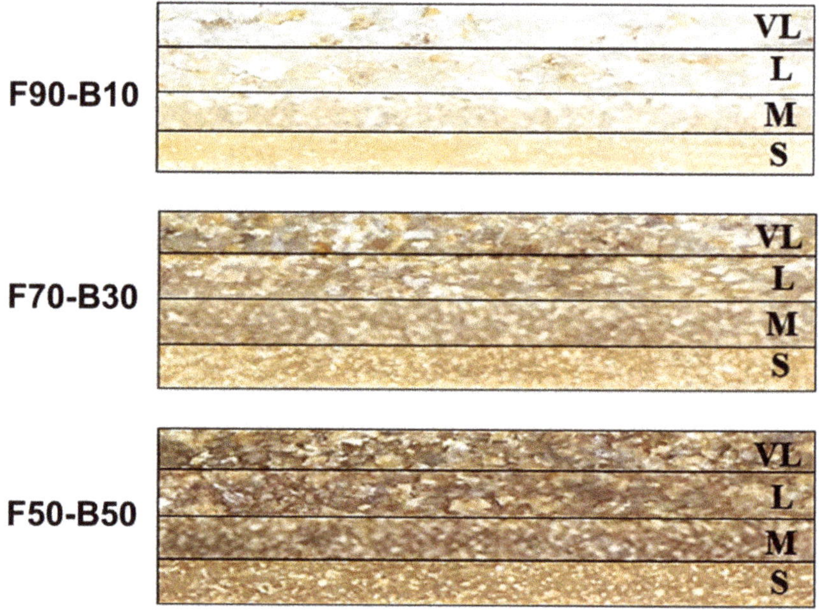

Figure 2. Appearance of fresh noodles prepared with flour-bran blends at different blending ratios and bran sizes. VL, L, M, and S indicate the brans with very large, large, medium, and small particle sizes.

The resistance to extension (R), extensibility (E), and R/E values of fresh noodles measured for textural properties are shown in Table 4. Gliadins contribute to dough viscosity and extensibility, and glutenin imparts elasticity. The R/E ratio of fresh noodles represents a balance between dough strength and extensibility, indicating gluten network strength [45]. An appropriate balance between dough viscosity and elasticity is essential for well-developed doughs [46]. The fresh noodles prepared with flour alone (F100-B0) exhibited 1.37 N resistance, 14.76 mm extensibility, and 0.093 R/E ratios. The resistance values of fresh noodles prepared with flour-bran blends with VL bran were 1.18 for F90-B10, 0.83 for F70-B30, and 0.51 N for F50-B50, decreasing significantly with increasing blending ratio. Within the same blending ratio, the resistance of the noodles increased (e.g., from 1.00 to 1.18 N for F70-B30) as bran size decreased from L to S ($p < 0.05$). The extensibility values of the blends with VL bran were 11.20 for F90-B10, 10.84 for F70-B30, and 10.49 mm for F50-B50, decreasing slightly with an increase in the bran blending ratio. Within the same blending ratio, the extensibility of noodles with blends decreased slightly or negligibly when bran size was reduced from L to S (e.g., from 9.67 to 9.58 mm for F70-B30). The R/E values of fresh noodles with blends containing VL bran were 0.106 for F90-B10, 0.077 for F70-B30, and 0.048 for F50-B50, decreasing significantly with an increase in the blending ratio ($p < 0.05$). Within the same blending ratio, the R/E ratios of noodles with blends increased significantly with decreasing bran size from L to S (e.g., from 0.103 to 0.123 for F70-B30) ($p < 0.05$).

Table 4. Resistance (R), extensibility (E), and R/E values of flour-bran blends with different bran sizes and blend ratios.

Flour-Bran Blend	Bran Size	Resistance (N)	Extensibility (mm)	R/E
F100-B0 [1]	-	1.37 ± 0.04 [e(2)]	14.76 ± 0.07 [f]	0.093 ± 0.003 [cd]
F90-B10	VL [3]	1.18 ± 0.05 [de]	11.20 ± 0.03 [e]	0.106 ± 0.006 [de]
	L	1.20 ± 0.03 [de]	10.08 ± 0.15 [bc]	0.119 ± 0.002 [ef]
	M	1.32 ± 0.03 [e]	9.81 ± 0.07 [ab]	0.134 ± 0.003 [fg]
	S	1.42 ± 0.03 [e]	9.78 ± 0.06 [ab]	0.145 ± 0.003 [g]
F70-B30	VL	0.83 ± 0.03 [bc]	10.84 ± 0.09 [de]	0.077 ± 0.003 [bc]
	L	1.00 ± 0.05 [cd]	9.67 ± 0.09 [ab]	0.103 ± 0.006 [de]
	M	1.16 ± 0.10 [de]	9.64 ± 0.09 [a]	0.120 ± 0.013 [ef]
	S	1.18 ± 0.08 [de]	9.58 ± 0.05 [a]	0.123 ± 0.011 [efg]
F50-B50	VL	0.51 ± 0.01 [a]	10.49 ± 0.01 [cd]	0.048 ± 0.001 [a]
	L	0.55 ± 0.02 [a]	9.50 ± 0.07 [a]	0.058 ± 0.002 [ab]
	M	0.71 ± 0.04 [ab]	9.53 ± 0.01 [a]	0.074 ± 0.006 [bc]
	S	0.74 ± 0.03 [abc]	9.52 ± 0.03 [a]	0.078 ± 0.004 [bc]

[1] F100-B0, Keumkang flour; F90-B10, F70-B30, and F50-B50: blend of Keumkang flour and Ariheuk bran at ratios of 9:1, 7:3, and 5:5, respectively. [2] Results are expressed as mean ± SD. Values with the different letters within the same column are significantly different ($p < 0.05$) according to Tukey's HSD test. [3] VL, L, M, and S indicate the brans with very large, large, medium, and small particle sizes.

Our results for resistance and extensibility of noodle dough sheets within the same ratio of flour-bran blend showed trends similar to other publications. Bran particle size distribution of WWF influences gluten network formation [47]. The effect of the bran particle size on the dough rheological property is somewhat contradictory due to the effects of the milling method and bran composition. Smaller bran particles resulted in higher resistance of bread dough than larger particles [48], and reduced particle size of wheat bran by micronization decreased dough extensibility [49]. The dough containing fine bran exhibited higher dough strength after resting for 180 min in the farinograph and extensograph evaluations, owing to the quick recovery of the gluten network of the dough [50]. Sudha et al. [51] reported that the dough extensibility of WWF decreased with an increase in the bran level. Overall, it can be speculated that the textural properties of fresh noodle sheets might reflect stronger gluten strength in the noodles with flour-bran blends containing the decreased size of bran. Özboy and Köksel [52] explained that coarse brans occupy more space and result in a more detrimental effect on the gluten network formation in dough by decreasing connectivity.

3.5. Turbidity of Cooking Water and Textural Characteristics of Cooked Noodles

The color values of the fresh dough sheet and cooked noodles prepared from flour-bran blends with different blend ratios and bran sizes are shown in Table 5. As the bran blending ratio increased, the fresh dough sheet showed a decrease in brightness (L*) and an increase in redness (a*), indicating the significant influence of purple-colored bran. Bran size also affected the color of the dough sheet, which slightly decreased in brightness and increased in redness and yellowness (b*) with size reduction.

Table 5. Color of fresh noodle sheets and cooked noodles prepared from flour-bran blends with different bran sizes and blend ratios.

Flour-Bran Blend	Bran Size	Uncooked			Cooked		
		L*	a*	b*	L*	a*	b*
F100-B0 [1]	-	86.2 ± 0.5 [f(2)]	2.7 ± 0.1 [a]	15.3 ± 0.8 [d]	74.4 ± 0.3 [h]	(−0.8 ± 0.1) [a]	12.0 ± 0.2 [bcde]
F90-B10	VL [3]	73.5 ± 1.8 [e]	3.8 ± 0.3 [ab]	9.6 ± 0.4 [a]	56.5 ± 1.1 [g]	4.6 ± 0.6 [b]	7.3 ± 1.1 [a]
	L	72.0 ± 1.2 [de]	4.8 ± 0.4 [bc]	10.7 ± 0.6 [ab]	54.8 ± 0.7 [fg]	5.8 ± 0.2 [bc]	8.9 ± 0.3 [ab]
	M	67.2 ± 1.9 [de]	6.0 ± 0.5 [cd]	11.6 ± 0.5 [abc]	52.2 ± 0.4 [f]	6.5 ± 0.3 [bcd]	10.4 ± 0.4 [abcd]
	S	65.9 ± 0.5 [d]	6.8 ± 0.2 [de]	12.8 ± 0.4 [bcd]	51.1 ± 0.7 [f]	7.2 ± 0.3 [cde]	10.9 ± 0.2 [abcd]
F70-B30	VL	57.8 ± 1.3 [c]	6.4 ± 0.4 [cde]	11.6 ± 0.7 [abc]	45.4 ± 1.4 [e]	6.7 ± 0.6 [cd]	9.3 ± 1.4 [abc]
	L	54.0 ± 1.7 [c]	8.0 ± 0.5 [ef]	13.2 ± 0.6 [bcd]	44.2 ± 0.4 [de]	7.4 ± 0.4 [cde]	10.1 ± 0.6 [abc]
	M	51.4 ± 1.0 [bc]	8.5 ± 0.1 [fg]	13.8 ± 0.3 [cd]	41.3 ± 0.4 [cd]	8.4 ± 0.4 [defg]	12.3 ± 1.1 [bcde]
	S	53.9 ± 0.6 [c]	9.1 ± 0.1 [fg]	14.7 ± 0.3 [cd]	40.4 ± 0.5 [bcd]	9.4 ± 0.3 [fg]	14.0 ± 0.6 [de]
F50-B50	VL	46.0 ± 2.1 [a]	8.9 ± 0.2 [g]	13.6 ± 0.2 [cd]	38.1 ± 1.0 [abc]	7.7 ± 0.2 [cdef]	9.7 ± 0.5 [ab]
	L	41.7 ± 0.6 [a]	9.7 ± 0.1 [g]	13.9 ± 0.2 [cd]	37.2 ± 0.3 [ab]	8.8 ± 0.3 [efg]	12.0 ± 0.4 [bcde]
	M	41.4 ± 0.6 [a]	9.5 ± 0.2 [fg]	13.9 ± 0.3 [cd]	35.3 ± 0.3 [a]	9.4 ± 0.2 [fg]	13.8 ± 0.4 [cde]
	S	43.6 ± 0.5 [ab]	9.9 ± 0.1 [g]	14.6 ± 0.2 [cd]	34.8 ± 0.2 [a]	10.2 ± 0.2 [g]	15.1 ± 0.3 [e]

[1] F100-B0, Keumkang flour; F90-B10, F70-B30, and F50-B50: blend of Keumkang flour and Ariheuk bran at ratios of 9:1, 7:3, and 5:5, respectively. [2] Results are expressed as mean ± SD. Values with the different letters within the same column are significantly different ($p < 0.05$) according to Tukey's HSD test. [3] VL, L, M, and S indicate the brans with very large, large, medium, and small particle sizes.

The turbidity of the cooking water for the noodles prepared from the flour-bran blends is presented in Table 6. The fresh noodle prepared with flour alone (F100-B0) exhibited turbidity of 0.78 ΔA hr^{-1} g flour^{-1} in the cooking water. The turbidity values of the cooking water for fresh noodles with the blends with VL bran were 1.09 for F90-B10, 1.30 for F70-B30, and 1.35 ΔA hr^{-1} g flour^{-1} for F50-B50, increasing significantly up to the 30% blending ratio and leveling off with a further increase in the blending ratio. Within the same blending ratio, the turbidity in cooking water for the noodles with blends decreased significantly (e.g., from 0.69 to 0.62 ΔA hr^{-1} g flour^{-1} for F70-B30) as bran size decreased from L to S ($p < 0.05$). The increased turbidity of the cooking water with an increasing blending ratio indicated that more solids leached from the noodles by loosening and weakening the gluten network structure. In contrast, the decreased turbidity of the cooking water with decreasing bran size reflected fewer solids leaching out from the noodles with a stronger gluten network structure. High cooking loss reflected the weaker gluten strength and lower degree of connectivity in the noodle structure [53].

The textural parameters of the cooked noodles made with the flour-bran blends are listed in Table 6. Cooked noodles prepared with flour alone exhibited 18.1 N firmness, 0.62 cohesiveness, 0.91 springiness, and 19.8 mJ chewiness. The firmness of cooked noodles prepared with the blends with VL bran decreased significantly from 13.9 N for F90-B10 to 4.4 N for F50-B50 with an increasing blending ratio ($p < 0.05$). Chen et al. [27] reported a decrease in hardness of cooked dry white Chinese noodles by increasing bran addition level, similar to our result. Within the same blending ratio of bran, the firmness of the noodles with blends increased significantly (e.g., from 8.8 to 22.0 N for F50-B50) as bran size decreased from L to S ($p < 0.05$). These increases were more significant in the noodles with a higher blending ratio than those with a lower blending ratio. Although the damaged starch content increased by decreasing bran size (Table 2), the noodles showed a decrease in turbidity of cooking water and an increase in firmness of cooked noodles, which might be a more significant contribution by gluten development than by damaged starch.

Table 6. Textural parameters of cooked noodles prepared from flour alone and flour-bran blends with different bran sizes and blend ratios.

| Flour-Bran Blend | Bran Size | Textural Parameter | | | | Turbidity of Cooked Water (ΔA hr^{-1} g flour^{-1}) |
		Firmness (N)	Cohesiveness	Springiness	Chewiness (mJ)	
F100-B0 [1]	-	18.1 ± 0.9 [def(2)]	0.62 ± 0.00 [c]	0.91 ± 0.00 [c]	19.8 ± 1.2 [e]	0.78 ± 0.06 [e]
F90-B10	VL [3]	13.9 ± 0.2 [bcd]	0.56 ± 0.02 [c]	0.84 ± 0.01 [bc]	15.5 ± 0.9 [de]	1.09 ± 0.01 [f]
	L	14.6 ± 0.7 [cde]	0.57 ± 0.02 [c]	0.86 ± 0.03 [bc]	16.0 ± 1.5 [de]	0.65 ± 0.01 [bcd]
	M	16.2 ± 1.1 [de]	0.57 ± 0.02 [c]	0.90 ± 0.01 [c]	16.6 ± 1.6 [de]	0.46 ± 0.02 [a]
	S	16.4 ± 0.9 [de]	0.59 ± 0.02 [c]	0.91 ± 0.02 [c]	17.1 ± 1.0 [de]	0.56 ± 0.00 [ab]
F70-B30	VL	9.5 ± 0.6 [abc]	0.43 ± 0.03 [abc]	0.74 ± 0.01 [abc]	7.7 ± 0.5 [abc]	1.30 ± 0.01 [g]
	L	14.7 ± 0.7 [cde]	0.49 ± 0.12 [bc]	0.83 ± 0.05 [bc]	13.1 ± 4.3 [cde]	0.69 ± 0.02 [cde]
	M	17.8 ± 0.5 [def]	0.46 ± 0.02 [abc]	0.85 ± 0.04 [bc]	13.7 ± 0.8 [cde]	0.59 ± 0.00 [bc]
	S	19.3 ± 1.1 [ef]	0.48 ± 0.02 [abc]	0.86 ± 0.02 [bc]	15.8 ± 0.6 [de]	0.62 ± 0.00 [bc]
F50-B50	VL	4.4 ± 0.5 [a]	0.24 ± 0.01 [a]	0.57 ± 0.02 [a]	2.0 ± 0.4 [a]	1.35 ± 0.01 [g]
	L	8.8 ± 0.7 [ab]	0.29 ± 0.06 [ab]	0.57 ± 0.09 [a]	3.8 ± 0.9 [ab]	0.74 ± 0.01 [de]
	M	14.5 ± 1.3 [cde]	0.31 ± 0.02 [ab]	0.67 ± 0.03 [ab]	7.0 ± 1.7 [abc]	0.65 ± 0.01 [bcd]
	S	22.0 ± 1.8 [f]	0.31 ± 0.06 [ab]	0.80 ± 0.02 [bc]	10.5 ± 1.9 [bcd]	0.63 ± 0.00 [bc]

[1] F100-B0, Keumkang flour; F90-B10, F70-B30, and F50-B50: blend of Keumkang flour and Ariheuk bran at ratios of 9:1, 7:3, and 5:5, respectively. [2] Results are expressed as mean ± SD. Values with the different letters within the same column are significantly different ($p < 0.05$) according to Tukey's HSD test. [3] VL, L, M, and S indicate the brans with very large, large, medium, and small particle sizes.

The cohesiveness values of cooked noodles prepared with flour-bran blends with VL bran were 0.56 for F90-B10, 0.43 for F70-B30, and 0.24 for F50-B50. The springiness values of cooked noodles prepared from blends with VL bran were 0.84 for F90-B10, 0.74 for F70-B30, and 0.57 for F50-B50. The cohesiveness and springiness of cooked noodles with the blends significantly decreased with an increase in the blending ratio ($p < 0.05$). Among noodles with same blending ratio, the cohesiveness of noodles increased negligibly (e.g., from 0.29 to 0.31 for F50-B50) with increasing bran size from L to S. However, springiness of noodles with blends increased significantly (e.g., from 0.57 to 0.80 for F50-B50) as bran size decreased from L to S ($p < 0.05$). Niu et al. [26] reported that WWF noodles increased in springiness and cohesiveness with decreasing millfeed particle size.

The chewiness of cooked noodles prepared from the blends containing VL bran were 15.5 for F90-B10, 7.7 for F70-B30, and 2.0 mJ for F50-B50. Similar to the results for firmness, the chewiness of cooked noodles with the blends significantly decreased with increasing blending ratio ($p < 0.05$). Chen et al. [27] also reported a decrease in chewiness of cooked dry white Chinese noodles by increasing bran addition level. Within the same blending ratio, the chewiness of the noodles with blends increased significantly (e.g., 3.8 to 10.5 mJ for F50-B50) as bran size decreased from L to S ($p < 0.05$).

Overall, as the bran size decreased, the firmness, cohesiveness, springiness, and chewiness of cooked noodles increased while the turbidity of the cooking water decreased, demonstrating improved desirable attributes of noodle. This result was similar to that reported by Sozer et al. [54], who reported that the reduction of bran particle size in biscuits showed the highest elastic modulus value because overall strength was improved by better incorporating bran particles into the biscuit matrix. Hatcher et al. [35] also reported excellent cooking quality of noodles with a fine particle size of flour. Ma et al. [55] explained that a small particle size of flour increases the change in the gluten index and wet gluten content of the flour and increases the hardness, elasticity, adhesiveness, and chewiness of the noodles by increasing the network structure and water-holding capacity of the noodles. Small bran particles enhanced noodle quality more significantly at higher blending ratios. In the future, a sensory evaluation will be needed to correlate with the textural data measured by the instrument to confirm the eating quality.

3.6. Total Phenolic Compound and Total Anthocyanin Contents of Fresh Noodles Prepared with Flour-Bran Blends

The total phenolic compound and total anthocyanin contents of fresh noodles prepared from the flour-bran blends with different blend ratios and bran sizes are shown in Figure 3. Fresh noodles prepared with flour alone exhibited a total phenolic content of 69.9 mg GAE/100 g and total anthocyanin content of 0.31 mg C3GE/100 g. The total phenolic content (TPC) of fresh noodles with the blend with VL bran was 158.7 for F90-B10, 302.8 for F70-B30, and 482.8 mg GAE/100 g for F50-B50, increasing significantly and proportionally as the blending ratio increased ($p < 0.05$). Within the same blending ratio of bran, the total phenolic content of the noodles with blends increased significantly (e.g., from 328.8 to 380.1 mg GAE/100 g for F70-B30, and from 504.2 to 546.3 mg GAE/100 g for F50-B50) as bran size decreased from L to S. The TPC increase is possibly due to an increase in surface area to mass by particle size reduction and the increased extraction of phytochemicals [56]. Black-purple wheat varieties have higher total phenolic content than white varieties [57]. Avarzed et al. [29] also reported that purple-colored wheat bran contains higher total phenolic content than normal wheat bran, which is a similar trend to our result.

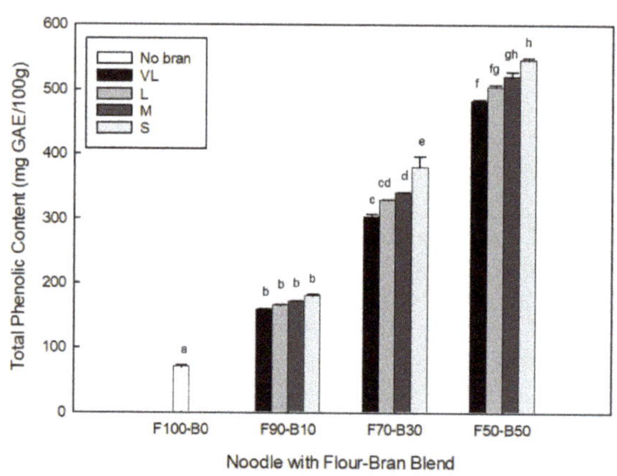
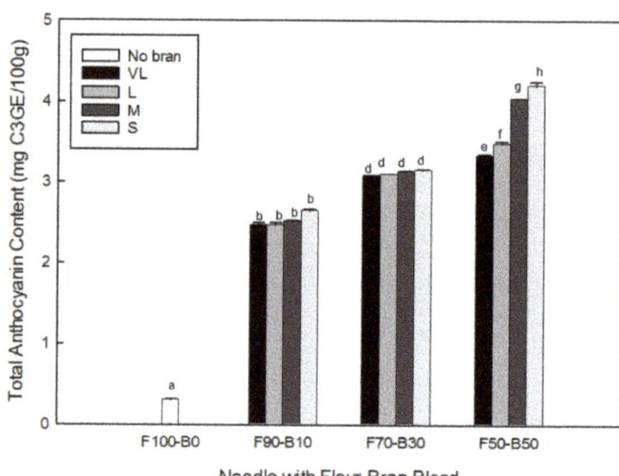

Figure 3. Total phenolic and anthocyanin content of fresh noodles made with flour-bran blends at different blending ratios and bran sizes. The different letters above the bars are significantly different ($p < 0.05$) according to Tukey's HSD test. VL, L, M, and S indicate the brans with very large, large, medium, and small particle sizes.

The total anthocyanin contents of the fresh noodles prepared with the blends with VL bran were 2.47 for F90-B10, 3.08 for F70-B30, and 3.33 mg C3GE/100 g for F50-B50, increasing significantly as the blending ratio increased ($p < 0.05$). Within the same blending ratio of bran, the total anthocyanin content of the noodles with blends increased slightly or significantly (e.g., from 3.10 to 3.15 mg C3GE/100 g for F70-B30 and from 3.48 to 4.20 mg C3GE/100 g for F50-B50) as bran size decreased from L to S ($p < 0.05$). Brewer et al. [56] reported that whole wheat bran showed an increase in phenolic acids and anthocyanins with decreasing particle size distribution compared to un-milled bran, showing a similar trend to the results of our study. The increase in total anthocyanin with decreasing bran size is also possibly explained by increasing surface area and extraction as the increase in TPC [56]. Like the total phenolic content, Avarzed et al. [29] reported higher total anthocyanin content in purple-colored wheat bran than in normal wheat bran.

3.7. Antioxidant Activity of Fresh Noodles Prepared with Flour-Bran Blends

The ABTS and DPPH radical-scavenging activities of fresh noodles prepared from flour-bran blends with different blend ratios and bran sizes are shown in Figure 4. Fresh noodles prepared with flour alone exhibited ABTS radical-scavenging activity of 78.7 μmol

TE/100 g. The ABTS radical-scavenging activity of the fresh noodles with the blends with VL bran was 245.5 for F90-B10, 315.3 for F70-B30, and 487.4 μmol TE/100 g for F50-B50, increasing significantly as the blending ratio increased ($p < 0.05$). Even with 10% blending of bran, the ABTS radical-scavenging activity of fresh noodles increased dramatically. Within the same blending ratio of bran, the ABTS radical-scavenging activity of the noodles with blends increased significantly (e.g., from 499.2 to 561.2 μmol TE/100 g for F50-B50) as the bran size decreased from L to S ($p < 0.05$). In addition, as the blending ratio increased, the increase in ABTS radical-scavenging activity of fresh noodles with reduced bran size increased. In particular, fresh noodles prepared with the blends at 50% blending ratio (F50-B50) from L to S bran showed the largest increase in ABTS radical-scavenging activity (Figure 4).

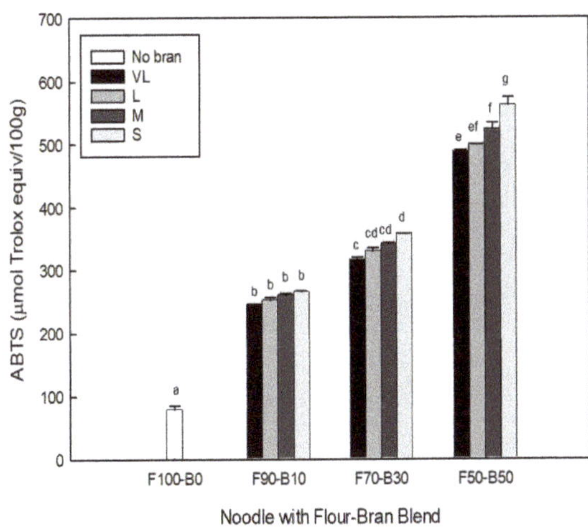

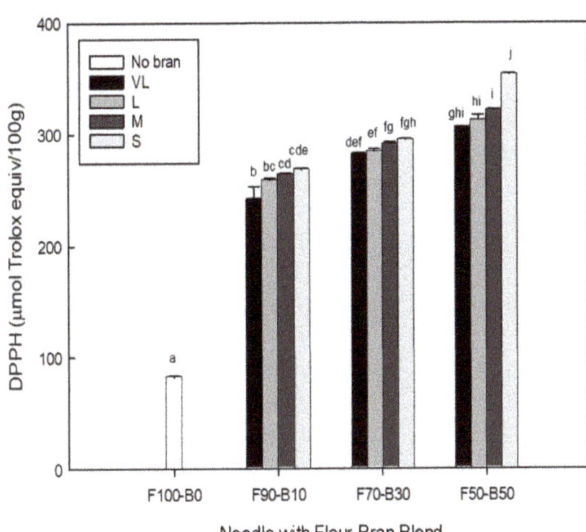

Figure 4. Antioxidant activity as indicated by ABTS and DPPH radical-scavenging activity of fresh noodles with flour-bran blends at different blending ratios and bran sizes. The different letters above the bars are significantly different ($p < 0.05$) according to Tukey's HSD test. VL, L, M, and S indicate the brans with very large, large, medium, and small particle sizes.

DPPH radical-scavenging activity of fresh noodles prepared with flour alone was 82.4 μmol TE/100 g. The noodles prepared with the blends containing VL bran showed the DPPH radical-scavenging activity was 242.0 for F90-B10, 282.1 for F70-B30, and 306.1 μmol TE/100 g for F50-B50, increasing significantly as the blending ratio increased ($p < 0.05$). The DPPH radical-scavenging activity of the noodles even with 10% blending of bran increased dramatically. Within the same blending ratio of bran, the DPPH radical-scavenging activity of the noodles with the blends increased slightly (e.g., from 312.5 to 353.8 μmol TE/100 g for F50-B50) as the bran size decreased from L to S ($p < 0.05$). Similar to the results of ABTS radical-scavenging activity, the extent of increase in DPPH radical-scavenging activity of fresh noodles with reducing bran size increased with increasing bran blending ratio (Figure 4). In particular, fresh noodles prepared with blends at 50% blending ratio (F50-B50) from L to S bran showed the largest increase in DPPH radical-scavenging activity. A positive correlation was found between antioxidant activity and the blending ratio of bran and bran size.

According to Avarzed et al. [29], purple-colored wheat bran has significantly higher antioxidant activity than normal wheat bran (1247 vs. 972 mg μmol TE/100 g for ABTS, 936 vs. 749 mg μmol TE/100 g for DPPH). Phenolic acids, pre-dominantly ferulic acid, are concentrated in the bran fraction of whole wheat and contribute to antioxidant properties [5,57,58]. Extracts with the highest TPC showed the greatest antioxidant properties [56]. Habuš et al. [39] also reported the highest TPC and AO in fine bran samples are possibly led by the enhanced release of phenolic compounds from the wheat bran matrix. In our study, the enhancing effect of small bran particles on antioxidant activity appeared more significant for higher blending ratios by a relatively increased cumulative release.

4. Conclusions

This study demonstrated that the particle size of wheat bran has a significant impact on fresh-noodle texture and antioxidant activity. Compared to VL wheat bran, the R/E ratios of fresh noodles increased significantly with decreasing bran size. The texture of cooked noodles showed an increase in firmness and springiness upon reducing the size of the bran particles. The antioxidant activity of fresh noodles increased significantly as the bran blending ratio and bran particle size decreased. Fine bran particles were embedded more closely to the noodle sheet, resulting in fewer bran particles leaching out of the cooking water. Noodle quality and antioxidant activity were more significantly enhanced by small bran particles at higher blending ratios. Thus, reducing the bran particle size could improve the textural properties of noodles with flour-bran blends.

Author Contributions: Conceptualization, K.K. and M.K.; methodology, G.P. and H.C.; formal analysis, G.P. and H.C.; investigation, G.P., H.C. and M.K.; data curation, G.P. and H.C.; writing—original draft, G.P. and M.K.; validation, K.K. and M.K.; visualization, K.K. and M.K.; writing—review and editing, K.K. and M.K.; resources, M.K.; supervision, M.K.; project administration, M.K.; funding acquisition, M.K. All authors have read and agreed to the published version of the manuscript.

Funding: This research was supported by the Cooperative Research Program for the Agriculture Science and Technology Department (Project No. PJ014543), funded by the Rural Development Administration (Korea).

Data Availability Statement: The data presented in this study are available on request from the corresponding author.

Conflicts of Interest: The authors declare no conflict of interest.

References

1. Vitaglione, P.; Napolitano, A.; Fogliano, V. Cereal dietary fibre: A natural functional ingredient to deliver phenolic compounds into the gut. *Trends Food Sci. Technol.* **2008**, *19*, 451–463. [CrossRef]
2. Junejo, S.H.; Geng, H.; Wang, N.; Wang, H.; Ding, Y.; Zhou, Y.; Rashid, A. Effects of particle size on physicochemical and in vitro digestion properties of durum wheat bran. *Int. J. Food Sci. Technol.* **2019**, *54*, 221–230. [CrossRef]
3. Coda, R.; Kärki, I.; Nordlund, E.; Heiniö, R.L.; Poutanen, K.; Katina, K. Influence of particle size on bioprocess induced changes on technological functionality of wheat bran. *Food Microbiol.* **2014**, *37*, 69–77. [CrossRef] [PubMed]
4. Deroover, L.; Tie, Y.; Verspreet, J.; Verspreet, J.; Courtin, C.M.; Verbeke, K. Modifying wheat bran to improve its health benefits. *Crit. Rev. Food Sci. Nutr.* **2020**, *60*, 1104–1122. [CrossRef] [PubMed]
5. Rico, D.; Villaverde, A.; Martinez-Villaluenga, C.; Gutierrez, A.L.; Caballero, P.A.; Ronda, F.; Penas, E.; Frias, J.; Diana Martin, A.B. Application of autoclave treatment for development of a natural wheat bran antioxidant ingredient. *Foods* **2020**, *9*, 781. [CrossRef] [PubMed]
6. Abdel-Aal, E.-S.M.; Hucl, P. Composition and stability of anthocyanins in blue-grained wheat. *J. Agric. Food Chem.* **2003**, *51*, 2174–2180. [CrossRef] [PubMed]
7. Hosseinian, F.S.; Li, W.; Beta, T. Measurement of anthocyanins and other physicochemicals in purple wheat. *Food Chem.* **2008**, *109*, 916–924. [CrossRef]
8. Sanz Penella, J.M.; Collar, C.; Haros, M. Effect of wheat bran and enzyme addition on dough functional performance and phytic acid levels in bread. *J. Cereal Sci.* **2008**, *48*, 715–721. [CrossRef]
9. Noort, M.W.J.; van Haaster, D.; Hemery, Y.; Schols, H.A.; Hamer, R.J. The effect of particle size of wheat bran fractions on bread quality—Evidence for fibre-protein interactions. *J. Cereal Sci.* **2010**, *52*, 59–64. [CrossRef]
10. Zhang, J.; Li, M.; Li, C.; Liu, Y. Effect of wheat bran insoluble dietary fiber with different particle size on the textural properties, protein secondary structure, and microstructure of noodles. *Grains Oil Sci. Technol.* **2019**, *2*, 97–102. [CrossRef]
11. Aravind, N.; Sissons, M.; Fellows, C.M. Effect of soluble fibre (guar gum and carboxymethylcellulose) addition on technological, sensory and structural properties of durum wheat spaghetti. *Food Chem.* **2012**, *131*, 893–900. [CrossRef]
12. Lang, R.; Jebb, S.A. Who consumes whole grains, and how much? *Proc. Nutr. Soc.* **2003**, *62*, 123–127. [CrossRef] [PubMed]
13. Bressiani, J.; Oro, T.; Santetti, G.S.; Almeida, J.L.; Bertolin, T.E.; Gómez, M.; Gutkoski, L.C. Properties of whole grain flour and performance in bakery products as a function of particle size. *J. Cereal Sci.* **2017**, *75*, 269–277. [CrossRef]
14. Fu, B.X. Asian noodles: History, classification, raw materials, and processing. *Food Res. Int.* **2008**, *41*, 888–902. [CrossRef]

15. Kartina, K.; Juvonen, R.; Laitila, A.; Flander, L.; Nordlund, E.; Kariluoto, S.; Piironen, V.; Poutanen, K. Fermented wheat bran as a functional ingredient in baking. *Cereal Chem.* **2012**, *89*, 126–134. [CrossRef]

16. Wu, T.; Li, Z.; Sui, W.; Zhang, M. Effect of extrusion, steam explosion and enzymatic hydrolysis on functional properties of wheat bran. *Food Sci. Technol. Res.* **2018**, *24*, 591–598. [CrossRef]

17. Park, E.Y.; Fuerst, E.P.; Baik, B.K. Effect of bran hydration with enzymes on functional properties of flour-bran blends. *Cereal Chem.* **2019**, *96*, 273–282. [CrossRef]

18. Seguchi, M.; Uozu, M.; Oneda, H.; Murayama, R.; Okusu, H. Effect of outer bran layers from germinated wheat grains on bread-making properties. *Cereal Chem.* **2010**, *87*, 231–236. [CrossRef]

19. Cai, L.; Choi, L.; Hyun, J.N.; Jeong, Y.K.; Baik, B.K. Influence of bran particle size on bread-baking quality of whole grain wheat flour and starch retrogradation. *Cereal Chem.* **2014**, *91*, 65–71. [CrossRef]

20. Jacobs, P.J.; Hemdane, S.; Dornez, E.; Delcour, J.A.; Courtin, C.M. Study of hydration properties of wheat brans as a function of particle size. *Food Chem.* **2015**, *179*, 296–304. [CrossRef]

21. Cai, L.; Choi, L.; Park, C.S.; Jeong, Y.K.; Baik, B.K. Bran hydration and physical treatment improve the bread-baking quality of whole grain wheat flour. *Cereal Chem.* **2015**, *92*, 557–564. [CrossRef]

22. Zhang, Y.; Gao, F.; He, Z. Effects of bran hydration and autoclaving on processing quality of Chinese steamed bread and noodles produced from whole grain wheat flour. *Cereal Chem.* **2019**, *96*, 104–114. [CrossRef]

23. Hemdane, S.; Leys, S.; Jacobs, P.J.; Dornez, E.; Delcour, J.A.; Courtin, C.M. Wheat milling by-products and their impact on bread making. *Food Chem.* **2015**, *187*, 280–289. [CrossRef] [PubMed]

24. De Bondt, Y.; Liberloo, I.; Roye, C.; Windhab, E.J.; Lamothe, L.; King, R.; Courtin, C.M. The effect of wet milling and cryogenic milling on the structure and physicochemical properties of wheat bran. *Foods* **2020**, *9*, 1755. [CrossRef]

25. Alzuwaid, N.T.; Pleming, D.; Fellows, C.M.; Laddomada, B.; Sissons, M. Influence of duram wheat bran particle size on phytochemical content and on leavened bread baking quality. *Foods* **2021**, *10*, 489. [CrossRef]

26. Niu, M.; Hou, G.G.; Wang, L.; Chen, Z. Effects of superfine grinding on the quality characteristics of whole-wheat flour and its raw noodle product. *J. Cereal Sci.* **2014**, *60*, 382–388. [CrossRef]

27. Chen, J.S.; Fei, M.J.; Shi, C.L.; Tian, J.C.; Sun, C.L.; Zhang, H.; Ma, Z.; Dong, H.X. Effect of particle size and addition level of wheat bran on quality of dry white Chinese noodles. *J. Cereal Sci.* **2011**, *53*, 217–224. [CrossRef]

28. Memon, A.A.; Mahar, I.; Memon, R.; Soomro, S.; Harnly, J.; Memon, M.; Bhangar, M.I.; Luthria, D.L. Impact of flour particle size on nutrient and phenolic acid composition of commercial wheat varieties. *J. Food Compost. Anal.* **2020**, *86*, 103358. [CrossRef]

29. Avarzed, E.; Wang, X.; Moon, Y.; Kim, K.H.; Kweon, M. Quality characteristics and antioxidant activities of cookies formulated with the blend of Korean domestic wheat flour and purple wheat bran. *Korean J. Food Cook Sci.* **2020**, *36*, 152–162. [CrossRef]

30. AACC. Method 44-15.02, 56-11.02, 76-31.01. In *Approved Methods of Analysis*, 11th ed.; AACC International: St. Paul, MN, USA, 2010.

31. Kweon, M.; Slade, L.; Levine, H. Solvent retention capacity (SRC) testing of wheat flour: Principles and value in predicting flour functionality in different wheat-based food processes and in wheat breeding–A review. *Cereal Chem.* **2011**, *88*, 537–552. [CrossRef]

32. Moon, Y.; Kim, K.H.; Kweon, M. Effects of flour quality and drying rates controlled by temperature, air circulation, and relative humidity on the quality of dried white-salted noodles. *Cereal Chem.* **2019**, *96*, 1011–1021. [CrossRef]

33. Wang, X.; Kweon, M. Quality of US soft red wheat flours and their suitability for making fresh noodles. *Korean J. Food Cook Sci.* **2021**, *37*, 134–143.

34. Yu, L.; Beta, T. Identification and antioxidant properties of phenolic compounds during production of bread from purple wheat grains. *Molecules* **2015**, *20*, 15525–15549. [CrossRef] [PubMed]

35. Hatcher, D.W.; Anderson, M.J.; Desjardins, R.G.; Edwards, N.M.; Dexter, J.E. Effects of flour particle size and starch damage on processing and quality of white salted noodles. *Cereal Chem.* **2002**, *79*, 64–71. [CrossRef]

36. Doblado-Maldonado, A.F.; Pike, O.A.; Sweley, J.C.; Rose, D.J. Key issues and challenges in whole wheat flour milling and storage. *J. Cereal Sci.* **2012**, *56*, 119–126. [CrossRef]

37. Okusu, H.; Otsubo, S.; Dexter, J. Wheat milling and flour quality analysis for noodles in Japan. In *Asian Noodles*; Hou, G.G., Ed.; John Wiley & Sons, Inc.: Hoboken, NJ, USA, 2010; pp. 57–73.

38. Zhu, K.; Huang, S.; Peng, W.; Qian, H.; Zhou, H. Effect of ultrafine grinding on hydration and antioxidant properties of wheat bran dietary fiber. *Food Res. Int.* **2010**, *43*, 943–948. [CrossRef]

39. Habuš, M.; Novotni, D.; Gregov, M.; Štifter, S.; Mistač, N.Č.; Voucko, B.; Ćuric, D. Influence of particle size reduction and high-intensity ultrasound on polyphenol oxidase, phenolics, and technological properties of wheat bran. *J. Food Process. Preserv.* **2021**, *45*, 15204. [CrossRef]

40. Hossain, K.; Ulven, C.; Glover, K.; Ghavami, F.; Simsek, S.; Alamri, M.; Kumar, A.; Mergoum, M. Interdependence of Cultivar and Environment on Fiber Composition in Wheat Bran. *Aust. J. Crop. Sci.* **2013**, *7*, 525–531.

41. Hemdane, S.; Jacobs, P.J.; Dornez, E.; Verspreet, J.; Delcour, J.A.; Courtin, C.M. Wheat (*Triticum aestivum* L.) bran in bread making: A critical review. *Compr. Rev. Food Sci. Food Saf.* **2016**, *15*, 28–42. [CrossRef]

42. Hemdane, S.; Langenaeken, N.A.; Jacobs, P.J.; Verspreet, J.; Delcour, J.A.; Courtin, C.M. Study of the intrinsic properties of wheat bran and pearlings obtained by sequential debranning and their role in bran-enriched bread making. *J. Cereal Sci.* **2016**, *71*, 78–85. [CrossRef]

43. Barak, S.; Mudgil, D.; Khatkar, B.S. Effect of flour particle size and damaged starch on the quality of cookies. *J. Food Sci. Technol.* **2012**, *51*, 1342–1348. [CrossRef] [PubMed]
44. Lu, L.; Cao, X.; Zhou, S.; Li, Y.; Wang, L.; Qian, H.; Zhang, H.; Qi, X. Effect of wheat bran particle size on the quality of whole wheat based instant fried noodles. *J. Food Nutr. Res.* **2018**, *6*, 295–301. [CrossRef]
45. Dhaka, V.; Khatkar, B.S. Effects of gliadin/glutenin and HMW-GS/LMW-GS ratio on dough rheological properties and bread-making potential of wheat varieties. *J. Food Qual.* **2015**, *38*, 71–82. [CrossRef]
46. Goesaert, H.; Brijs, K.; Veraverbeke, W.S.; Courtin, C.M.; Gebruers, K.; Delcour, J.A. Wheat flour constituents: How they impact bread quality, and how to impact their functionality. *Trends Food Sci. Technol.* **2005**, *16*, 12–30. [CrossRef]
47. Xiong, L.; Zhang, B.; Niu, M.; Zhao, S. Protein polymerization and water mobility in whole-wheat dough influenced by bran particle size distribution. *LWT-Food Sci. Technol.* **2017**, *82*, 396–403. [CrossRef]
48. Wang, N.; Hou, G.G.; Dubat, A. Effects of flour particle size on the quality attributes of reconstituted whole-wheat flour and Chinese southern-type steamed bread. *LWT-Food Sci. Technol.* **2017**, *82*, 147–153. [CrossRef]
49. Lin, S.; Jin, X.; Gao, Z.; Ying, J.; Wang, Y.; Dong, Z.; Zhou, W. Impact of wheat bran micronization on dough properties and bread quality: Part I–Bran functionality and dough properties. *Food Chem.* **2021**, *353*, 129407. [CrossRef]
50. Zhang, D.; Moore, W.R. Effect of wheat bran particle size on dough rheological properties. *J. Sci. Food Agric.* **1997**, *74*, 490–496. [CrossRef]
51. Sudha, M.L.; Vetrimani, R.; Leelavathi, K. Influence of fibre from different cereals on the rheological characteristics of wheat flour dough and on biscuit quality. *Food Chem.* **2007**, *100*, 1365–1370. [CrossRef]
52. Özboy, Ö.; Köksel, H. Unexpected strengthening effects of a coarse wheat bran on dough rheological properties and baking quality. *J. Cereal Sci.* **1997**, *25*, 77–82. [CrossRef]
53. Niu, M.; Hou, G.G.; Kindelspire, J.; Krishnan, P.; Zhao, S. Microstructural. Textural, and sensory properties of whole wheat noodle modified by enzymes and emulsifiers. *Food Chem.* **2017**, *223*, 16–24. [CrossRef] [PubMed]
54. Sozer, N.; Cicerelli, L.; Heiniö, R.L.; Poutanen, K. Effect of wheat bran addition on in vitro starch digestibility, physico-mechanical and sensory properties of biscuits. *J. Cereal Sci.* **2014**, *60*, 105–113. [CrossRef]
55. Ma, S.; Wang, C.; Li, L.; Wang, X. Effects of particle size on the quality attributes of wheat flour made by the milling process. *Cereal Chem.* **2020**, *97*, 172–182. [CrossRef]
56. Brewer, L.R.; Kubola, J.; Siriamornpun, S.; Herald, T.J.; Shi, Y.C. Wheat bran particle size influence on phytochemical extractability and antioxidant properties. *Food Chem.* **2014**, *152*, 483–490. [CrossRef]
57. Li, W.D.; Shan, F.; Sun, S.C.; Corke, H.; Beta, T. Free radical scavenging properties and phenolic content of Chinese black-grained wheat. *J. Agric. Food Chem.* **2005**, *53*, 8533–8536. [CrossRef] [PubMed]
58. Zhou, K.; Su, L.; Yu, L. Phytochemicals and antioxidant properties in wheat bran. *J. Agric. Food Chem.* **2004**, *52*, 6108–6114. [CrossRef] [PubMed]

Article

Evaluation of Physical Characteristics of Typical Maize Seeds in a Cold Area of North China Based on Principal Component Analysis

Han Tang [1,2], **Changsu Xu** [1], **Yeming Jiang** [1], **Jinwu Wang** [1,*], **Zhenhua Wang** [3,*] and **Liquan Tian** [2,4,*]

[1] College of Engineering, Northeast Agricultural University, Harbin 150030, China; tanghan@neau.edu.cn (H.T.); ChangsuXu@neau.edu.cn (C.X.); jiangyeming951120@163.com (Y.J.)
[2] Key Laboratory of Crop Harvesting Equipment Technology of Zhejiang Province, Jinhua 321007, China
[3] College of Agriculture, Northeast Agricultural University, Harbin 150030, China
[4] School of Mechanical and Electrical Engineering, Jinhua Polytechnic, Jinhua 321007, China
* Correspondence: jinwuw@neau.edu.cn (J.W.); Zhenhuawang_2006@163.com (Z.W.); tlqbuct@foxmail.com (L.T.); Tel.: +86-0451-55190950 (J.W. & Z.W. & L.T.)

Abstract: The physical properties of maize seeds are closely related to food processing and production. To study and evaluate the characteristics of maize seeds, typical maize seeds in a cold region of North China were used as test varieties. A variety of agricultural material test benches were built to measure the maize seeds' physical parameters, such as thousand-grain weight, moisture content, triaxial arithmetic mean particle size, coefficient of static friction, coefficient of rolling friction, angle of natural repose, coefficient of restitution, and stiffness coefficient. Principal component and cluster comprehensive analyses were used to simplify the characteristic parameter index used to judge the comprehensive score of maize seeds. The results showed that there were significant differences in the main physical characteristics parameters of the typical maize varieties in this cold area, and there were different degrees of correlation among the physical characteristics. Principal component analysis was used to extract the first three principal component factors, whose cumulative contribution rate was over 80%, representing most of the information of the original eight physical characteristic parameters, and had good representativeness and objectivity. According to the test results, the classification standard of the evaluation of the physical characteristics of 15 kinds of maize seeds were determined, and appropriate evaluations were conducted. The 15 kinds of maize seeds were clustered into four groups by cluster analysis, and the physical characteristics of each groups were different. This study provides a new idea for the evaluation and analysis of the physical properties of agricultural materials, and provides a new method for the screening and classification of food processing raw materials.

Keywords: northern cold area; maize seeds; physical characteristics; principal component analysis; cluster comprehensive analysis

Citation: Tang, H.; Xu, C.; Jiang, Y.; Wang, J.; Wang, Z.; Tian, L. Evaluation of Physical Characteristics of Typical Maize Seeds in a Cold Area of North China Based on Principal Component Analysis. *Processes* **2021**, *9*, 1167. https://doi.org/10.3390/pr9071167

Academic Editors: Yonghui Li and Shawn/Xiaorong Wu

Received: 2 June 2021
Accepted: 2 July 2021
Published: 5 July 2021

Publisher's Note: MDPI stays neutral with regard to jurisdictional claims in published maps and institutional affiliations.

1. Introduction

Maize is the main grain crop in China, contributing grain and feed to the economy. Maize is also an important raw material for food processing. Effective evaluation and screening of raw materials is important to enhance the nutrition and taste of raw food [1–3]. The processing parameters of drying, cooking, and baking maize with different physical properties also differ, and the evaluation of physical properties of maize seeds is also crucial for guiding food processing [4–6].

Maize can be made into many kinds of food and additives after deep or primary processing. Deep processing can extract crude amino acid and starch from corn seeds to form liquor or yeast [7]. The primary processed products are widely used in daily life, and can be made into corn flour or corn paste for flushing [8]. The comminution degree of maize in primary processing is closely related to its physical properties. The main factors influencing the comminution degree are the moisture content, particle surface area, friction characteristics, and mechanical properties of maize grain. Kyttä et al. [9] found that the friction coefficient is an important parameter reflecting the influence of raw materials on the wear of the pressed film. The higher the friction coefficient of the raw materials, the lower the service life of the pressed film. Córdova-Noboa et al. [10] found that physical properties can affect the flow of maize grain in food processing machinery, and then affect the binding process, with effects between different grains. The research on the physical characteristics of maize grain has important guiding value for the development and production of related food processing machinery and equipment.

The study of the characteristics of agricultural materials is a basic discipline formed in conjunction with the development of agricultural engineering [11]. The physical parameters of maize seed, as a typical agricultural material, mainly include basic characteristic parameters, friction characteristic parameters, mechanical characteristic parameters, and electrical characteristic parameters. [12]. Through study of maize seeds' physical characteristics, varieties with similar physical characteristics can be comprehensively evaluated and selected for statistical analysis, which can be combined with clustering methods to extract a reasonable range of physical parameters. Scholars have researched the determination of the physical characteristics parameters of maize seeds [13–15]. Zhang et al. [16] proposed a method for calibrating the interspecies contact parameters of the maize seed particle model based on the discrete element method, which provided effective boundary conditions for its virtual simulation. Wang et al. [17] proposed a method for measuring the rolling friction coefficient and collision recovery coefficient of maize seeds based on the energy conservation law and the principle of specular reflection, which provided a new method for obtaining the mechanical parameters of granular agricultural materials. However, the corresponding planting varieties that are suitable in different areas of China are different, and their physical characteristics may also be significantly different. There are few reports on the measurement of physical parameters of maize varieties suitable for planting in the cold areas of Northern China. The literature data and experience do not fully provide the data needed for machine development.

At present, the evaluation of the parameters of agricultural materials is mainly based on conventional appearance quality and sensory evaluations, chemical composition evaluation, and single-factor evaluation [18,19]. Appearance quality and sensory evaluations are vulnerable to subjective influence, so have poor generalization and application value. Single-factor evaluation is vulnerable to the influence of single indexes, which reduces the overall evaluation accuracy, so it is impossible to conduct a systematic and comprehensive quantitative evaluation. Therefore, objective, scientific, and effective methods are needed for this type of evaluation [20]. Chemical composition evaluation methods are mainly achieved through weight analysis or bulk density analysis to extract products from samples. Related research mainly focused on the determination of polysaccharide, ash, crude fat, and protein contents in maize, wheat, rice, and other staple crops; pineapple; apple; and other baked fruits and vegetables [21,22]. Based on a chemical composition evaluation method, Salvador-Reyes et al. [23] studied different varieties of maize from different areas and found that there was no significant differences in ash, crude fat, and protein contents. Principal component analysis reduces the loss of information from raw data, simplifies the data structure, and avoids subjective randomness. It is widely used in comprehensive evaluation in various fields. Yang et al. [24] classified the characteristics of various potato varieties and their processing suitability based on the principal component analysis, and provided a reference for the division of potato processing uses. Mu et al. [25] used principal component analysis and cluster analysis to evaluate and analyze multiple agronomic traits of multiple peanut germplasm resources, which provided a reference for the selection of excellent peanut varieties and multi-functional applications. Lu, Dai, Luo, and Liu [26–29] measured and evaluated the physical characteristics of rice, soil, earthworm dejecta substrate, and potato, respectively. The above research mainly classified and evaluated some kinds of crops, but the principal component analysis and the evaluation of the physical characteristics of maize, especially the maize in cold areas of North China, have not been reported.

Given this background, we used typical maize varieties in the cold areas of Northern China to determine their thousand-grain weight, moisture content, triaxial arithmetic mean particle size, coefficient of static friction, coefficient of rolling friction, angle of natural repose, coefficient of restitution, and stiffness coefficient. Principal component analysis was used to simplify the parameter index, and the comprehensive score of the various maize varieties was judged in order to construct a scientific and reasonable evaluation system. Our findings provide a new idea for the evaluation and analysis of the physical properties of agricultural materials, and provides a new method for the screening and classification of raw food processing materials.

2. Materials and Methods

2.1. Principal Component Analysis of Physical Maize Seed Characteristics

The basic idea for the evaluation of the physical characteristics of maize seeds using principal component analysis mainly involved constructing a linear combination of the characteristic physical parameters and calculating new variables that are irrelevant and contain the information of the original variables. Through the new variables that were determined to replace the complex original variables to analyze and solve the problem, complex problems can be simplified [30]. The steps are were follows: (1) The indexes needed to evaluate the physical properties of maize seeds were determined, including thousand-seed weight, moisture content, triaxial arithmetic mean particle size, natural angle of repose, static friction coefficient, rolling friction coefficient, impact recovery coefficient, and stiffness coefficient. (2) A bench test of material characteristics was conducted to obtain the test data of physical parameters. (3) We conducted correlation analysis and a preliminary evaluation of the characteristic physical parameters. (4) We conducted principal component analysis of the characteristic physical parameters. (5) Through the cumulative index contribution rate of the principal components and the bottom lithotripsy map, the number of principal components was selected to achieve index dimensionality reduction. (6) The test data the physical parameters were substituted into the principal component formula to obtain a principal component score, and then the contribution rate of each principal component was used as the weight value to produce a comprehensive score, and then we conducted the final quantitative evaluation.

In step 2, during the bench test of material characteristics, the test data were standardized.

$$x_i{'} = \frac{x_i - \overline{x}_i}{\sqrt{\frac{1}{n-1}\sum\limits_{i=1}^{n}(x_i - \overline{x}_i)^2}} \tag{1}$$

In step 3, correlation analysis was conducted, and the correlation coefficient matrix R of each index was solved. The Pearson product moment formula was adopted for the correlation coefficient between indexes x and y.

$$r_{xy} = \frac{\sum\limits_{i=1}^{n}(x_i - \overline{x})(y_i - \overline{y})}{\sqrt{\sum\limits_{i=1}^{n}(x_i - \overline{x})^2 \sum\limits_{i=1}^{n}(y_i - \overline{y})^2}} \tag{2}$$

where $\overline{x}$ and $\overline{y}$ are the mean values of indexes x and y, respectively; and x_i and y_i are the ith measured values of indexes x and y, respectively.

The characteristics values of the correlation coefficient matrix R and the corresponding normalized eigenvectors were solved. In the process of principal component analysis and number selection in steps 4 and 5, the characteristics values were ranked as $\{\lambda_1, \lambda_1, \lambda_1, \ldots\}$

from small to large, and several larger characteristics values were selected, and their contribution rate was calculated.

$$\mu_{(p)} = \frac{\lambda_1 + \lambda_2 + \ldots + \lambda_p}{m} \tag{3}$$

where m is the number of principal components.

2.2. Test Materials for Aize Seeds Characteristics

We focused on the measurement and analysis of the characteristic physical parameters of typical maize seeds in a cold area of North China and selected 15 kinds of maize varieties as the test materials. We focused on the investigation of different maize varieties in different temperature zones in Heilongjiang Province, China. The accumulated temperature zone is calculated as the sum of the daily average temperature during periods with a daily average temperature $\geq 10\,°C$ in a year, that is, the sum of active temperature, is referred to as accumulated temperature. It is an index used to study the relationship between temperature and the growth and development speed of biological organisms from two aspects: intensity and action time [31]. The first accumulated temperature zone ($\geq 2700\,°C$) included Xianyu335, Zhongdan909 and Xiangyu998. The second accumulated temperature zone (2500–2700 °C) included Zhengxianda408, Jingnongke728, and Xianyu696. The third accumulated temperature zone (2300–2500 °C) included Suiyu23, Fuer116, and Dongnong259. The forth accumulated temperature zone (2100–2300 °C) included Dongnong254, Demeiya3, and Longfuyu9; and the fifth and sixth accumulated temperature zones ($\leq 2100\,°C$) included Demeiya1, Xinkeyu1, and Keyu16. The above varieties were provided by the Northeast Agricultural University, Heilongjiang Academy of Agricultural Sciences and Beidahuang Seed Industry Group Co., Ltd. Through manual grading and cleaning treatment, the shape of the tested maize seeds was uniform, full, and free of damage. The distribution of each temperate zone and variety number are shown in Table 1.

Table 1. Selection and number of maize varieties in a cold area of China.

No.	Accumulated Temperature Zone in Cold Area	Variety Name
1		Xianyu335
2	First accumulated temperature zone	Zhongdan909
3	($\geq 2700\,°C$)	Xiangyu998
4		Xianzhengda408
5	Second accumulated temperature zone	Jingnongke728
6	(2500–2700 °C)	Xianyu696
7		Suiyu23
8	Third accumulated temperature zone	Fuer116
9	(2300–2500 °C)	Dongnong259
10		Dongnong254
11	Forth accumulated temperature zone	Demeiya3
12	(2100–2300 °C)	Longfuyu9
13		Demeiya1
14	Fifth and sixth accumulated temperature zones	Xinkeyu1
15	($\leq 2100\,°C$)	Keyu16

2.3. Selection of Physical Characteristics of Maize Seeds and Test Instrument

The basic parameters of physical maize seeds characteristics mainly include thousand grain weight, moisture content, density, and geometry size. The tribological parameters include the coefficient of static friction, coefficient of rolling friction, angle of natural repose, etc. The mechanical parameters include the coefficient of restitution, stiffness coefficient, and the modulus of elasticity. The physical properties of maize seeds also include their rheological, thermal, optical, electrical, and comminution properties [32]. Based on the comprehensive analysis of the correlation between the characteristics and the design of mechanical components, eight physical characteristics indexes, i.e., the thousand-grain weight, moisture content, triaxial arithmetic mean particle size, coefficient of static friction, coefficient of rolling friction, angle of natural repose, coefficient of restitution, and stiffness coefficient, were selected as indexes in this study.

1. Thousand-grain weight (g) is an important index used to measure the weight and plumpness of 1000 maize seeds [33]. It is also an important factor that affects the mechanical characteristics of materials, mainly related to variety, shape, size, plumpness, bulk density, and moisture content.
2. Moisture content (%) is used to assess the quantity of water in maize seeds. Wet basis representation was applied in this study.
3. Triaxial arithmetic mean particle size (mm): Large differences exist in the geometrical dimensions along the three axes of a seed. The axial dimension method was mainly used to determine the shape characteristics seeds; establish the three-dimensional coordinate system; define the length, width, and thickness; and measurement of the maize seeds. The calculation of the triaxial arithmetic mean particle size comprehensively reflects the length, width, and thickness of maize seed.
4. The coefficient of static friction reflects the friction characteristics between a maize seed and the contact surface, and is the main parameter used to characterize the friction and scattering characteristics. It is mainly related to the surface roughness of the contact body and directly affects the movement trend of maize seeds [34]. A test bench for measuring the coefficient of static friction was set up using the inclined plane method.
5. The coefficient of rolling friction shows that, when maize seeds roll or have the tendency to roll relative to the surface of the contact body, the rolling is hindered by the deformation of the contact part under pressure [35]. Based on the law of energy conservation, high-speed camera technology was used to build a test bench for measuring the coefficient of rolling friction between the maize seeds.
6. The angle of natural repose (°) is also an important parameter reflecting the friction between maize seeds and the flow characteristics of maize seeds. The larger the value, the greater the friction resistance between seeds, and the smaller the tendency toward free dispersion [36]. Injection method was used to determine the angle of natural repose of maize seeds.
7. The coefficient of restitution reflects the ability of maize seeds to return to their original shape after collision and deformation. The larger the value, the stronger the ability to restore shape after deformation and the higher the elasticity of maize seeds after collision [37]. A test bench was built to measure the coefficient of restitution.
8. Stiffness coefficient (N/mm): Under the action of external force, the basic parameters of elastic deformation behavior of maize seeds reflect their resistance to elastic deformation, which also characterizes the mechanical damage limit of maize seeds [38]. The stiffness coefficients of maize seeds in horizontal, lateral, and vertical laying were measured by an electronic texture analyzer.

The experiment instruments in this study mainly included an electronic analytical balance (FC204, Shanghai Precision Scientific Instrument Co., Ltd., Shanghai, China. accuracy of 0.001 g), a halogen moisture tester automatic (DHS-16, Shanghai JINGHAI Instrument Co., Ltd., Shanghai, China. accuracy of 0.02%), an automatic microcomputer seed counter (SLY-C, Zhejiang TOP Instrument Co., Ltd., Zhejiang, China. count speed $\geq 1000/\text{min}$), a Vernier caliper (Shanghai SHOUFENG Precision Instrument Co., Ltd., Shanghai, China. accuracy of 0.01 mm), a digital inclinometer (SANHE, Resolution of 0.05°, accuracy of $\pm 0.2°$), a high-speed camera (Vision research Co., Ltd., Wayne, New Jersey, USA), a computer (Hewlett-Packard), an electric blast drying oven (DHG-9053A, Shanghai YIHENG Scientific Instrument Co., Ltd. Palo Alto, CA, USA), a micromaterial crusher (FZ102, Tianjin TAISITE Instrument Co., Ltd., Tianjin, China), a microcomputer-controlled electronic texture analyzer (TA.XT.plus, Stable Micro systems Co., Ltd., London, UK), a static-friction test bench, a rolling-friction test bench, a collision-recovery test bench, and a free-injection test bench. The principles, methods, and instruments used for measuring the physical characteristics of maize seeds are shown in Table 2.

Table 2. Principles, methods, and instruments used for measuring the physical characteristics of maize seeds.

No.	Project	Test Principles and Methods	Instrument and Test Bench
1	Thousand-grain weight	The maize seeds were scattered on an imaging disk and placed on an electronic balance with an RS232 communication line. When it was stable, the weight data were sent to the computer, and the number of seeds in the viewing area was analyzed synchronously. Then, the thousand-grain weight of maize seeds was obtained (GB/T 5519-2008 Cereals and pulses—Determination of the mass of 1000 grains).	
2	Moisture content	High temperature drying method: We selected maize seeds to measure the total mass before drying, and used the micromaterial grinder to crush and grind the maize seeds, we placed the crushed maize seeds into the electric blast drying oven, and adjusted it to 103 °C for drying for 3–4 h, and then weighed it again to determine the total mass after drying (GB/T 10362-2008 Inspection of grain and oils—Determination of moisture content of maize).	
3	Triaxial arithmetic mean particle size	We selected more than 500 maize seeds at random, and used a Vernier caliper to measure the geometric dimensions (length, width, and thickness) of the maize seeds, with an accuracy of 0.01 mm.	
4	Coefficient of static friction	We constructed the maize seeds material platform, adjusted the platform surface to the horizontal position, placed one maize seed on the surface, ensured the same direction, lifted the platform slowly until the maize seed had a sliding trend, and recorded the inclination of the platform surface [39].	
5	Coefficient of rolling friction	We constructed the maize seeds material roller and platform, reduced the energy loss caused by relative sliding and friction, approximately adjusted the platform to 45° by rolling, and collected the coordinate value and instantaneous speed of the maize roller center point in the process of movement with a high-speed camera test [40].	
6	Angle of natural repose	Discharge method: We placed the cylinder on the flat plate vertically, placed maize seeds in the cylinder, and the cylinder moved at a slow speed perpendicular to the plate. We captured photos of the maize seeds in three-dimensional space. MATLAB software was used to process image noise, gray scale, and binary, and fit the envelope curve equation (GB/T 11986-1989 Surface active agents-Powders and granules—Measurement of the angle of repose).	
7	Coefficient of restitution	Speed definition method: We constructed the maize seeds material platform, adjusted it to 45° with the horizontal plane; we dropped one maize seed freely to the platform at a certain height, which collided with the platform, forming a parabola movement after rebound, and finally falling to the ground. We measured parameters such as the horizontal and vertical displacements of the maize seed falling to the ground [41].	

Table 2. *Cont.*

No.	Project	Test Principles and Methods	Instrument and Test Bench
8	Stiffness coefficient	The maize seeds were placed on the platform in all directions to ensure that the center of the maize seed was aligned with the center of the plate indenter of the analyzer. The pressure head was set to decrease steadily and compress the maize seeds. The load and displacement parameters at all times were automatically collected and recorded, and the curve of the compression load displacement relationship of maize seeds was drawn in real time [42].	

During the test, the methods and instruments in Table 2 were used to determine and analyze the physical characteristics of 15 maize varieties. Excel 2013 and SPSS 22.0 software were used to analyze the data, and principal component analysis and cluster analysis were conducted. Through the analysis and simplification of the physical characteristics index, an effective and scientific evaluation system was constructed, and the comprehensive scores of each variety were obtained. The maize varieties with excellent comprehensive physical characteristics and suitable for mechanized planting were selected.

3. Results

3.1. Results of Physical Maize Seeds Characteristics

Eight physical characteristic parameters, such as thousand-grain weight, moisture content, triaxial arithmetic mean particle size, and coefficient of static friction, of the 15 kinds of selected maize seeds from a cold region are shown in Table 3.

According to the analysis of the test results in Table 3, the average thousand-grain weight of each maize variety was 353.19 g, ranging from 295.92–418.9 g, with larger variation. The average value of moisture content was 12.30%, with a range of 11.20–13.05%, with small variation. The average triaxial arithmetic mean particle size was 8.62 mm (7.84–9.80 mm), with large variation. The average coefficient of static friction was 0.29, stable at 0.24–0.34, with large variation. The average coefficient of rolling friction was 0.07, stable at 0.053–0.083, with large variation. The average natural repose angle was 23.11°, stable at 20.37–24.59°, with large variation. The average coefficient of restitution was 0.40, stable at 0.332–0.471, with large variation. The average value of stiffness coefficient was 89.85 N/mm, stable at 78.5–102.23 N/mm, with large variation.

Table 3. Parameters of physical characteristics of typical maize seeds in cold area.

Index	Statistical Analysis	Typical Maize Seeds in Cold Region														
		1	2	3	4	5	6	7	8	9	10	11	12	13	14	15
Thousand grain weight, g	Mean	342.80	339.21	402.68	357.13	313.83	370.49	348.69	418.90	360.80	366.46	370.43	340.63	331.08	295.92	338.75
	SD	1.32	1.04	1.42	0.69	0.56	1.52	0.34	0.81	1.55	2.43	0.15	1.23	1.89	0.55	1.59
	CV, %	0.39	0.31	0.35	0.19	0.18	0.41	0.98	0.19	0.43	0.67	0.40	0.36	0.57	0.19	0.47
Moisture content, %	Mean	12.13	11.94	13.02	12.47	12.42	12.88	11.74	13.05	12.61	11.42	12.90	12.37	11.60	11.20	12.74
	SD	0.66	0.58	0.64	0.52	0.60	0.99	0.47	1.29	0.34	0.48	0.35	1.44	0.83	0.27	0.12
	CV, %	5.44	4.86	4.92	4.17	4.83	7.69	4.00	9.89	0.03	4.20	2.71	11.64	7.16	2.41	0.94
Triaxial arithmetic mean particle size, mm	Mean	8.33	8.35	9.16	8.45	8.45	9.46	8.20	9.80	8.37	8.63	9.10	8.56	8.32	7.84	8.34
	SD	0.31	0.52	0.32	0.21	0.23	0.51	0.09	0.34	0.42	0.23	0.32	0.06	0.47	0.20	0.23
	CV/%	3.72	6.22	3.49	2.49	2.72	5.39	1.10	3.47	5.02	2.67	3.52	0.70	5.65	2.55	2.76
Coefficient of static friction	Mean	0.27	0.30	0.29	0.34	0.24	0.25	0.32	0.29	0.30	0.28	0.29	0.26	0.28	0.25	0.33
	SD	0.05	0.03	0.07	0.03	0.01	0.06	0.03	0.03	0.05	0.02	0.03	0.05	0.05	0.09	0.08
	CV, %	18.52	10.00	24.14	8.82	4.17	24.00	9.38	10.34	16.67	7.14	10.34	19.23	25.00	36.00	24.24
Coefficient of rolling friction	Mean	0.069	0.071	0.083	0.075	0.067	0.078	0.075	0.072	0.080	0.080	0.081	0.067	0.062	0.053	0.078
	SD	0.02	0.02	0.03	0.02	0.01	0.03	0.02	0.01	0.01	0.01	0.02	0.01	0.01	0.02	0.03
	CV, %	28.98	28.17	36.14	26.67	14.93	38.46	26.67	13.89	12.50	12.50	24.69	14.93	16.13	37.74	38.46
Angle of natural repose, °	Mean	24.21	23.03	24.59	23.18	22.31	23.73	23.05	23.82	24.20	23.84	24.23	22.63	22.20	21.32	20.37
	SD	0.43	1.62	1.59	2.31	1.96	2.90	1.48	1.94	2.04	0.59	1.00	0.94	1.54	1.56	1.88
	CV, %	1.78	7.03	6.47	9.97	8.79	12.22	6.42	8.14	8.43	2.47	4.13	4.15	6.94	7.32	9.23
Coefficient of restitution	Mean	0.471	0.406	0.428	0.390	0.332	0.403	0.415	0.425	0.372	0.385	0.455	0.374	0.382	0.398	0.390
	SD	0.03	0.01	0.01	0.03	0.01	0.02	0.01	0.02	0.01	0.01	0.02	0.04	0.03	0.02	0.01
	CV, %	6.37	2.46	2.34	7.69	3.01	4.96	2.41	4.71	2.69	2.60	4.40	10.70	7.85	5.03	2.56
Stiffness coefficient, N/mm	Mean	79.83	98.26	100.11	89.25	94.36	102.23	84.35	97.26	80.05	81.06	98.41	83.74	98.37	78.50	81.92
	SD	0.22	0.21	0.33	0.10	0.08	0.21	0.41	0.06	0.20	0.42	0.09	0.22	0.31	0.29	0.12
	CV, %	0.28	0.21	0.33	0.11	0.08	0.21	0.49	0.06	0.25	0.52	0.09	0.26	0.36	0.37	0.15

"Mean" is the mean value of each parameter; "SD" is the standard deviation; "CV" is the coefficient of variation.

3.2. Principal Component Analysis of Physical Maize Seeds Characteristics

The SPSS 22.0 software was used to perform the principal component analysis of the physical characteristics of maize seeds [43]. The process was as follows: 1. index data normalization and standardization, 2. using SPSS software, the Correlation Matrix module, to judge the correlation; 3. determining the number of principal components, using the Total Variance Explained module in the SPSS to calculate the cumulative contribution rate of the principal component variance $\geq 80\%$, and combining the Component Matrix module with no loss of variables to determine the number of principal components m; 4. determining principal component Z_i expression, dividing the ith column vector in the Component Matrix module of SPSS by the open root of the ith characteristic root to obtain the variable coefficient vector of the ith principal component Z_i (in the Transform-compute module), obtaining the principal component Z_i expression; 5. naming principal component Z_i and naming the corresponding variable with a large absolute value of the coefficient in the ith column of the Component Matrix in SPSS; 6. The principal component and integrated principal $Z_t = \sum\limits_{i=1}^{m} (\lambda_i/p)Z_i$, λ_i/p is in the % of variance of the Initial Eigenvalues in the Total Variance Explained modules of SPSS software, $VarZ_t = (\sum\limits_{i=1}^{m} \lambda_i{}^3)p^2$; 7. examination, combining the actual results and experience of the comprehensive principal component evaluation value to test the original data.

SPSS 22.0 software was used to obtained the correlation coefficient matrix, variance contribution analysis table, principal component load matrix, and eigenvector of each index parameter, as shown in Tables 4–6.

Table 4. Correlation coefficient matrix between the main physical characteristic parameters of maize seeds.

Index	Thousand-Grain Weight	Moisture Content	Triaxial Arithmetic Mean Particle Size	Coefficient of Static Friction	Coefficient of Rolling Friction	Angle of Natural Repose	Coefficient of Restitution	Stiffness Coefficient
Thousand grain weight	1.000							
Moisture content	0.657 *	1.000						
Triaxial arithmetic mean particle size	0.856 **	0.733 **	1.000					
Coefficient of static friction	0.284	0.164	−0.095	1.000				
Coefficient of rolling friction	0.700 **	0.604 *	0.500	0.473	1.000			
Angle of natural repose	0.691 *	0.349	0.545	−0.034	0.550	1.000		
Coefficient of restitution	0.423	0.188	0.302	0.175	0.225	0.476	1.000	
Stiffness coefficient	0.429	0.469	0.655*	−0.112	0.209	0.293	0.106	1.000

Note: * and ** indicate significant differences at the 0.05 and 0.01 levels, respectively.

Table 5. Variance contribution analysis table.

Component	Initial Characteristics Value			Extract Square Sum Loading		
	Total	Variance, %	Cumulative, %	Total	Variance, %	Cumulative, %
1	4.018	50.222	50.222	4.018	50.222	50.222
2	1.406	17.579	67.801	1.406	17.579	67.801
3	1.035	12.935	80.736	1.035	12.935	80.736
4	0.631	7.881	88.617			
5	0.443	5.543	94.160			
6	0.283	3.531	97.692			
7	0.165	2.065	99.757			
8	0.019	0.243	100.000			

Table 6. Load matrix and eigenvector of each principal component.

Test Index	Principal Component Load Matrix			Principal Component Eigenvector		
	Z_1	Z_2	Z_3	Z_1	Z_2	Z_3
Thousand-grain weight	0.938	0.087	0.024	0.468	0.073	0.024
Moisture content	0.798	−0.071	−0.372	0.398	−0.060	−0.366
Triaxial arithmetic mean particle size	0.883	−0.359	−0.071	0.441	−0.303	−0.070
Coefficient of static friction	0.221	0.859	−0.283	0.110	0.724	−0.278
Coefficient of rolling friction	0.766	0.423	−0.201	0.382	0.357	−0.198
Angle of natural repose	0.734	−0.020	0.472	0.366	−0.017	0.464
Coefficient of restitution	0.475	0.229	0.705	0.237	0.193	0.693
Stiffness coefficient	0.580	−0.544	−0.225	0.289	−0.459	−0.221

We concluded that there were obvious differences in the physical characteristic parameters of the maize varieties, and there were different degrees of correlation. We found a significant difference in the thousand-grain weight among the different varieties of maize. The higher the moisture content, the higher the thousand-grain weight. Triaxial arithmetic mean particle size was positively correlated with thousand-grain weight and moisture content. The coefficient of static friction was positively correlated with the thousand-grain weight and moisture content and was negatively correlated with the triaxial arithmetic mean particle size. The coefficient of rolling friction was positively correlated with the thousand-grain weight, moisture content, triaxial arithmetic mean particle size, and coefficient of static friction. Both the angle of natural repose and the stiffness coefficient were negatively correlated with the coefficient of static friction. We identified different degrees of correlation between maize seeds physical characteristics, which showed that the information reflected overlaps and interweaves, and each single index parameter had a different effect on the physical maize seed characteristics, so we were unable to directly and comprehensively evaluate the maize seeds' physical characteristics using the above indexes.

On this basis, principal component analysis was conducted for the eight physical parameters of the 15 maize varieties. The variance of the principal component was taken as the characteristic's value, indicating how much the corresponding component can describe the original information. The larger the characteristic value of the principal component, the more information contained by the variable [44]. The scree plot of principal component analysis is shown in Figure 1. The characteristic value can reflect the corresponding relationship between the characteristic value of the correlation matrix of the index and the principal component number. Combined with the principal component load matrix and the eigenvector, the first three characteristic values of the principal component were all greater than one, and the cumulative contribution rate was more than 80%, which represented the information from the raw data of the physical characteristics of maize. According to the bottom lithotripsy map, there is an inflection point at the third principal component. After the fourth principal component, the characteristic values are small and close to each other. According to the cumulative contribution rate, the cumulative contribution rate of the first three principal components was 80.736%. Therefore, the first three principal components were selected to comprehensively evaluate the physical characteristics of maize seeds.

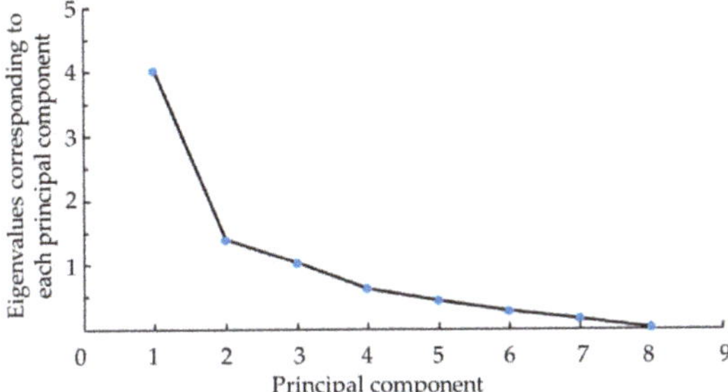

Figure 1. Scree plot of principal component analysis: The X axis is the principal component; The Y axis is the eigenvalues corresponding to each principal component—the amount of each principal component contributed. Generally, the principal component with characteristic value greater than one is selected for evaluation.

Based on a series of processing and analyses, and combined with the load matrix and eigenvector of each principal component in Table 6, the relationships between the principal component and the corresponding variable were obtained, and the linear equations between each principal component and the physical characteristics indexes of maize seeds was constructed.

1. The first principal component was:

$$Z_1 = 0.468X_1 + 0.398X_2 + 0.441X_3 + 0.110X_4 + 0.382X_5 + 0.366X_6 + 0.237X_7 + 0.289X_8 \tag{4}$$

2. The second principal component was:

$$Z_2 = 0.073X_1 - 0.060X_2 - 0.303X_3 + 0.724X_4 + 0.357X_5 - 0.017X_6 + 0.193X_7 - 0.459X_8 \tag{5}$$

3. The third principal component was:

$$Z_3 = 0.024X_1 - 0.366X_2 - 0.070X_3 - 0.278X_4 - 0.198X_5 + 0.464X_6 + 0.693X_7 - 0.221X_8 \tag{6}$$

According to the analysis of variance contribution, the first principal component accounted for 50.222%, which contained a large amount of information. It mainly included the thousand grain weight, triaxial arithmetic mean particle size, moisture content, coefficient of rolling friction, and the angle of natural repose. The second principal component accounted 17.579%, which mainly included the coefficient of static friction and stiffness coefficient. The third principal component accounted for 12.935%, which mainly included the coefficient of restitution. Combining the principal component coefficient and its corresponding variance contribution rate, the comprehensive evaluation formula was established as $Z = 0.502Z_1 + 0.176Z_2 + 0.129Z_3$. The comprehensive scores of the physical maize seeds characteristics in this cold region could be obtained through the evaluation formula.

According to the comprehensive scores, the maize varieties in the cold area were sorted, as shown in Table 7. Combining the analysis results with the performance of maize varieties in the mechanized planting process, the evaluation standard of the eight physical characteristics of 15 maize varieties in a cold region were determined. A comprehensive score $\geq$110 was level 1, 105–110 was level 2, 100–105 was level 3, 95–100 was level 4, and a comprehensive score $\leq$95 was level 5. So, the order of maize varieties was: Fuer116 > Xiangyu998 > Xianyu696 > Demeiya3 > Dongnong254 > Dongnong259 > Xianzhengda408 > Suiyu23 > Xianyu335 > Zhongdan909 > Longfuyu9 > Keyu16 > Demeiya1 > Jingnongke728 > Xinkeyu1.

Table 7. Principal component scores and composite scores.

Variety	Principal Component			Composite Score	Rating Level
	Z_1	Z_2	Z_3		
Fuer116	242.548	−17.899	−5.652	117.880	1
Xiangyu998	235.772	−20.208	−6.257	113.994	1
Xianyu696	221.070	−23.636	−7.871	105.802	2
Demeiya3	219.988	−21.747	−6.755	105.735	2
Dongnong254	212.159	−13.852	−2.669	103.722	3
Dongnong259	209.708	−13.790	−2.846	102.479	3
Xianzhengda408	210.261	−18.252	−5.390	101.643	3
Suiyu23	204.450	−16.504	−4.264	99.179	4
Xianyu335	201.034	−14.971	−2.965	97.902	4
Zhongdan909	204.167	−23.654	−7.660	97.341	4
Longfuyu9	200.739	−17.008	−4.784	97.161	4
Keyu16	198.572	−16.168	−5.606	96.115	4
Demeiya1	199.930	−24.282	−8.145	95.04	4
Jingnongke728	191.107	−23.827	−7.956	90.716	5
Xinkeyu1	177.037	−17.563	−4.806	85.162	5

3.3. Cluster Analysis of Typical Maize in Cold Area

In this context, we conducted a cluster analysis of the 15 maize varieties. According to the comprehensive physical characteristics parameters, the cluster results can be used to judge the intimate relationship according to the comprehensive physical characteristic parameters to better understand the nature of the data [45]. First, the physical characteristics of maize seeds were analyzed by principal component analysis, and the high-dimensional data were transformed into low-dimensional data. The Euclidean distance showed a good effect on the status data, so we used the Euclidean distance [46]. Using the discrete sum of squares method, all varieties of maize were divided into four categories at the Euclidean distance of 5.0, as shown in Figure 2. Xianyu696 and Demeiya3 were classified into one group; Dongnong259, Dongnong254, Xianzhengda408, Zhongdan909, Demeiya1, Longfuyu9, Keyu16, Xianyu335, and Suiyu23 were classified into another group; Jingnongke728 and Xinkeyu1 were classified into a third group; and Xiangyu998 and Fuer116 were classified into a fourth group. Through principal component and cluster analyses, the eight physical characteristics parameters of typical cold-area maize were analyzed. The results showed that the eight physical parameters belong to three principal components, representing 80.736% of the total variance in the information. The 15 maize varieties were categorized into four groups. Due to the wide differences in the physical characteristics and the Euclidean distance between groups, the principal component complementation and selection of Euclidean distance should be considered in the evaluation of physical characteristics.

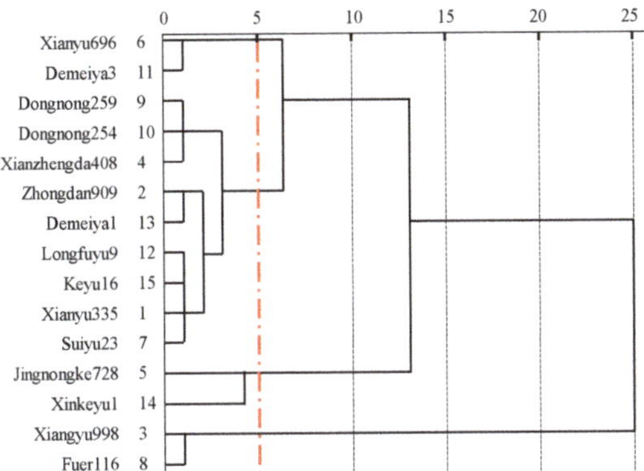

Figure 2. Dendrogram of cluster analysis of the maize seeds of 15 different varieties: 0–25 indicates the range of Euclidean distance, and different Euclidean distances lead to different categories. The red line in the figure indicates that Euclidean distance 5.0 is selected, and the number of branches passed by the red line represents the clustering results of maize seed varieties.

3.4. Discussion

In this study, we measured the thousand-grain weight, moisture content, triaxial arithmetic mean particle size, coefficient of static friction, coefficient of rolling friction, angle of natural repose, coefficient of restitution, and stiffness coefficient of 15 varieties of maize seeds. Principal component evaluation and cluster comprehensive analysis were used to simplify the characteristic parameter index, which was then used to judge the comprehensive score of maize seeds. The 15 varieties of maize seeds were clustered into four groups by cluster analysis. The physical characteristics of the different groups were different, and the physical characteristics of the same group showed strong similarity and correlation. Therefore, during food processing, maize seeds of the same group can be processed with the same equipment or control parameters, whereas maize seeds of different groups showed considerable differences, so it would be necessary to adjust the control parameters of the equipment to process maize and produce the same quality product. In addition, if the seeds of the same group are mixed to produce the same taste and quality, it is necessary to analyze the chemical components of different varieties of maize seeds, such as starch, sugar, crude fat, protein, and ash contents. The thermal characteristics of different varieties of maize seeds should also be analyzed to explore the specific heat value changes in different varieties of maize at different temperatures to provide a theoretical basis for the processing and production of maize baking food. Then, it will also be necessary to explore the differences in the gelatinization characteristics of different varieties of maize seeds to control the degree of gelatinization of maize seeds in the process of preparation and improve the forming characteristics of maize-based food. In the future, we will conduct determination research combining physical analysis and chemical analysis to comprehensively, objectively, and systematically evaluate the characteristics of maize seeds, and provide reliable data support for the screening and classification of maize seeds in food processing and production.

With the development of China's modern maize industry, adjusting and optimizing the structure of maize varieties directly affect the application of maize germplasm resources, cultivation technology, protection technology, and post-production processing and utilization of system links, which is crucial for promoting sustainable agricultural development and the moderate-scale management of grain [47]. In this study, the principal component analysis method was used to combine multiple indicators into a few comprehensive indicators. The principal components reflect most of the original variable information, and the information contained is complementary and repeated. In future research, the selection capacity of typical maize seed samples should be continuously increased, and the evaluation basis of the physical characteristics of maize seed samples suitable for planting in northern cold areas and the scientific quality of the evaluation should be improved.

4. Conclusions

In this study, we focused on the determination of physical characteristics parameters, principal component evaluation, and cluster analysis of typical cold-region maize varieties; judged the comprehensive scores of maize varieties; constructed a scientific and reasonable evaluation system; and obtained the following conclusions:

1. The main physical characteristics parameters of typical cold-region maize varieties differ significantly, and the physical characteristics indicators show different degrees of correlation.
2. Principal component analysis was used to select the first three principal component factors, with a cumulative contribution rate to the total variance of 80.736%, which represented most of the information of the original eight physical characteristics indicators, with good representativeness and objectivity. The comprehensive evaluation formula is $Z = 0.502Z_1 + 0.176Z_2 + 0.129Z_3$, and the comprehensive score and grade classification of the 15 maize varieties in a cold region were determined.
3. The maize species in a cold region were clustered into four groups by cluster analysis, and the physical characteristics of each group were different. Our findings provide a new idea for the evaluation and analysis of the physical properties of agricultural materials, and provide a new method for the screening and classification of food processing raw materials.

Author Contributions: Conceptualization, J.W., H.T., and C.X.; methodology, H.T. and C.X.; software, Y.J. and Z.W.; validation, L.T.; investigation, C.X. and Z.W.; resources, H.T.; data curation, H.T. and L.T.; writing—original draft preparation, J.W., H.T., and C.X.; writing—review and editing, J.W. and H.T.; visualization, H.T. and C.X.; supervision, J.W. and Z.W.; funding acquisition, H.T. All authors have read and agreed to the published version of the manuscript.

Funding: This work was funded by the National Natural Science Foundation of China (NSFC), grant number: 31901414; the Postdoctoral Science Foundation of Heilongjiang Province, grant number: LBH-Z19007; and the Young Talents Project of Northeast Agricultural University, grant number: 19QC41.

Institutional Review Board Statement: Not applicable.

Informed Consent Statement: Not applicable.

Data Availability Statement: All data are presented in this article in the form of figures and tables.

Acknowledgments: The authors would like to acknowledge the College of Engineering, Northeast Agricultural University; the College of Agriculture, Northeast Agricultural University; and the Key Laboratory of Crop Harvesting Equipment Technology of Zhejiang Province.

Conflicts of Interest: The authors declare no conflict of interest.

References

1. Li, Y.; Bingxin, Y.; Yiming, Y.; Xiantao, H.; Quanwei, L.; Zhijie, L.; Xiaowei, Y.; Tao, C.; Dongxing, Z. Global overview of research progress and development of precision maize planters. *Int. J. Agric. Biol. Eng.* **2016**, *9*, 9–26. [CrossRef]
2. Wang, J.; Tang, H.; Wang, J. Comprehensive Utilization Status and Development Analysis of Crop Straw Resource in Northeast China. *Nongye Jixie Xuebao/Trans. Chin. Soc. Agric. Mach.* **2017**, *48*, 1–21. [CrossRef]
3. Gao, Y.; Zhai, C.; Yang, S.; Zhao, X.; Wang, X.; Zhao, C. Development of CAN-based Downforce and Sowing Depth Monitoring and Evaluation System for Precision Planter. *Nongye Jixie Xuebao/Trans. Chin. Soc. Agric. Mach.* **2020**, *51*, 15–28. [CrossRef]
4. Wang, J.; Tang, H.; Zhou, W.; Yang, W.; Wang, Q. Improved design and experiment on pickup finger precision seed metering device. *Nongye Jixie Xuebao/Trans. Chin. Soc. Agric. Mach.* **2015**, *46*, 68–76. [CrossRef]
5. Tang, H.; Xu, C.; Wang, J.; Wang, Z.; Han, H.; Li, J. Design and experiment of finger-clip maize no-tillage precision planter. *Int. Agric. Eng. J.* **2020**, *29*, 86–97.
6. Yatskul, A.; Lemiere, J.P.; Cointault, F. Influence of the divider head functioning conditions and geometry on the seed's distribution accuracy of the air-seeder. *Biosyst. Eng.* **2017**, *161*, 120–134. [CrossRef]

7. Brown, W.E.; Bradford, B.J. Effects of a high-protein corn product compared with soy and canola protein sources on nutrient digestibility and production responses in mid-lactation dairy cows. *J. Dairy Sci.* **2020**, *103*, 6233–6243. [CrossRef]

8. Aboagye, G.; Gbolonyo-Cass, S.; Kortei, N.K.; Annan, T. Microbial evaluation and some proposed good manufacturing practices of locally prepared malted corn drink ("asaana") and Hibiscus sabdarifa calyxes extract ("sobolo") beverages sold at a university cafeteria in Ghana. *Sci. Afr.* **2020**, *8*, 1–16. [CrossRef]

9. Kyttä, K.M.; Lakio, S.; Wikström, H.; Sulemanji, A.; Fransson, M.; Ketolainen, J.; Tajarobi, P. Comparison between twin-screw and high-shear granulation-The effect of filler and active pharmaceutical ingredient on the granule and tablet properties. *Powder Technol.* **2020**, *376*, 187–198. [CrossRef]

10. Córdova-Noboa, H.A.; Oviedo-Rondón, E.O.; Ortiz, A.; Matta, Y.; Hoyos, J.S.; Buitrago, G.D.; Martinez, J.D.; Yanquen, J.J.; Chico, M.; Martin, V.E.S.; et al. Effects of corn kernel hardness and grain drying temperature on particle size and pellet durability when grinding using a roller mill or hammermill. *Anim. Feed Sci. Technol.* **2021**, *271*, 114715. [CrossRef]

11. Han, Y.; Jia, F.; Li, G.; Liu, H.; Li, J.; Chen, P. Numerical analysis of flow pattern transition in a conical silo with ellipsoid particles. *Adv. Powder Technol.* **2019**, *30*, 1870–1881. [CrossRef]

12. Gürsoy, S.; Güzel, E. Determination of physical properties of some agricultural grains. *Res. J. Appl. Sci. Eng. Technol.* **2010**, *2*, 492–498.

13. Kelley, D.S.; Adkins, Y.; Reddy, A.; Woodhouse, L.R.; Mackey, B.E.; Erickson, K.L. Sweet bing cherries lower circulating concentrations of markers for chronic inflammatory diseases in healthy humans1-4. *J. Nutr.* **2013**, *143*, 340–344. [CrossRef]

14. Liu, C.; Wang, Y.; Song, J.; Li, Y.; Ma, T. Experiment and discrete element model of rice seed based on 3D laser scanning. *Nongye Gongcheng Xuebao/Trans. Chin. Soc. Agric. Eng.* **2016**, *32*, 294–300. [CrossRef]

15. Abdolahzare, Z.; Abdanan Mehdizadeh, S. Real time laboratory and field monitoring of the effect of the operational parameters on seed falling speed and trajectory of pneumatic planter. *Comput. Electron. Agric.* **2018**, *145*, 187–198. [CrossRef]

16. Wang, Y.; Liang, Z.; Zhang, D.; Cui, T.; Shi, S.; Li, K.; Yang, L. Calibration method of contact characteristic parameters for corn seeds based on EDEM. *Nongye Gongcheng Xuebao/Trans. Chin. Soc. Agric. Eng.* **2016**, *32*, 36–42. [CrossRef]

17. Wang, J.W.; Han, T.; Wang, J.F.; Jiang, D.X.; Li, X. Measurement and analysis of restitution coefficient between maize seed and soil based on high-speed photography. *Int. J. Agric. Biol. Eng.* **2017**, *10*, 102–114. [CrossRef]

18. Ozturk, I.; Kara, M.; Uygan, F.; Kalkan, F. Restitution coefficient of chick pea and lentil seeds. *Int. Agrophysics* **2010**, *24*, 209–211.

19. Chen, L.; Ma, X.; Cao, X.; Wen, Z.; Ji, C.; Li, H. Evaluation research of physical characteristics of hybrid rice buds based on principal component analysis. *Nongye Gongcheng Xuebao/Trans. Chin. Soc. Agric. Eng.* **2019**, *35*, 334–342. [CrossRef]

20. Ganiyu, S.A.; Badmus, B.S.; Olurin, O.T.; Ojekunle, Z.O. Evaluation of seasonal variation of water quality using multivariate statistical analysis and irrigation parameter indices in Ajakanga area, Ibadan, Nigeria. *Appl. Water Sci.* **2018**, *8*, 1–15. [CrossRef]

21. Mounjouenpou, P.; Ngono Eyenga, S.N.N.; Kamsu, E.J.; Bongseh Kari, P.; Ehabe, E.E.; Ndjouenkeu, R. Effect of fortification with baobab (Adansonia digitata L.)pulp flour on sensorial acceptability and nutrient composition of rice cookies. *Sci. Afr.* **2018**, *1*, e00002. [CrossRef]

22. Liu, G.; Chen, Y.; He, X.; Yao, F.; Guan, G.; Zhong, B.; Zhou, G. Seasonal changes of mineral nutrients in the fruit of navel orange plants grafted on trifoliate orange and citrange. *Sci. Hortic.* **2020**, *264*, 109156. [CrossRef]

23. Salvador-Reyes, R.; Rebellato, A.P.; Lima Pallone, J.A.; Ferrari, R.A.; Clerici, M.T.P.S. Kernel characterization and starch morphology in five varieties of Peruvian Andean maize. *Food Res. Int.* **2021**, *140*, 110044. [CrossRef]

24. Yang, B.; Zhang, X.; Zhao, F.; Yang, Y.; Liu, W.; Li, S. Suitability evaluation of different potato cultivars for processing products. *Nongye Gongcheng Xuebao/Trans. Chin. Soc. Agric. Eng.* **2015**, *31*, 301–308. [CrossRef]

25. Sood, S.; Khulbe, R.K.; Arun Kumar, A.K.R.; Agrawal, P.K.; Upadhyaya, H.D. Barnyard millet global core collection evaluation in the submontane Himalayan region of India using multivariate analysis. *Crop. J.* **2015**, *3*, 517–525. [CrossRef]

26. Lu, F.; Ma, X.; Tan, S.; Chen, L.; Zeng, L.; An, P. Simulative Calibration and Experiment on Main Contact Parameters of Discrete Elements for Rice Bud Seeds. *Nongye Jixie Xuebao/Trans. Chin. Soc. Agric. Mach.* **2018**, *49*, 93–99. [CrossRef]

27. Dai, F.; Song, X.; Zhao, W.; Zhang, F.; Ma, H.; Ma, M. Simulative Calibration on Contact Parameters of Discrete Elements for Covering Soil on Whole Plastic Film Mulching on Double Ridges. *Nongye Jixie Xuebao/Trans. Chin. Soc. Agric. Mach.* **2019**, *50*, 49–56. [CrossRef]

28. Luo, S.; Yuan, Q.; Gouda, S.; Yang, L. Parameters Calibration of Vermicomposting Nursery Substrate with Discrete Element Method Based on JKR Contact Model. *Nongye Jixie Xuebao/Trans. Chin. Soc. Agric. Mach.* **2018**, *49*, 343–350. [CrossRef]

29. Liu, W.; He, J.; Li, H.; Li, X.; Zheng, K.; Wei, Z. Calibration of Simulation Parameters for Potato Minituber Based on EDEM. *Nongye Jixie Xuebao/Trans. Chin. Soc. Agric. Mach.* **2018**, *49*, 125–135. [CrossRef]

30. Jones, J.R.; Lawrence, H.G.; Yule, I.J. A statistical comparison of international fertiliser spreader test methods-Confidence in bout width calculations. *Powder Technol.* **2008**, *184*, 337–351. [CrossRef]

31. Qiu-ju, W.; Feng, L.; Pan, G.; Zhong-chao, G.; Ben-chao, C.; Yan-xia, L.; Li-li, Z. Effects of Rice Yield and Quality Across Accumulated Temperature Zone Planting in Cold Area. *J. Northeast. Agric. Univ.* **2015**, *22*, 1–7. [CrossRef]

32. Wu, P.; Jia, C.; Fan, S.; Sun, Y. Principal component analysis and fuzzy comprehensive evaluation of fruit quality in cultivars of cherry. *Nongye Gongcheng Xuebao/Trans. Chin. Soc. Agric. Eng.* **2018**, *34*, 291–300. [CrossRef]

33. Ma, S.; Xu, L.; Yuan, Q.; Niu, C.; Zeng, J.; Chen, C.; Wang, S.; Yuan, X. Calibration of discrete element simulation parameters of grapevine antifreezing soil and its interaction with soil-cleaning components. *Nongye Gongcheng Xuebao/Trans. Chin. Soc. Agric. Eng.* **2020**, *36*, 40–49. [CrossRef]

34. He, X.; Cui, T.; Zhang, D.; Wei, J.; Wang, M.; Yu, Y.; Liu, Q.; Yan, B.; Zhao, D.; Yang, L. Development of an electric-driven control system for a precision planter based on a closed-loop PID algorithm. *Comput. Electron. Agric.* **2017**, *136*, 184–192. [CrossRef]
35. Tian, L.; Tang, H.; Wang, J.; Li, S.; Zhou, W.; Yan, D. Design and Experiment of Rebound Dipper Hill-drop Precision Direct Seed-metering Device for Rice. *Nongye Jixie Xuebao/Trans. Chin. Soc. Agric. Mach.* **2017**, *48*, 65–72. [CrossRef]
36. Shi, L.; Sun, W.; Zhao, W.; Yang, X.; Feng, B. Parameter determination and validation of discrete element model of seed potato mechanical seeding. *Nongye Gongcheng Xuebao/Trans. Chin. Soc. Agric. Eng.* **2018**, *34*, 35–42. [CrossRef]
37. Lü, J.; Feng, X.; Yang, X.; Li, Z.; Zou, F. Design and experiment of supply and suction device on air-suction potato planter. *Int. Agric. Eng. J.* **2019**, *28*, 89–100.
38. Liu, Y.; Zong, W.; Ma, L.; Huang, X.; Li, M.; Tang, C. Determination of three-dimensional collision restitution coefficient of oil sunflower grain by high-speed photography. *Nongye Gongcheng Xuebao/Trans. Chin. Soc. Agric. Eng.* **2020**, *36*, 44–53. [CrossRef]
39. Aspelmeier, T.; Zippelius, A. Dynamics of a one-dimensional granular gas with a stochastic coefficient of restitution. *Phys. Stat. Mech. Appl.* **2000**, *282*, 450–474. [CrossRef]
40. Tang, H.; Wang, J.; Wang, F.; Li, X.; Li, J. Measurement and analysis of rolling friction coefficient of maize seed based on high-speed photography. *Int. Agric. Eng. J.* **2018**, *27*, 185–198.
41. Leblicq, T.; Smeets, B.; Ramon, H.; Saeys, W. A discrete element approach for modelling the compression of crop stems. *Comput. Electron. Agric.* **2016**, *123*, 80–88. [CrossRef]
42. Hastie, D.B. Experimental measurement of the coefficient of restitution of irregular shaped particles impacting on horizontal surfaces. *Chem. Eng. Sci.* **2013**, *101*, 828–836. [CrossRef]
43. Zhan, Z.; Yaoming, L.; Jin, C.; Lizhang, X. Numerical analysis and laboratory testing of seed spacing uniformity performance for vacuum-cylinder precision seeder. *Biosyst. Eng.* **2010**, *106*, 344–351. [CrossRef]
44. Xu, D.; Li, W. Effects of no-tillage opening seedbed on maize growth and yield in Northeast China. *Int. Agric. Eng. J.* **2018**, *27*, 25–32.
45. Gaitani, N.; Lehmann, C.; Santamouris, M.; Mihalakakou, G.; Patargias, P. Using principal component and cluster analysis in the heating evaluation of the school building sector. *Appl. Energy* **2010**, *87*, 2079–2086. [CrossRef]
46. Galhano dos Santos, R.; Bordado, J.C.; Mateus, M.M. Estimation of HHV of lignocellulosic biomass towards hierarchical cluster analysis by Euclidean's distance method. *Fuel* **2018**, *221*, 72–77. [CrossRef]
47. Guo, Y.; Lu, X.; Peng, G. Investigation and analysis of the building technology of low temperature germplasm genebanks of China. *Nongye Gongcheng Xuebao/Trans. Chin. Soc. Agric. Eng.* **2005**, *21*, 186–190.

Article

Effect of Heat Resource Effectiveness Change on Rice Potential Yield in Southern China

Qing Ye [1], Xiaoguang Yang [2,*], Wenjuan Xie [3], Junmeng Yao [4] and Zhe Cai [4]

1 College of Forestry, Jiangxi Agricultural University, Nanchang 330045, China; yeqing@jxau.edu.cn
2 College of Resources and Environmental Sciences, China Agricultural University, Beijing 100193, China
3 Hebei Meteorological Disaster Prevention Center, Shijiazhuang 050021, China; xiewenjuanwork@gmail.com
4 Jiangxi Agricultural Meteorological Center, Nanchang 330096, China; junmengyao86@gmail.com (J.Y.); caizheread@gmail.com (Z.C.)
* Correspondence: yangxg@cau.edu.cn

Abstract: During the rice growing season, farmers' decisions about cropping systems and seed varieties directly affect the utilization of heat resource, and eventually affect the potential yield. In this study, we used the hourly accumulated temperature model to calculate the available heat resource as well as the effective heat resource in southern China. We conducted a spatiotemporal analysis of the heat resource effectiveness during rice growing season and an impact assessment of heat resource effectiveness on rice potential yield and cereal yield reduction. The results showed that, during the period of 1951–2015, heat resource effectiveness generally declined in the rice cropping area of southern China. And this decrease worsened during the most recent three decades compared with the period of 1951–1980. A strong correlation was detected between heat resource effectiveness and rice potential yield in the study area. When the effective heat resource during the growing season increased by 1 °C·d, rice potential yield would increase by 14 kg ha^{-1}. For each percentage increase in heat resource effectiveness, the rice potential yield reduction rate would go down by 0.65%. This agro-climatological study aims to offer a scientific basis for rice production decisions in southern China, such as when to plant, which varieties to choose and so on.

Keywords: climate change; heat resource effectiveness; hourly accumulated temperature simulation; rice potential yield

Citation: Ye, Q.; Yang, X.; Xie, W.; Yao, J.; Cai, Z. Effect of Heat Resource Effectiveness Change on Rice Potential Yield in Southern China. *Processes* **2021**, *9*, 896. https://doi.org/10.3390/pr9050896

Academic Editor: Yonghui Li

Received: 4 March 2021
Accepted: 15 May 2021
Published: 19 May 2021

Publisher's Note: MDPI stays neutral with regard to jurisdictional claims in published maps and institutional affiliations.

1. Introduction

Rice is one of the three main food crops in China. China's rice yield accounts for 30% of the total global rice yield. The fluctuations of rice yield in China affect the national or even global food security. Against the background of climate change, such impact could be amplified by the changes in effective heat resources during the rice growing period.

Heat resources (temperature, herein unless otherwise specified) plays a major role in the geological distribution of crops [1–4]. Given sufficient available water, the heat resource determines the crop mixture structure and seed variety allocation [5,6], crop growing season length [7,8], and eventually the potential yield [9]. Conversely, farming decisions of crop mixture, cropping system, and seed varieties could affect the utilization of heat resource during the crop growing season.

Previous scholars had discovered a clear relationship between crop development and thermal time [7,10–13]. The application of thermal time has been widely used to predict crop phenology [14–20] and crop yield [21–23] in crop models. Recently, the application of thermal time has been expanded to evaluate the effective heat resource during the crop growing season. The spatiotemporal analyses on the effective heat resources for different crops in different regions of China have been reported. Nevertheless, those studies did not consider the effect of temperatures greater than the upper threshold beyond which temperatures accelerate crop growth or crop development. Similar to the temperatures less than the lower threshold of crop growth requirement, temperatures greater than the upper threshold of crop growth requirement are useless or even detrimental to crop development [24]. Therefore, some researchers came up with three-point temperatures to measure the effective temperature for crop development, and further recognized that only temperatures greater than the lower threshold and less than the upper threshold of crop growth requirement are seen as effective [25,26]. The ratio of accumulated effective temperature to accumulated available temperature is defined as the accumulated temperature effectiveness and is used to quantify the effectiveness of the heat resource [27]. In addition, the ratio of heat resource effectiveness has been used to analyze the effect of temperature change on crop yield [28].

In China, three different methods are primarily used to calculate the accumulated effective temperature during the growing season: (a) the average temperature-based method [26,29–32], (b) the adjusted maximum and minimum temperatures method [33–36]; and (c) the hourly temperature-based method [37]. Jiang and Wen (2013) have pointed out that the hourly temperature-based method is more accurate than the others [38]. In this study, we used the hourly temperature-based method to calculate the available and effective heat resources during rice growing season in the rice cropping area of southern China. We investigated the spatiotemporal characteristics of heat resource effectiveness in the southern rice cropping area and further discussed the relationships between heat resource effectiveness and rice potential yield, as well as grain yield reduction rate. This study aims to offer scientific support for the optimal cropping system arrangement in the rice cropping area of southern China.

2. Materials and Methods

2.1. Study Region and Data

In this study, the research area is southern China (99°–123° E and 18°–34° N) where national rice production is concentrated. The boundaries of the research area followed the suggestions in Liu and Han [39], where natural resources (e.g., terrain factors, radiation and water resources), socio-economic conditions, agricultural background (e.g., crop varieties, crop mixtures, and maturity types), and the integrity of county-level administrative divisions were taken into consideration. In the research area, rice-based cropping systems include double-cropping systems rotated with middle rice and winter wheat and triple-cropping systems rotated with early or middle rice and winter wheat or rape. The regional paddy land area accounts for 83.52% of the national paddy land area [40]. Within the regional area, the double-cropping system accounts for 66% of the paddy land area, and produces 61.3% of the national rice grain yield [41]. Therefore, improved understanding of the effects of temperature change on rice grain yields and production for multiple rice-based cropping systems in southern China is critically important for the country's food security.

According to [42], the study area can be further divided into four sub-regions: single rice-cropping system (SRCS), early double rice-cropping system (EDRCS), middle double rice-cropping system (MDRCS) and late double rice-cropping system (LDRCS) (Figure 1B,C), and rice varieties for the four main rice-based cropping systems are shown in Table 1. The division of these sub-regions is based on cropping suitability for different rice-based cropping systems from a climatological perspective [43].

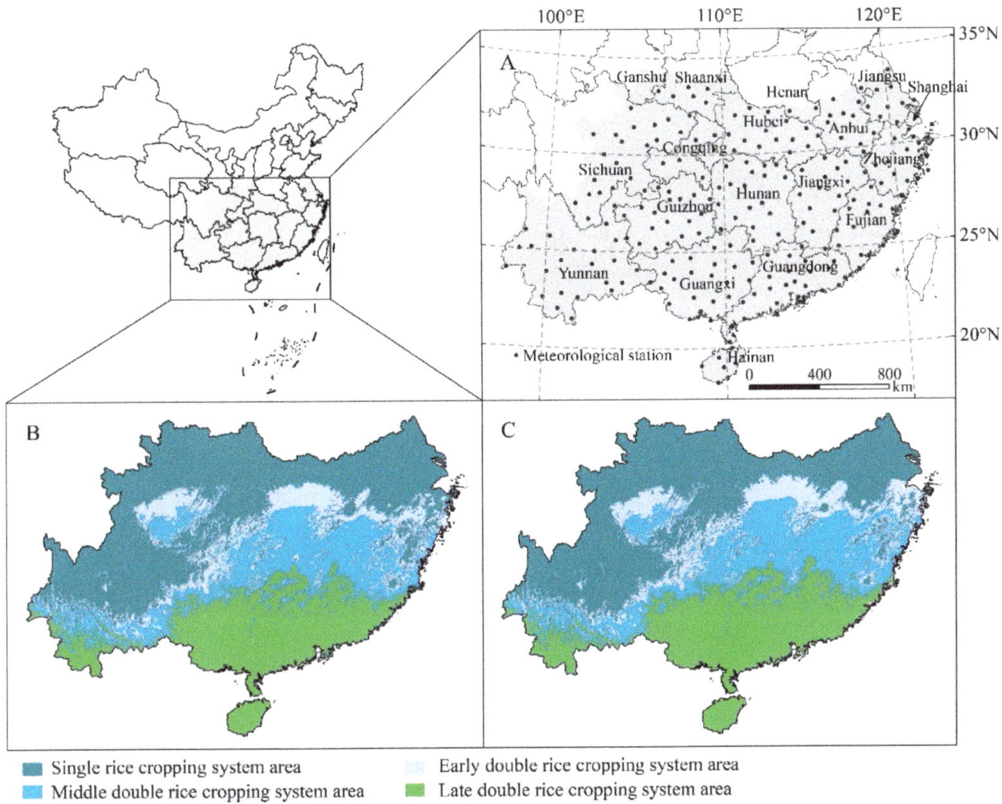

Figure 1. Overview of the study area. (**A**) Distribution of the 254 meteorological stations in southern China; (**B**) suitable planting area for the four rice-based cropping systems during the period of 1951–1980; (**C**) suitable planting area for the four rice-based cropping systems during the period of 1981–2015.

Table 1. Rice varieties for the four main rice-based cropping systems.

Rice-Based Cropping System	Rice Variety
Single rice	Hybrid rice
Early double rice	Medium maturity early indica + medium maturity japonica
Middle double rice	Late maturity early indica + hybrid rice
Late double rice	Hybrid rice + hybrid rice

The historical climate data from 1951 to 2015 for 254 meteorological stations (Figure 1) were obtained through the China Meteorological Science Data Sharing Service (http://cdc.cma.gov.cn, accessed on 8 March 2018), including atmospheric pressure, maximum, minimum, and average temperatures, relative humidity, precipitation, wind speed, and sunshine hours. The crop phenology data are acquired from China Meteorological Bureau agricultural meteorological observations (http://cdc.cma.gov.cn, accessed on 8 March 2018).

2.2. Determination of Rice Growing Season

According to [44,45], we divided the entire rice growing season into three stages: early stage (from sowing to booting), middle stage (from booting to flowering), and late stage (from flowering to maturity). The climatic-ecological model proposed by Gao, et al. [46] was adopted to calculate potential growing season length for different rice varieties during the two periods of 1951–1980 and 1981–2015 (Table 2). For early and middle maturity rice varieties, the sowing date was specified as the date when the possibility of average daily temperature equal to or greater than 10 °C reached 80%. For late maturity rice varieties, sowing date (the beginning date of seedling) was specified as 30 days earlier than the maturity date of early maturity rice varieties, and transplanting date was specified as 5 days (typical time needed for harvest activities) after the maturity of early maturity rice varieties. The maturity date was also calculated based on the climatic-ecological model. As suggested by FAO [44], we used the mean phenology stage length during the period of 1981–2015 to interpolate the booting and heading dates. The mean phenology dates for different rice cropping systems during the two study periods of 1951–1980 and 1981–2015 are presented in Table 3.

Table 2. The potential growing season length for different rice varieties.

Rice Variety	Rice Type	Potential Growing Season Length (day)
Hybrid rice	Shanyou II	$N = 101.56 - 3.52\Delta T + 0.16(\Delta T)^2 + 0.16\Delta D + 3.28\Delta\Phi$
Medium maturity early indica	Yuanfengzao	$N = 71.82 - 2.42\Delta T + 0.14\Delta D + 1.49\Delta\Phi$
Late maturity early indica	Guangsi	$N = 71.73 - 3.826\Delta T + 0.088\Delta D + 1.856\Delta\Phi$
Medium maturity japonica	Nanjing 34	$N = 122.64 - 3.13\Delta T + 0.39\Delta D + 1.09\Delta\Phi$

Note: N means the potential growing season length (day); ΔT means the difference between 25 °C and the mean temperature from the safe sowing date to the safe fully heading date (°C); ΔD means the number of days between April 1 and the sowing date; and $\Delta\Phi$ means the difference between station latitude and 30° N.

Table 3. Simulated rice phenology (day of the year) of different varieties for the four main rice-based cropping systems in southern China.

Period	Cropping System	Rice Variety	Sowing Date	Transplant Date	Booting Date	Flowering Date	Mature Date	Length of Growing Season (days)
1951–1980	SRCS	MM	84	115	196	206	247	163
	EDRCS	EM	78	108	151	160	187	109
		LM	162	192	251	245	304	142
	MDRCS	EM	68	98	144	153	182	114
		LM	157	187	215	244	275	118
	LDRCS	EM	2	50	106	116	149	147
		LM	124	154	205	216	244	120
1981–2015	SRCS	MM	84	114	190	199	238	154
	EDRCS	EM	83	113	156	165	193	110
		LM	168	198	256	268	309	141
	MDRCS	EM	72	102	149	159	187	115
		LM	162	192	248	260	300	138
	LDRCS	EM	2	50	106	116	149	147
		LM	124	154	205	216	252	128

Note: MM: middle maturity; EM: early maturity, and LM: late maturity.

In the crop model, the potential growing season length for the four representative cultivars was calibrated based on the actual crop phenology data from China Meteorological Bureau agricultural meteorological observations. The validation results show that the correlation coefficient (R^2) between the observed and simulated growing season length was greater than 0.80 for all four representative rice cultivars (Figure 2).

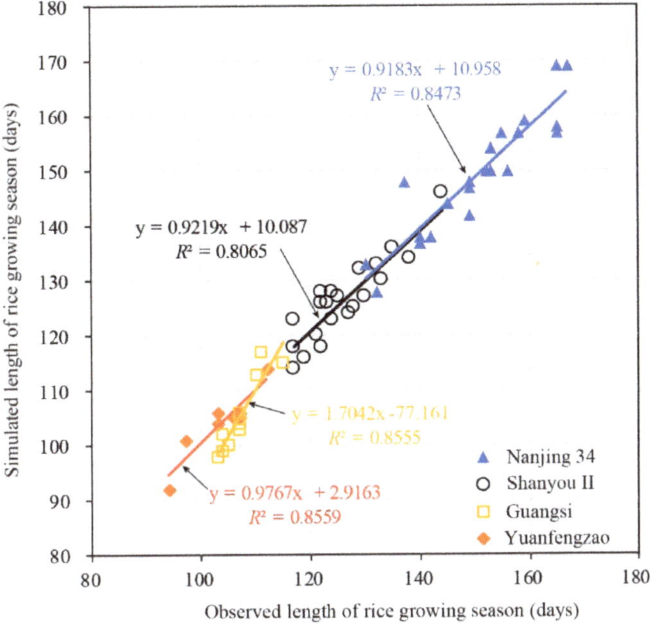

Figure 2. Simulated vs. observed growing season length for the four study rice cultivars.

2.3. Available Heat Resources (AHR)

Temperatures below the lower threshold of crop growth requirement can be detrimental [47,48]; likewise temperatures above the upper threshold of crop growth requirement can be detrimental [23,49,50]. In this study, we used the lower threshold of 10 °C, and the upper threshold of 35 °C to evaluate the heat resources during the rice growing season [51–53]. The calculation equations are

$$DH_i = \begin{cases} 0 & T_i < T_b \\ T_i - T_b & T_b \leq T_i \leq T_u \\ 0 & T_i > T_u \end{cases} \tag{1}$$

$$DD = \left(\sum_{i=1}^{24} DH_i \right) / 24 \tag{2}$$

$$GDD = \sum_{k=1}^{n} DD_k \tag{3}$$

where DH_i is the available heat resource in the ith hour of a day (°C·h); T_i is the average temperature (°C); T_b is the lower threshold (°C); T_u is the upper threshold (°C); DD is the available heat resource in the day (°C·d); GDD is the available heat resource during the growing season (°C·d); and n is the number of days in the growing season (unitless).

2.4. Effective Heat Resource (EHR)

The effective heat resource is defined as the accumulated temperatures that are above the lower threshold of crop growth [32,54–57]. It has been pointed out that rice growth rate is exponentially related to the daily mean temperature [53,58]. The response of growth rate to hourly temperature can be described by the bilinear model [59]. Hodges [60] found that daily mean temperature between base and optimum temperature has a positive relationship with the crop growth rate, while temperature between optimum temperature and upper threshold has a negative relationship with the crop growth rate. This implies that the effectiveness of temperature to crop growth varies among different ranges. Based on the "cardinal" temperature theory, we modified the effective temperature model brought up by Bouman et al. (2001) [61] as below:

$$EHR_i = \begin{cases} \frac{1}{24}\left[T_{ou} - \frac{(T_i - T_{ou})(T_{ou} - T_b)}{T_u - T_{ou}}\right] & T_{ou} < T_i < T_u \\ 0 & T_i \leq T_b, T_i \geq T_u \\ \frac{T_{ob} - T_b}{24} & T_{ob} \leq T_i \leq T_{ou} \\ \left(\frac{T_i - T_b}{24}\right) & T_b < T_i < T_{ob} \end{cases} \tag{4}$$

$$DEHR = \sum_{i=1}^{24} EHR_i \tag{5}$$

$$GEHR = \sum_{k=1}^{3} \sum_{j=1}^{n} \sum_{i=1}^{24} EHR_{ijk} \tag{6}$$

where EHR_i is the EHR for the ith hour (°C·h); $DEHR$ is the daily EHR (°C·d); T_{ob} is the lower range of the optimum temperature (°C); T_{ou} is the upper range of the optimum temperature (°C); T_i is the average temperature for the ith hour (°C); T_u is the upper threshold (°C); T_b is the lower threshold (°C); $GEHR$ is the sum of total growing season EHR for different rice cropping systems (°C·d); EHR_{ijk} is the EHR at the ith hour on the jth day at the kth crop stage; and n is the number of days in the growing season (unitless).

In Table 4, we present the three "cardinal" temperatures for different crop development stages of rice [24,51,52,62].

Table 4. The cardinal (i.e., the base, maximum and optimum) temperatures for different crop development stages of rice.

Development Stages	Three Cardinal Temperature (°C)			
	Lower Threshold (T$_b$)	Upper Threshold (T$_u$)	Lower Optimum Range (T$_{ob}$)	Upper Optimum Range (T$_{ou}$)
Sowing to booting	10	35	25	30
Booting to flowering	22	35	30	33
Flowering to maturity	15	35	20	29

We calculated the hourly temperatures based on the daily minimum and maximum temperatures by using the method from Bouman, Kropff, Tuong, Wopereis, ten Berge and van Laar [61].

$$T = \begin{cases} \frac{T_{max} + T_{min}}{2} + \frac{T_{max} - T_{min}}{2} \cdot cos\left(\frac{h_i - 14}{14 - h_r} \cdot \pi\right) & h_r \leq h_i \leq 14 \\ \frac{T_{max} + T'_{min}}{2} + \frac{T_{max} - T'_{min}}{2} \cdot cos\left(\frac{h_i - 14}{10 + h_s} \cdot \pi\right) & 14 \leq h_i \leq 24 + h_s \end{cases} \tag{7}$$

where T_i is the hourly mean temperature (°C); T_{max} is the daily maximum temperature (°C); T_{min} is the daily minimum temperature (°C); h_i is the ith hour; h_r is the sunrise time in a day; and h_s is the sunset time in a day. The calculations for sunrise time and sunset time are cited from Allen, et al. [63].

2.5. Heat Resource Effectiveness (HRE)

The term "resource effectiveness" was coined by Liu and Zhong (1996) to indicate the effectiveness of environmental resources on crop development [64]. Furthermore, heat resource effectiveness was defined as the percentage ratio of actually utilized heat resource (effective heat resource) to available heat resource during the crop growing stages or the entire crop growing season [65]. By investigating heat resource effectiveness during the entire crop growing season, we could quantify the utilization of heat resource by the plants. Based on the investigation results, we can make better decisions about seed varieties, crop mixture, and farming management to make better use of the heat resource.

Heat resource effectiveness is the percentage ratio of effective heat resource to available heat resource (%). The *HRE* during growing season for a crop type can be calculated as:

$$HRE = \frac{GEHR}{GDD} \times 100 \tag{8}$$

where *HRE* is the heat resource effectiveness during the growing season (%); *GDD* is the total available heat resource during the growing season (°C·d); and *GEHR* is the total effective heat resource during the growing season (°C·d).

Without considering the blank time between different crops, the gross EHR of a cropping system is the sum of the EHR for all crops. The gross HRE (*GHRE*, %) for a cropping system can be calculated as:

$$GHRE = \frac{\sum_{i=1}^{n} GEHR_i}{\sum_{i=1}^{n} GDD_i} \times 100 \tag{9}$$

where $GEHR_i$ is the total EHR for the *i*th crop during the growing season (°C·d); and GDD_i is the total AHR for the *i*th crop during the growing season (°C·d).

2.6. Calculation of Rice Potential Yield and Yield Reduction Rate

Crop potential yield is the maximum yield when crops grow under an ideal environment, without any limitations in soil fertility farming technology, and it is solely determined by climate elements of solar radiation, temperature, and precipitation. In the research area, irrigation is typically used for rice production. Therefore, we define the crop potential yield without any water limitation. This is similar to the so-called photo-thermal potential [66–68]. The crop potential yield (Y_T) can be calculated as

$$Y_T = Y_P \cdot f(T) \tag{10}$$

$$Y_P = \frac{s \cdot \Omega \cdot \varepsilon \cdot \phi \cdot (1-\alpha)(1-\beta)(1-\rho)(1-\gamma)(1-\omega) \cdot f(L)}{q \cdot (1-\eta)(1-\delta)} \sum_{i=1}^{n} Q_i \tag{11}$$

where Y_P is the daily photosynthetic potential, kg ha^{-1}; Y_T is the daily photo-thermal potential, kg ha^{-1}; *f(T)* is the temperature correction coefficient and can be calculated with Equation (12); and $\sum_{i=1}^{n} Q_i$ is the total solar radiation during the growing season, MJ·m^{-2}. The rest of the parameters are listed in Table 5 together with values from earlier studies [66,67,69,70].

Table 5. Parameters and values for rice potential yield calculation.

Parameter	Value	Description
s	0.45	The harvest index of rice
Ω	0.9	The capacity of crop photosynthetic CO_2 fixation
ε	0.49	The ratio of photosynthetic active radiation to total solar radiation on the ground
φ	0.224	The conversion efficiency from light to energy
α	0.06	The reflectance of crop canopy
β	0.08	The transmittance of crop canopy
ρ	0.1	The ineffective absorption rate (absorption rate of non-photosynthetic organs)
γ	0.05	The rate of solar above photosynthetic saturation point
ω	0.3	The respiration rate of rice
q	17.8	The energy requirement for one-kilogram dry matter formation (MJ/kg)
η	0.14	The standard water content of rice grain
δ	0.08	The mineral content of rice plant
$f(L)$	0.56	The correction index of daily leaf area index (LAI)

Since the critical temperatures are different among the rice growing stages, the effects of temperature on crop development vary. And we use the equations below to revise the temperature thresholds for different crop development stages:

$$f(T) = \begin{cases} 0 & T \leq T_b, \ T \geq T_u \\ \frac{T-T_b}{T_{ob}-T_b} & T_b < T < T_{ob} \\ 1 & T_{ob} < T < T_{ou} \\ \frac{T_u-T}{T_u-T_{ou}} & T_{ou} < T < T_u \end{cases} \tag{12}$$

where $f(T)$ is the temperature correction coefficient (unitless); T is the daily average temperature (°C); T_b is the lowest temperature (°C); T_u is the upper temperature limit (°C); T_{ob} is the lower range of optimal temperature (°C); T_{ou} is the upper range of optimal temperature (°C). In this study, critical temperatures for different rice growing stages (Table 4) are drawn from some previous studies [24,51,52,62].

The reduction rate of rice potential yield is calculated as:

$$RPY = \frac{Y_P - Y_T}{Y_P} \times 100 \tag{13}$$

where RPY is the reduction rate of crop potential yield (%).

3. Results

3.1. Temporal Variations in Major Heat Resource Indices

We analyzed the temporal trends in the three elements of AHR, EHR and HRE for five spatial scales: the entire study area, SRCS area, EDRCS area, MDRCS area and LDRCS area. We detected that in the areas of EDRCS, MDRCS, and LDRCS, both AHR and EHR showed a slightly decreasing trend during the period of 1951–1980, then a significantly increasing trend during the period of 1981–2015. In the entire study area, both AHR and EHR showed a slightly increasing trend during the study period of 1951–2015. Meanwhile, both AHR and HRE showed a slightly decreasing trend in SRCS area during the period of 1951–2015 (Figure 3A,B).

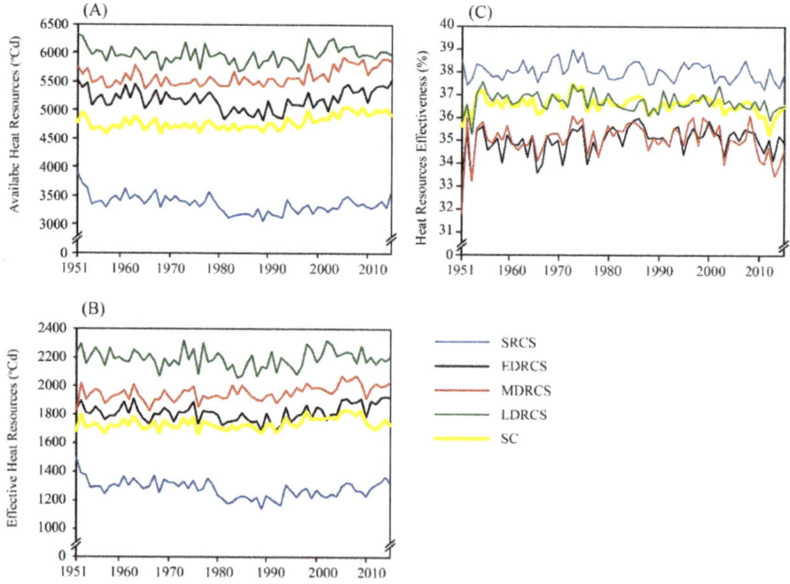

Figure 3. Temporal variations in AHR (**A**), EHR (**B**), HRE (**C**) for different rice cropping system areas in southern China during 1951–2015.

As seen in Figure 3C, HRE failed to show a clear trend for the entire study area or for the areas of four rice-based cropping systems during the period of 1951–2015. However, within the study period, there was a slightly increasing trend in HRE during the period of 1951–1980, but a slightly decreasing trend during the period of 1981–2015.

Overall, in the southern China rice cropping area, both AHR and EHR increased, while HRE decreased. The heat resource effectiveness decreased during the study period of 1951–2015.

3.2. Spatial Patterns of Major Heat Resources Indices

3.2.1. Available Heat Resource

During the period of 1951–2015, AHR during the rice growing season decreased from southeast to northwest in the study area (Figure 4A,C). The areas with AHR lower than 3000 °C·d were located in the southwestern part of the study area (i.e., Yunnan province), while the areas with AHR greater than 5500 °C·d were located in the central (i.e., Poyang Lake Plain and Dongting Lake Plain) and southeastern (e.g., Fujian and Guangdong provinces) parts of the study area.

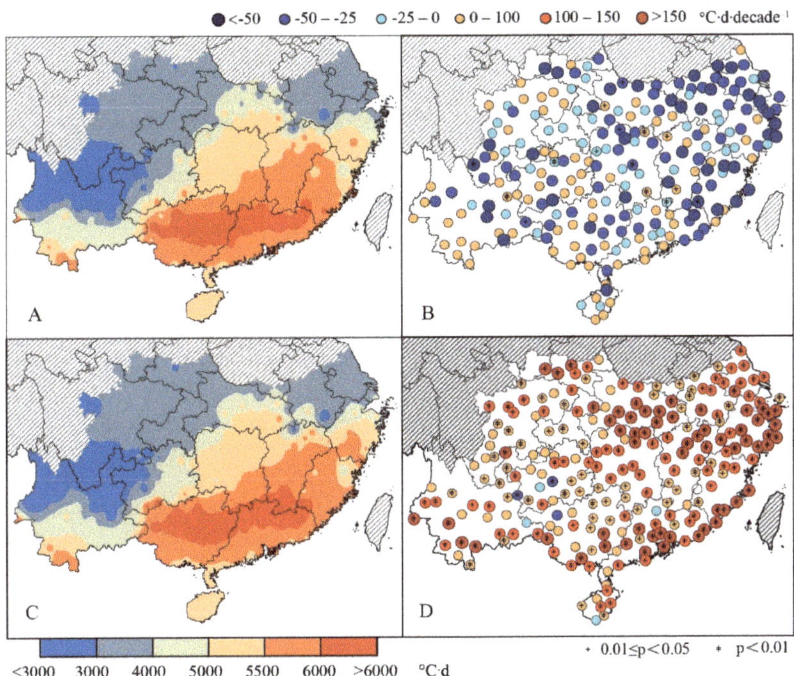

Figure 4. The amount of growing season AHR during the two periods of 1951–1980 (**A**) and 1981–2015 (**C**), and the trend in growing season AHR during the two periods of 1951–1980 (**B**) and 1981–2015 (**D**) in southern China.

On average, the total growing season AHR for the entire study area was 4609 °C·d and 4628 °C·d during the periods of 1951–1980 and 1981–2015, respectively. During the period of 1951–1980, a total of 66% of the research locations showed a decreasing trend in AHR (6% were statistically significant ($p < 0.05$), and were concentrated in the southeastern part of the study area), and the composite trend was -16.9 °C·d·decade^{-1} (Figure 4B). However, during the period of 1981–2015, a total of 96.5% of the research locations showed an increasing trend in growing season AHR (86% were statistically significant ($p < 0.05$)), and the composite trend was 114.5 °C·d·decade^{-1}.

For the same rice varieties, growing season AHR was greater during the period of 1951–1980 than 1981–2015 for different rice-based cropping systems. During the period of 1951–1980, total growing season AHR for SRCS, EDRCS, MDRCS and LDRCS was 3308 °C·d, 5072 °C·d, 5377 °C·d, and 5810 °C·d, respectively. And during the period of 1981–2015, total growing season AGR for SRCS, EDRCS, MDRCS and LDRCS was 3105 °C·d, 4884 °C·d, 5407 °C·d, and 5736 °C·d, respectively. This is because potential growing season was shortened for different rice cropping systems under the background of climate change (Table 6).

Table 6. Growing season AHR for different rice-based cropping systems in southern China during the two periods of 1951–1980 and 1981–2015 (unit: °C·d).

Rice Cropping System	1951–1980	1981–2015
SRCS	3308	3105
EDRCS	5072	4884
MDRCS	5377	5407
LDRCS	5810	5763

3.2.2. Effective Heat Resource

From 1951 to 2010, during the rice growing season, EHR also decreased from southeast to northwest in the study area (Figure 5A,C). During the period of 1951–1980, growing season EHR was 1668 °C·d on average in the study area, the areas with growing season EHR greater than 2000 °C·d were located in Guangdong, Guangxi, and southern Yunnan; the areas with growing season EHR lower than 1000 °C·d were located in northern Yunnan and southern Sichuan. During the period of 1981–2015, growing season EHR was 1676 °C·d on average in the study area; the spatial distribution characteristics were similar with the 1951–1980 period, but the areas with growing season EHR greater than 2000 °C·d expanded.

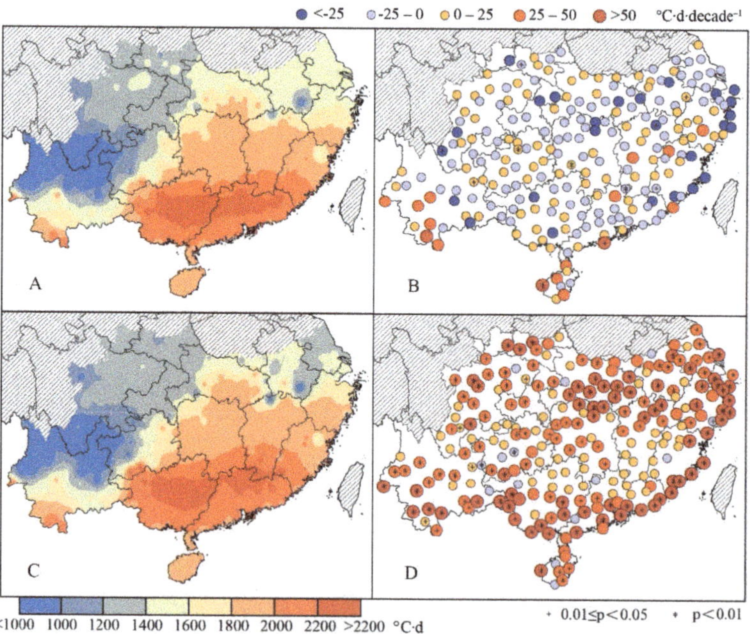

Figure 5. The amount of growing season EHR during the two periods of 1951–1980 (**A**) and 1981–2015 (**C**) and the trend in growing season AHR during the two periods of 1951–1980 (**B**) and 1981–2015 (**D**) in southern China.

During the period of 1951–1980, the composite trend in growing season EHR was −4.2 °C·d·decade^{-1}, with a total of 55% research locations showing decreasing trends (only 5% were statistically significant at $p < 0.05$) (Figure 5B). During the period of 1981–2015, the composite trend in growing season EHR was 37.6 °C·d·decade^{-1}, with a total of 95% of the research locations showing increasing trends (63% were statistically significant at $p < 0.05$).

During the study period of 1951–2015, the rice growing season was shortened against the background of climate warming; therefore, growing season EHR had been decreasing. Compared with the period of 1951–1980, growing season EHR for the entire study area had decreased by 27 °C·d during the period of 1981–2015. During the period of 1981–2015, the order of growing season EHR for different rice-based cropping systems from high to low was: LTCRS, MTCRS, ETCRS, and SRCS (Table 7).

Table 7. The amount of growing season EHR for different rice-based cropping systems in southern China during the two periods of 1951–1980 and 1981–2015 (unit: °C·d).

Rice Cropping System	1951–1980	1981–2015
SRCS	1260	1177
EDRCS	1747	1719
MDRCS	1873	1898
LDRCS	2115	2093

3.2.3. Heat Resource Effectiveness

During the period of 1951–2015, growing season HRE increased from southeast to northwest in the study area (Figure 6). On average, growing season HRE for the entire study area was 36% and 35.8% during the period of 1951–1980 and 1981–2015, respectively. During the period of 1951–1980, the single rice planting area had relatively higher growing season HRE (>37%), e.g., Sichuan, Shaanxi, Anhui and Jiangsu area, while the main double rice planting area had relatively lower growing season HRE (<34%), e.g., Hu'nan and Jiangxi area. During the period of 1981–2015, the area with relatively higher growing season HRE had decreased, while the area with relatively lower growing season HRE had expanded.

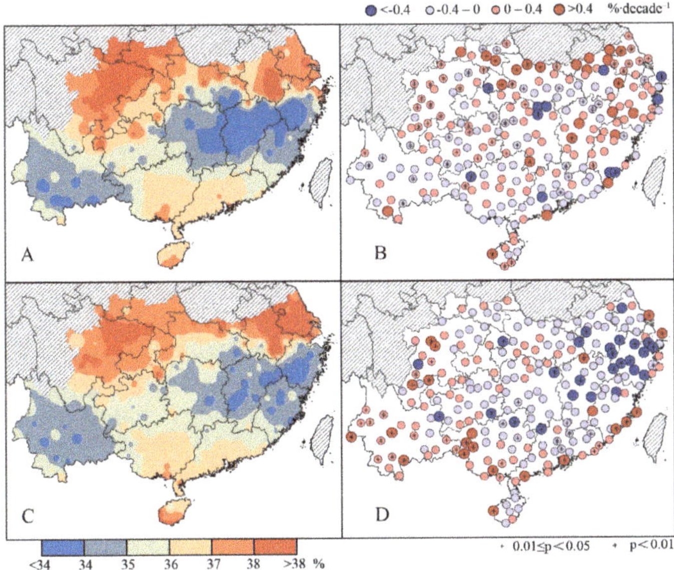

Figure 6. The amount of growing season HRE during the two periods of 1951–1980 (**A**) and 1981–2015 (**C**) and the trend in growing season AHR during the two periods of 1951–1980 (**B**) and 1981–2015 (**D**) in southern China.

During the period of 1951–1980, growing season HRE showed a slightly increasing trend at 58% of the research locations, with a composite trend of 0.01%·decade^{-1}, but none of them were statistically significant. During the period of 1981–2015, a total of 57% of the research locations showed a decreasing trend in growing season HRE (22% are statistically significant at $p < 0.01$, and they were located in Zhejiang and Fujian, Figure 6D), with a composite trend of -0.03%·decade^{-1}.

During the period of 1981–2015, growing season HRE has decreased by 0.1% on average compared with 1951–1980. In general, single rice cropping system has higher growing season HRE than double rice cropping systems. More specifically, the order of growing season HRE from high to low is: SRCS, LDRCS, MDRCS, and EDRCS (Table 8).

Table 8. Growing season HRE in southern China during the two periods of 1951–1980 and 1981–2015 (unit: %).

Rice Cropping System	1951–1980	1981–2015
SRCS	37.3	37.2
EDRCS	33.9	34.7
MDRCS	34.3	34.5
LDRCS	36.2	36.1

3.3. The Effect of EHR on Photo-Thermal Potential for Different Rice Cropping Systems

We detected a significant linear relationship ($R^2 > 0.90$) between growing season GEHR and photo-thermal potential yield (Figure 7), which implies a strong correlation between growing season heat resource and rice potential yield. With a 1 °C·d increase in growing season GEHR, rice photo-thermal potential yield would increase by 14.3 kg·ha^{-1} and 13.7 kg·ha^{-1} during the period of 1951–1980 and 1981–2015, respectively.

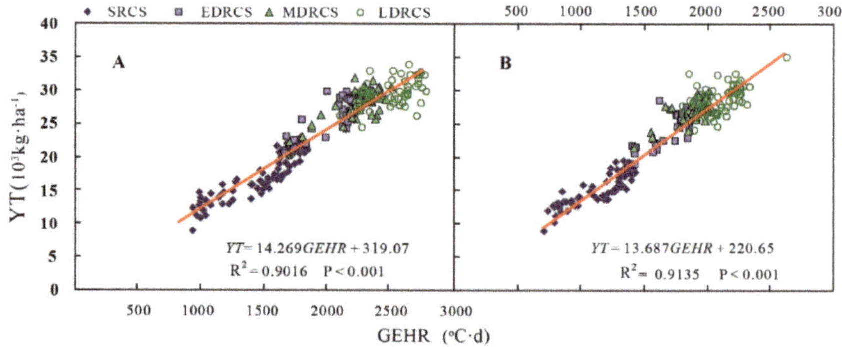

Figure 7. Correlation between growing season GEHR and rice photo-thermal potential yield in southern China during the two periods of 1951–1980 (**A**) and 1981–2015 (**B**).

3.4. The Effect of Growing Season HRE on Rice Potential Yield

We detected a negative relationship between HRE and RPY ($R^2 > 0.30$, $p < 0.01$) during the two periods of 1951–1980 and 1981–2015. For per unit of increase in growing season HRE, the temperature-induced potential yield reduction rate would be offset by 0.6% and 0.7% during the periods 1951–1980 and 1981–2015, respectively (Figure 8A,B).

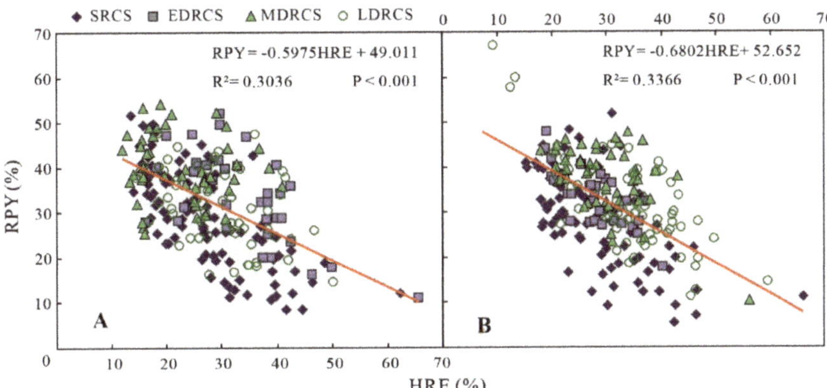

Figure 8. Correlation between growing season HRE and temperature-induced potential yield reduction rate in southern China during the two periods of 1951–1980 (**A**) and 1981–2015 (**B**).

4. Discussion

Agro-thermal resources are the main factors affecting the layout, structure, and yield of crops. Previous related studies have used agricultural limit temperature, accumulated temperature, and the growing season length [42,71–73] or the temperature suitability indices based on the development characteristics during various growth stages [74]. However, few studies have focused on quantifying the effectiveness of heat resource or heat resource use efficiency.

Traditional methods used to evaluate heat resource during the crop growing season (e.g., accumulated temperature) fail to consider the changing temperature demands of the crop within different growth stages. In this study, we adopted the three "cardinal" temperatures for rice in different developmental stages to calculate the growing season EHR. In addition, we used hourly temperature simulated from the daily maximum and minimum temperatures during the computing process [38]. Therefore, the growing season EHR can better represent the heat resource demands during different growth stages for rice crops.

Previous studies have shown that growing season thermal time has changed over the last six decades due to a warming climate [31,32,75,76]. However the changes in growing season HRE for rice production in southern China have rarely been documented. As pointed out by Bouman et al. (2001) effective temperature is what really matters for rice development; thus, we analyzed the changes in growing season HRE in this study. In general, growing season HRE for rice production had decreased during the research period in southern China. This decrease might have been related to the shortening rice growing season [77] and the increase in the number of days with daily maximum temperature $\geq 35\,^{\circ}\mathrm{C}$ [78,79].

Statistical methods and crop models are commonly used to predict the potential growing season for rice. In this study, we selected the meteorological–biological model for the rice growing season that was brought up by Gao (1983) [46] because it has relatively fewer parameters and is simple to run. This model is superior to the trending rice models that require many parameters to simulate the growing season and which are also complicated and difficult to use.

This study was based on the assumption that no changes had been made to the rice varieties for the four main cropping systems in southern China during the study period [62]. In practice, the choices of seed varieties could be influenced by breeding technology, governmental policy, cereal grain market, weather extremes (e.g., low temperatures and waterlogging), etc. By using the historical rice cultivars to analyze the effects of climate change on rice potential yield, we omitted the factor that farmers would switch to newer hybrids to combat climate change. Hence, our results might exaggerate the effects of climate change on rice potential yield.

5. Conclusions

During the period of 1951–1980, both growing season AHR and EHR showed a slightly decreasing trend in southern China. By contrast, during the period of 1981–2015, the trends in growing season AHR and EHR significantly increased.

In the past six decades, growing season AHR and EHR decreased from southeast to northwest in the southern rice cropping area. The areas with relatively shorter growing season AHR and EHR are in the southwestern part of the study area (e.g., Yunnan), where the single rice-cropping system is typical. The areas with relatively longer growing season AHR and EHR are in the central (e.g., Poyang Lake plain and Dongting Lake plain) and southeastern (e.g., Fujian province, Guangdong province) parts of the study area, where the double rice-cropping systems are typical. Nevertheless, the growing season HRE for the single rice-cropping system is longer than that for the double rice-cropping systems.

With a 1 °C·d increase in growing season GEHR in the southern rice cropping area, the photo-thermal potential yield for rice would increase by 14.3 kg·ha^{-1} and 13.7 kg·ha^{-1} during the period of 1951–1980 and 1981–2015, respectively. For each percentage of increase in growing season HRE, the temperature-induced rice potential yield reduction rate would decrease by 0.6% and 0.7% during the period of 1951–1980 and 1981–2015, respectively.

Author Contributions: Conceptualization, Q.Y. and X.Y.; methodology, Q.Y. and W.X.; software, Q.Y. and J.Y.; validation, W.X., J.Y. and Z.C.; formal analysis, Q.Y. and W.X.; resources, Q.Y., X.Y. and Z.C.; data curation, Q.Y. and J.Y.; writing—original draft preparation, Q.Y.; writing—review and editing, X.Y. and Z.C.; supervision, X.Y.; project administration, Q.Y. and X.Y.; funding acquisition, Q.Y., X.Y. and Z.C. All authors have read and agreed to the published version of the manuscript.

Funding: This research was funded by the Ministry of Science and Technology of China, grant number 2016YFD0300101; the National Natural Science Foundation of China, grant number 31560337 and China Meteorological Administration, grant number CMA-CLYBZX(2019JX07).

Conflicts of Interest: The authors have no conflict of interest to declare.

References

1. Chen, T.B.; Zhang, B.C.; Huang, Z.C.; Ru, L.Y.; Zheng, Y.M.; Lei, M.; Liao, X.Y.; Piao, S.J. Geographical distribution and characteristics Of habitat of As-hyperaccumulator *Pteris vittata* L. in China. *Geogr. Res.* **2005**, *24*, 825–833.
2. Parmesan, C. Influences of species, latitudes and methodologies on estimates of phenological response to global warming. *Glob. Chang. Biol.* **2007**, *13*, 1860–1872. [CrossRef]
3. Song, Y.; Liu, B.; Zhong, H. Impact of Global Warming on the Rice Cultivable Area in Southern China in 1961-2009. *Adv. Clim. Chang. Res.* **2011**, *7*, 259–264.
4. Martin, Y.; Van Dyck, H.; Dendoncker, N.; Titeux, N. Testing instead of assuming the importance of land use change scenarios to model species distributions under climate change. *Glob. Ecol. Biogeogr.* **2013**, *22*, 1204–1216. [CrossRef]
5. Yang, X.G.; Liu, Z.J.; Chen, F. The Possible Effect of Climate Warming on Northern Limits of Cropping System and Crop Yield in China. *Agric. Sci. China* **2011**, *10*, 585–594. [CrossRef]
6. Dong, J.; Liu, J.; Tao, F.; Xu, X.; Wang, J. Spatio-temporal changes in annual accumulated temperature in China and the effects on cropping systems, 1980s to 2000. *Clim. Res.* **2009**, *40*, 37–48. [CrossRef]
7. Bonhomme, R. Bases and limits to using 'degree. day'units. *Eur. J. Agron.* **2000**, *13*, 1–10. [CrossRef]
8. Song, Y.; Linderholm, H.W.; Chen, D.; Walther, A. Trends of the thermal growing season in China, 1951–2007. *Int. J. Climatol.* **2010**, *30*, 33–43. [CrossRef]
9. Ma, S.; An, G.; Wang, Q.; Xi, Z.; Liu, Y. Study on the variation laws of the thermal resources in maize-growing belt of northeast China. *Resour. Sci.* **2000**, *22*, 41–45.
10. Hartz, T.; Moore, F. Prediction of potato yield using temperature and insolation data. *Am. Potato J.* **1978**, *55*, 431–436. [CrossRef]
11. Huang, Y.; Gao, L.; Jin, Z.; Chen, H. Simulating the optimal growing season of rice in the Yangtze River Valley and its adjacent area, China. *Agric. For. Meteorol.* **1998**, *91*, 251–262. [CrossRef]
12. Caton, B.; Foin, T.; Gibson, K.; Hill, J. A temperature-based model of direct-, water-seeded rice (*Oryza sativa*) stand establishment in California. *Agric. For. Meteorol.* **1998**, *90*, 91–102. [CrossRef]
13. Liu, D.; Kingston, G.; Bull, T. A new technique for determining the thermal parameters of phenological development in sugarcane, including suboptimum and supra-optimum temperature regimes. *Agric. For. Meteorol.* **1998**, *90*, 119–139. [CrossRef]

14. Leong, S.; Ong, C. The influence of temperature and soil water deficit on the development and morphology of groundnut (*Arachis hypogaea* L.). *J. Exp. Bot.* **1983**, *34*, 1551–1561. [CrossRef]
15. Dufault, R.J. Determining heat unit requirements for broccoli harvest in coastal South Carolina. *J. Am. Soc. Hortic. Sci.* **1997**, *122*, 169–174. [CrossRef]
16. McMaster, G.S.; Wilhelm, W. Growing degree-days: One equation, two interpretations. *Agric. For. Meteorol.* **1997**, *87*, 291–300. [CrossRef]
17. Black, C.; Ong, C. Utilisation of light and water in tropical agriculture. *Agric. For. Meteorol.* **2000**, *104*, 25–47. [CrossRef]
18. Butler, T.J.; Evers, G.W.; Hussey, M.A.; Ringer, L.J. Flowering in crimson clover as affected by planting date. *Crop Sci.* **2002**, *42*, 242–247. [CrossRef]
19. Caliskan, S.; Caliskan, M.; Arslan, M.; Arioglu, H. Effects of sowing date and growth duration on growth and yield of groundnut in a Mediterranean-type environment in Turkey. *Field Crops Res.* **2008**, *105*, 131–140. [CrossRef]
20. Caliskan, S.; Caliskan, M.; Erturk, E.; Arslan, M.; Arioglu, H. Growth and development of Virginia type groundnut cultivars under Mediterranean conditions. *Acta Agric. Scand. Sect. B Soil Plant Sci.* **2008**, *58*, 105–113. [CrossRef]
21. Idso, S.; Hatfield, J.; Jackson, R.; Reginato, R. Grain yield prediction: Extending the stress-degree-day approach to accommodate climatic variability. *Remote Sens. Environ.* **1979**, *8*, 267–272. [CrossRef]
22. Teal, R.; Tubana, B.; Girma, K.; Freeman, K.; Arnall, D.; Walsh, O.; Raun, W. In-season prediction of corn grain yield potential using normalized difference vegetation index. *Agron. J.* **2006**, *98*, 1488–1494. [CrossRef]
23. Challinor, A.; Wheeler, T.; Craufurd, P.; Slingo, J. Simulation of the impact of high temperature stress on annual crop yields. *Agric. For. Meteorol.* **2005**, *135*, 180–189. [CrossRef]
24. Han, X.L.; Qu, M.L. *Crop Ecology*; China Meteorological Press: Beijing, China, 1991.
25. Challinor, A.; Wheeler, T.; Craufurd, P.; Slingo, J.; Grimes, D. Design and optimisation of a large-area process-based model for annual crops. *Agric. For. Meteorol.* **2004**, *124*, 99–120. [CrossRef]
26. Tao, F.; Yokozawa, M.; Zhang, Z. Modelling the impacts of weather and climate variability on crop productivity over a large area: A new process-based model development, optimization, and uncertainties analysis. *Agric. For. Meteorol.* **2009**, *149*, 831–850. [CrossRef]
27. Yu, Y.; Ge, B.; Ren, S. Study on the effectiveness of accumulated tempreature in the sub-tropical western moutain areas in China. *Meteorol. Mon.* **1991**, *17*, 21–25.
28. Ravindra, G.M.; Sridhara, S.; Girijesh, G.K.; Nanjappa, H.V. Weed biology and growth analysis of Celosia argentea L., a weed associated with groundnut and finger millet crops in southern India. *Commun. Biom. Crop Sci.* **2008**, *3*, 80–87.
29. Zhang, L.; Lou, W. Impact of climate warming on the distribution of thermal resources in the lower-middle reaches of the Changjiang River. *J. Nat. Resour.* **2013**, *28*, 1361–1372.
30. Yue, W.; Cao, W.; Yang, T.; Wu, W.; Chen, G. Variable characteristics heat resources of single-season rice growing seasons in Anhui province and its impacts on rice yield. *Chin. Agric. Sci. Bull.* **2014**, *30*, 222–228.
31. Hu, Q.; Pan, X.; Shao, C.; Zhang, D.; Wang, X.; Wei, X. Distribution and variation of China agricultural heat resources in 1961–2010. *Chin. J. Agrometeorol.* **2014**, *35*, 119–127.
32. Miao, Q.; Ding, Y.; Wang, Y.; Duan, C. Impact of climate warming on the distribution of China's thermal resources. *J. Nat. Resour.* **2009**, *24*, 934–944.
33. Baskerville, G.; Emin, P. Rapid estimation of heat accumulation from maximum and minimum temperatures. *Ecology* **1969**, *50*, 514–517. [CrossRef]
34. DeGaetano, A.T.; Knapp, W.W. Standardization of weekly growing degree day accumulations based on differences in temperature observation time and method. *Agric. For. Meteorol.* **1993**, *66*, 1–19. [CrossRef]
35. Sacks, W.J.; Kucharik, C.J. Crop management and phenology trends in the US Corn Belt: Impacts on yields, evapotranspiration and energy balance. *Agric. For. Meteorol.* **2011**, *151*, 882–894. [CrossRef]
36. Lobell, D.B.; Bänziger, M.; Magorokosho, C.; Vivek, B. Nonlinear heat effects on African maize as evidenced by historical yield trials. *Nat. Clim. Chang.* **2011**, *1*, 42–45. [CrossRef]
37. Rodríguez-Rajo, F.J.; Frenguelli, G.; Jato, M. Effect of air temperature on forecasting the start of the Betula pollen season at two contrasting sites in the south of Europe (1995–2001). *Int. J. Biometeorol.* **2003**, *47*, 117–125. [CrossRef]
38. Jiang, H.; Wen, D. Methods of calculating growing degree-day based on LR assumption and daily extreme temperatures. *J. China Agric. Univ.* **2013**, *18*, 82–87.
39. Liu, X.; Han, X. *Regionalization of Cropping System in China*; Beijing Agricultural University Press: Beijing, China, 1987.
40. Hu, Z.X. Analysis of the situation of rice production in China. *Hybrid Rice* **2009**, *24*, 1–7.
41. Xin, L.; Li, X. Changes of Multiple Cropping in Double Cropping Rice Area of Southern China and Its Policy Implications. *J. Nat. Resour.* **2009**, *24*, 58–65.
42. Ye, Q.; Yang, X.G.; Liu, Z.J.; Dai, S.W.; Li, Y.; Xie, W.J.; Chen, F. The Effects of Climate Change on the Planting Boundary and Potential Yield for Different Rice Cropping Systems in Southern China. *J. Integr. Agric.* **2014**, *13*, 1546–1554. [CrossRef]
43. Ye, Q.; Yang, X.; Dai, S.; Chen, G.; Li, Y.; Zhang, C. Effects of climate change on suitable rice cropping areas, cropping systems and crop water requirements in southern China. *Agric. Water Manag.* **2015**, *159*, 35–44. [CrossRef]
44. Allen, R.G.; Pereira, L.S.; Raes, D.; Smith, M. *Crop Evapotranspiration Guidelines for Computing Crop Water Requirements-Irrigation and Drainage Paper 56*; Food and Agriculture Organization of the United Station: Rome, Italy, 1998.

45. Li, Y.; Yang, X.G.; Dai, S.W.; Wang, W.F. Spatiotemporal change characteristics of agricultural climate resources in middle and lower reaches of Yangtze River. *Chin. J. Appl. Ecol.* **2010**, *21*, 2912–2921.

46. Gao, L.; Li, L.; Guo, P. The length of growing season and climatic and ecologic regionalization for rice in China. *Chin. J. Agrometeorol.* **1983**, *4*, 50–55.

47. Thakur, P.; Kumar, S.; Malik, J.A.; Berger, J.D.; Nayyar, H. Cold stress effects on reproductive development in grain crops: An overview. *Environ. Exp. Bot.* **2010**, *67*, 429–443. [CrossRef]

48. Shimono, H.; Okada, M.; Kanda, E.; Arakawa, I. Low temperature-induced sterility in rice: Evidence for the effects of temperature before panicle initiation. *Field Crops Res.* **2007**, *101*, 221–231. [CrossRef]

49. Gooding, M.; Ellis, R.; Shewry, P.; Schofield, J. Effects of restricted water availability and increased temperature on the grain filling, drying and quality of winter wheat. *J. Cereal Sci.* **2003**, *37*, 295–309. [CrossRef]

50. Jagadish, S.; Craufurd, P.; Wheeler, T. High temperature stress and spikelet fertility in rice (*Oryza sativa* L.). *J. Exp. Bot.* **2007**, *58*, 1627–1635. [CrossRef]

51. De Datta, S.K. *Principles and Practices of Rice Production*; International Rice Research Institute: Los BanÄos, PI, USA, 1981.

52. Yoshida, S. *Fundamentals of Rice Crop Science*; International Rice Research Institute: Los BanÄos, PI, USA, 1981.

53. Gao, L.; Jin, Z.; Huang, Y.; Zhang, L. Rice clock model—A computer model to simulate rice development. *Agric. For. Meteorol.* **1992**, *60*, 1–16. [CrossRef]

54. Liu, Y. The impact of the contemporary climatic change on China's heat resource. *J. Nat. Resour.* **1993**, *8*, 166–175.

55. Wang, F. On the variation of accumulated temperature during the last 100 years and crop yield in China. *Acta Geogr. Sin.* **1982**, *37*, 272–280.

56. Zhang, H.; Zhang, Y. Perliminary discussion on the response of active accumulated temperature of China to climate warming. *Acta Geogr. Sin.* **1994**, *49*, 27–36.

57. Dong, H.; Deng, Z. *The Utilization of Agoclimatic Resources in Multiple Cropping Area*; China Meteorological Press: Beijing, China, 1988.

58. Yin, X.; Kropff, M.J.; McLaren, G.; Visperas, R.M. A nonlinear model for crop development as a function of temperature. *Agric. For. Meteorol.* **1995**, *77*, 1–16. [CrossRef]

59. Matthews, R.B.; Hunt, L.A. GUMCAS: A model describing the growth of cassava (*Manihot esculenta* L. Crantz). *Field Crops Res.* **1994**, *36*, 69–84. [CrossRef]

60. Hodges, T. *Predicting Crop Phenology*; CRC Press: Boca Raton, FL, USA, 1990.

61. Bouman, B.A.M.; Kropff, M.J.; Tuong, T.P.; Wopereis, M.C.S.; ten Berge, H.F.M.; van Laar, H.H. *ORYZA2000: Modeling Lowland Rice*; International Rice Research Institute: Los BanÄos, PI, USA; Wageningen University and Reserarch Centre: Wageningen, The Netherlands, 2001; Volume 1.

62. Gao, L.; Li, L. *Meteorological Ecology for Rice*; China Agricultural Press: Beijing, China, 1992.

63. Allen, R.G.; Pereira, L.S.; Raes, D.; Smith, M. *Crop Evapotranspiration: Guidelines for Computing Crop Water Requirements*; FAO Irrigation and Drainage Papers; FAO: Rome, Italy, 1998; Volume 56, p. 300.

64. Liu, Q.; Zhong, Z. The effects of water resources supply on clonal growth in Pleioblastus Maculata population. *Acta Phytoecol. Sin.* **1996**, *20*, 245–254.

65. Ye, Q.; Yang, X.G.; Dai, S.W.; Li, Y.; Guo, J.P. Variation Characteristics of Hydrothermal Resources Effectiveness Under the Background of Climate Change in Southern Rice Production Area of China. *J. Integr. Agric.* **2013**, *12*, 2260–2279. [CrossRef]

66. Hou, G.L.; Liu, Y.F. Climate potential productivity and regionalization in China. *J. Nat. Resour.* **1985**, *5*, 60–65.

67. Huang, B.W. Chinese agricultural potential productivity-photosynthetic potential productivity. *Ann. Geogr.* **1985**, *17*, 15–22.

68. Jiang, X.J.; Tang, L.; Liu, X.J.; Cao, W.X.; Zhu, Y. Spatial and Temporal Characteristics of Rice Potential Productivity and Potential Yield Increment in Main Production Regions of China. *J. Integr. Agric.* **2013**, *12*, 45–56. [CrossRef]

69. Yu, H.N.; Zhao, F.S. On the light and thermal resources and the crop potential productivity-taking Luancheng county of Hebei province as an example. *Acta Meteorol. Sin.* **1982**, *40*, 327–334.

70. Guo, J.P.; Gao, S.H.; Pan, Y.R. Agroclimatic Potentiality Development Application and Countermeasures in Northeast China. *Meteorol. Mon.* **1995**, *21*, 3–9.

71. Zhang, L.; Li, S.; Tan, F.; Guo, A.; Huo, Z. Potential agro-thermal resources dynamic for double-season rice cultivation across China under greenhouse gas emission scenarios. *Theor. Appl. Climatol.* **2021**, *144*, 67–75. [CrossRef]

72. Liu, M.; Liu, A.; Deng, A.; Wang, S.; Liu, Z. Changing characteristics of heat resources of rice growing seasons in Hubei Province and its impacts on rice production. *J. HuaZhong Agric. Univ.* **2011**, *30*, 746–752.

73. Ye, Q.; Yang, X.G.; Xie, W.j.; Li, Y.; Liu, Z.Q.; Dong, C.Y.; Sun, S. Tendency of use efficiency of rice growth season in southern China under the background of global warming. *Sci. Agric. Sin.* **2013**, *46*, 4399–4415.

74. Lai, C.; Qian, H.; Duan, H.; Song, Q.; Yu, F.; Zhang, Y.; Zhang, J. Climate suitability of wheat-rice double cropping system in Huaihe watershed. *Sci. Agric. Sin.* **2011**, *44*, 2868–2875.

75. Jiang, X.; Zhang, J.; Gao, J.; Chen, P.; Zang, Q.; Jiang, M. Characteristics of heat resources during crop growth season in Shenyang region, Liaoning province. *J. Meteorol. Environ.* **2011**, *27*, 19–24.

76. Ye, Q.; Yang, X.; Li, Y.; Dai, S.; Xiao, J. Changes of China agricultural climate resources under the background of climate change. VIII. Change characteristics of heat resources during the growth period of double cropping rice in Jiangxi Province. *Chin. J. Appl. Ecol.* **2011**, *22*, 2021–2030.

77. Cui, D.C. The scenario analyses of possible effect of warming climate on rice growing period. *Q. J. Appl. Meteorol.* **1995**, *6*, 361–365.
78. Jiang, M.; Jin, Z.Q.; Shi, C.L.; Ge, D.K.; Zhuo, D.W. Occurrence patterns of high temperature at booting and flowering stages of rice in the middle and lower reaches of Yangtze River and their impacts on rice yield. *Chin. J. Ecol.* **2010**, *29*, 649–656.
79. Gao, R.; Wang, L.; Gao, G. The trend of variation in high temperature days during 1956–2006 in China. *Adv. Clim. Chang. Res.* **2008**, *4*, 177–181.

Article

Assessing Functionality of Alternative Sweeteners in Rolled "Sugar" Cookies

Melanie L. Heermann [1], Janae Brown [1], Kelly J. K. Getty [1,2,*] and Umut Yucel [1,2]

[1] Food Science Institute, Kansas State University, 216 Call Hall, 1530 Mid-Campus Drive N, Manhattan, KS 66506, USA; mheermann95@gmail.com (M.L.H.); jeb98@ksu.edu (J.B.); yucel@ksu.edu (U.Y.)

[2] Department of Animal Sciences and Industry, Kansas State University, 232 Weber Hall, 1424 Claflin Road, Manhattan, KS 66506, USA

* Correspondence: kgetty@ksu.edu; Tel.: +1-(785)-532-2203

Abstract: Sucrose contributes to the key physical and sensory characteristics of cookies. Due to the negative health effects associated with excess sucrose consumption, the replacement of sucrose in baking applications is of interest. In this study, nine variations of rolled cookies were prepared ($n = 3$) using a sucrose control (C), Splenda for baking (SB), Equal for baking (EB), Truvia (TR), Sweet'N Low (SNL), and 1:1 (wt%) mixtures of sweeteners and sucrose (S). The cookies were characterized by a width-to-thickness (W/T) ratio, moisture loss, color, hardness, and fracturability. The W/T ratios of TR (5.7) and TR + sucrose (6.6) were similar, the closest to C (7.7), and bigger than ($p < 0.05$) all other treatments. Color was not affected ($p > 0.05$) by the sugar type or concentration. C showed the greatest hardness (5268 N), and SNL had the greatest fracturability (8667 N). Overall, regarding physiochemical characteristics, TR + sucrose (1:1 replacement) and SB (100% replacement) were the closest to the control.

Keywords: alternative sweeteners; sucrose; cookies; baking; sugar reduction

Citation: Heermann, M.L.; Brown, J.; Getty, K.J.K.; Yucel, U. Assessing Functionality of Alternative Sweeteners in Rolled "Sugar" Cookies. *Processes* **2022**, *10*, 868. https://doi.org/10.3390/pr10050868

Academic Editor: Ofelia Anjos

Received: 7 April 2022
Accepted: 26 April 2022
Published: 28 April 2022

Publisher's Note: MDPI stays neutral with regard to jurisdictional claims in published maps and institutional affiliations.

1. Introduction

Sugar reduction is a current challenge being addressed in the food industry [1]. In 2015, the World Health Organization recommended that the daily consumption of free sugars be reduced to less than 10% of total energy intake to reduce the risk of obesity and dental decay [2]. Current trends show that the average consumption of added sugars is above 10%, except for in the elderly and infants [3]. It is projected that, by the year 2030, 1.12 billion people around the world will be obese if current trends are sustained [4].

As of January 1, 2020, the Food and Drug administration (FDA) requires added sugars to be included in the nutrition facts panel. As defined by the FDA, "added sugars are either added during the processing of foods, or are packaged as such, and include sugars (free, mono and disaccharides), sugars from syrups and honey, and sugars from concentrated fruit or vegetable juices" [5]. Therefore, many consumers are looking to lower the sugar content of their food [6]. In 2021, the International Food Information Council (IFIC) conducted a Food and Health survey that found that 72% of consumers are attempting to reduce their sugar intake [7].

Reducing the sugar content of bakery products presents challenges related to the functionalities provided by sucrose, such as the sensory attributes related to sweetness and mouthfeel, and the physical attributes related to the texture, color, and processing parameters of the dough [1]. Cookies typically have a high sugar content comprising 30–40% of the formula. Sugar is a key ingredient in rolled cookie formulations, with sucrose commonly being used in traditional recipes [8]. Sucrose granules in cookie dough dissolve with the heat of baking. This fluidity causes the cookies to spread in the oven, with higher concentrations of sucrose correlating with a greater spread [9]. Cookies expand in the oven, followed by a structural collapse and the recrystallization of dissolved sucrose upon drying and cooling. These phenomena are believed to give sugar snap cookies their characteristic cracked appearance on the top [9,10]. Sucrose also contributes to the formation of a golden brown color through browning reactions [11]. These characteristics determine the final product quality of sugar-snap cookies, including their crisp texture and storage stability [10].

The commercially relevant non-caloric sweeteners used as sugar alternatives include sucralose, aspartame, acesulfame potassium (Ace-K), saccharine, and plant-based sweeteners (such as stevia) [1]. Maltodextrin, sugar alcohols, and fibers can be used as bulking agents in the preparation of high-intensity sweeteners as a method to reduce sugar in baked goods [12]. In a study by Ho and Pulsawat, 2020, sugar was partially replaced at the 50% level with either maltitol, sorbitol, or isomalt [13]. It was noted that all treatments could reduce sugar and daily calorie intake, with maltitol and isomalt yielding similar sensory attributes to the control. In a review by Luo et al., 2019, maltitol served as a single substitute for sucrose in baked goods [14]. Natural products also have the benefit of serving as bulking agents or providing sweetness in baked goods. Dates may be added to baked goods to provide both sweetness and fiber as a bulking agent [15]. Apple pomace flour (APF) has been studied as a product high in dietary fibers that could be added to cookies with an acceptable flavor and texture profile. Therefore, APF may serve as a fiber/bulking agent for a low-sugar cookie [16]. Another natural approach to incorporating bulking agents in low-sugar cookies could be through the addition of high fiber flours. In 2021, Pavičić et al. found that the inclusion of carob, oat, or rye flours reduced the spread of 3D printed cookies with olive oil [17].

It is suggested that non-nutritive sweeteners in combination with polyols and bulking agents can be used to reduce sugar in baked goods. More research is needed to determine the minimum sugar level and/or sugar substitution levels to maintain similar functional properties in baked goods [14]. Therefore, the objective of our study was to determine the effect of different alternative sweeteners that contained sugar alcohols or bulking agents at 100% and 50% replacement on the physiochemical properties of rolled "sugar" cookies.

2. Materials and Methods

2.1. Experimental Design

The experiment consisted of nine treatments: sucrose control (C), Splenda for baking (SB), Equal for baking (EB), Truvia (TR), Sweet'N Low (SNL), and mixtures of Splenda for baking and sucrose (SB + S), Equal for baking and sucrose (EB + S), Truvia and sucrose (TR + S), and SNL and sucrose (SNL + S). The ingredients in each commercially available artificial sweetener are given in Tables 1 and 2. The individual and mixture amounts were determined based on relative sweetness values. The Baker's Percentages are given in Tables 1 and 2. The experiment was performed in triplicate ($n = 3$).

Alternative sweeteners were chosen from commercially available sources. Splenda for baking is made up of sucralose, a popular sugar alternative in baked goods, and maltodextrin, a bulking agent. Equal for baking includes aspartame, acesulfame potassium (Ace-K), and maltodextrin for bulking. Sweet'N Low for Baking was selected as aspartame is a popular sugar-free sweetener for beverages, but it is not heat stable and, therefore, has limited applications in baked foods. To improve functionality in baked goods, maltodextrin is added for bulking and acesulfame potassium is added to assist with heat stability. Truvia, composed of stevia and erythritol, is an all-natural sugar alternative with a sugar alcohol added for functional benefits. Sweet'N Low is made up of saccharin and nutritive dextrose as the bulking agent.

Table 1. Formulation of cookies with 100% sucrose replacement based on relative sweetness level.

Ingredient and Supplier	Baker's Percent (Flour Weight Basis)				
	Control (C)	Splenda for Baking (SB)	Equal for Baking (EB)	Truvia (TR)	Sweet'N Low (SNL)
All-Purpose Flour (bleached wheat flour, malted barley flour, niacin, iron (reduced), thiamine mononitrate, riboflavin, folic acid) (Great Value, Walt-Mart Stores, Inc., Bentonville, AR, USA)	100	100	100	100	100
Shortening (soybean oil, hydrogenated palm oil, palm oil, mono and diglycerides, TBHQ, citric acid) (The Kroger Co., Cincinnati, OH, USA)	28.4	28.4	28.4	28.4	28.4
Sugar (Great Value, Wal-Mart Stores, Inc., Bentonville, AR, USA)	57.8	-	-	-	-
Splenda for Baking (maltodextrin, sucralose) (McNeil Nutritionals LLC, Fort Washington, PA, USA)	-	7.2	-	-	-
Equal for Baking (maltodextrin, aspartame, acesulfame potassium) (Merisant US, Inc., Chicago, IL, USA)	-	-	6.8	-	-
Truvia (erythritol, stevia leaf extract, natural flavors) (Cargill, Inc., Minneapolis, MN, USA)	-	-	-	26.7	-
Sweet'N Low (nutritive dextrose, saccharin, cream of tartar, calcium silicate) (Cumberland Packing Corp., Brooklyn, NY, USA)	-	-	-	-	7.1
Iodized Salt (salt, calcium silicate, dextrose, potassium iodide) The Kroger Co., Cincinnati, OH, USA	0.9	0.9	0.9	0.9	0.9
Sodium Bicarbonate (Arm & Hammer, Church & Dwight, Co., Inc., Ewing, NJ, USA)	1.1	1.1	1.1	1.1	1.1
Dextrose Solution (8.9 g dextrose in 150 mL water) (STALEYDEX 333, Tate & Lyle, Decatur, IL, USA)	14.7	14.7	14.7	14.7	14.7
Water (high concentration) Municipal water, Manhattan, KS, USA	-	28.4	28.4	28.4	28.4
Water (low concentration) Municipal water, Manhattan, KS, USA	7.1	-	-	-	-

Table 2. Formulation of cookies with 1:1 sucrose to sweetener based on relative sweetness level.

Ingredient and Supplier	Baker's Percent				
	Control (C)	Splenda for Baking + Sucrose (SB + S)	Equal for Baking + Sucrose (EB + S)	Truvia + Sucrose (TR + S)	Sweet'N Low + Sucrose (SNL + S)
All-Purpose Flour (bleached wheat flour, malted barley flour, niacin, iron (reduced), thiamine mononitrate, riboflavin, folic acid) (Great Value, Walt-Mart Stores, Inc., Bentonville, AR, USA)	100	100	100	100	100
Shortening (soybean oil, hydrogenated palm oil, palm oil, mono and diglycerides, TBHQ, citric acid) (The Kroger Co., Cincinnati, OH, USA)	28.4	28.4	28.4	28.4	28.4
Sugar (Great Value, Wal-Mart Stores, Inc., Bentonville, AR, USA)	57.8	28.9	28.9	28.9	28.9
Splenda for Baking (maltodextrin, sucralose) (McNeil Nutritionals LLC, Fort Washington, PA, USA)	-	3.6	-	-	-
Equal for Baking (maltodextrin, aspartame, acesulfame potassium) (Merisant US, Inc., Chicago, IL, USA)	-	-	3.4	-	-
Truvia (erythritol, stevia leaf extract, natural flavors) (Cargill, Inc., Minneapolis, MN, USA)	-	-	-	13.3	-
Sweet'N Low (nutritive dextrose, saccharin, cream of tartar, calcium silicate) (Cumberland Packing Corp., Brooklyn, NY, USA)	-	-	-	-	3.6
Iodized Salt (salt, calcium silicate, dextrose, potassium iodide) The Kroger Co., Cincinnati, OH, USA	0.9	0.9	0.9	0.9	0.9
Sodium Bicarbonate (Arm & Hammer, Church & Dwight, Co., Inc., Ewing, NJ, USA)	1.1	1.1	1.1	1.1	1.1
Dextrose Solution (8.9 g dextrose in 150 mL water) (STALEYDEX 333, Tate & Lyle, Decatur, IL, USA)	14.7	14.7	14.7	14.7	14.7
Water (high concentration) Municipal water, Manhattan, KS, USA	-	28.4	28.4	28.4	28.4
Water (low concentration) Municipal water, Manhattan, KS, USA	7.1	-	-	-	-

2.2. Materials

All ingredients were purchased from a local store. Procedures and formulations were adapted from the American Association of Cereal Chemists (AACC) Method 10-50.05: Baking Quality of Cookie Flour (AACC International, 1999). Ingredients, suppliers, and formulations can be found below (Tables 1 and 2). Two water concentrations were used in the cookie formulations. A low concentration was used for the control, resulting in a non-sticky dough. Extra water was required for the alternative sweetener treatments to allow for the desired consistency for the rolling of the modified cookie dough. Without the extra addition of water, alternative sweetener formulations formed crumbly mixtures that could not be formed into cookies.

2.3. Cookie Preparation

The equipment needed to prepare the cookies included scales (Ohaus, Parsippany, NJ, USA), weigh boats, spoons, disposable pipettes, spatulas, a stand mixer with a flat paddle attachment (KitchenAid Artisan, Model No. KSM150PSWH, St. Joseph, MI, USA), a rolling pin, a 7.5 cm circular cookie cutter, 7 mm-diameter wooden dowel rods, parchment paper (Reynolds Consumer Products LLC, Lake Forest, IL, USA), plastic wrap (Handi-Film, Handi-foil of America Inc., Wheeling, IL, USA), baking sheets (The Vollrath Co., L.L.C., Sheboygan, WI; USA PAN, Crescent, PA, USA), ovens (Whirlpool Model RF367LXSB, Benton Harbor, MI, USA), cooling racks, and low-density polyethylene (LDPE) bags (Ziploc, SC Johnson, Racine, WI, USA).

The analytical equipment needed to collect data included digital calipers (Tool Shop, Eau Claire, WI, USA), standard rulers with cm/mm, a portable HunterLab MiniScan colorimeter (Model No. 4500L, Reston, VA, USA), a Texture Technologies Corp texture analyzer with a 30 kg load cell and a flat blade attachment (TA-XT2, Scarsdale, NY, USA), and Exponent Stable Micro Systems texture analysis software (Godalming, UK).

Shortening, sweetener(s), salt, and sodium bicarbonate were creamed in a stand mixer (KitchenAid Artisan, Model No. KSM150PSWH, St. Joseph, MI) for 3 min on Speed 2, pausing to scrape the sides of the bowl every 1 min. Dextrose solution and water were added and mixed on Speed 1 for 1 min and then for 1 min more on Speed 4, stopping in between to scrape the bowl. Flour was then added and mixed for 2 min on Speed 2, scraping the bowl every 30 s.

The dough was removed from the bowl and placed between two pieces of plastic wrap. Two 7 mm-diameter wooden dowel rods were taped to a clean countertop 18 cm apart so that the entire width of the rolling pin rested on top of them. The dough was rolled to a 7 mm thickness, using the dowel rods as guides. The cookies were cut using a 7.5 cm circular cookie cutter. The dough was rerolled up to two times to obtain as many cookies as possible from the dough to prevent overworking. The cookies were baked for 10 min in a preheated oven at 204.4 °C. The cookies were allowed to cool at room temperature approximately 30 min before they were placed in LDPE bags and frozen at −20 °C overnight so that all samples could be analyzed at the same time. The cookies were allowed to thaw approximately 1 h before analyzing.

2.4. Physical Analyses

2.4.1. Spread

The cookie spread was measured using a method adapted from the AACC Method 10-50.05: Baking Quality of Cookie Flour (AACC International, 1999). Three cookies were placed edge to edge, and the width was measured to the nearest 0.1 mm using digital calipers (Tool Shop, Eau Claire, WI, USA). The cookies were then rotated 90°, and the width was measured again. The average width was calculated. The same three cookies were then stacked on top of one another, and the thickness (height) of the stack was measured in mm. The stack of cookies was shuffled, and the thickness was measured again. The average thickness was calculated from the two thickness values. Then, the average width and average thickness were used to calculate the width-to-thickness (W/T) ratio (Equation (1)).

$$W/T = \frac{\text{Average width of 3 cookies}}{\text{Average thickness of 3 cookies}} \tag{1}$$

2.4.2. Moisture Loss

Moisture loss was determined using a method adapted from the AACC Method 44-01.01: Calculation of Percent Moisture (AACC International, 1999) to ensure that the amount of water added to the raw dough in all treatments did not significantly affect the final cooked dough product. The baking sheet was first tared on a scale (Ohaus I-20W Model B10AS, Parsippany, NJ, USA). The cut cookies were placed on the baking sheet, and their weight was recorded. After baking, the cookies were allowed to cool slightly before their final weight was recorded. These values were used to calculate the percent moisture loss that occurred during baking (Equation (2)).

$$\text{Percent Moisture loss} = \frac{\text{Weight raw cookies} - \text{Weight baked cookies}}{\text{Weight raw cookies}} \times 100 \tag{2}$$

2.4.3. Color

A portable HunterLab MiniScan colorimeter (Model No. 4500L, Reston, VA, USA) was calibrated using the standardized methods provided by Hunter Associates Laboratory, Inc., (Reston, VA, USA) using a black glass and white tile. Using the colorimeter, the top L*a*b* color of one cookie from each replication was collected. Using the collected L*a*b* values, the total color change (ΔE) was calculated (Equation (3)).

$$\Delta E = \sqrt{\left(L^{*2}a^{*2}b^{*2}\right)} \tag{3}$$

2.4.4. Texture Profile Analysis (TPA)

Texture profile analysis was performed on two cookies per treatment from each replication. The cookies were analyzed for their hardness and three-point fracturability with a calibrated Texture Technologies Corp texture analyzer with a 1 kg load cell and a flat blade attachment (TA-XT2, Scarsdale, NY, USA) and Exponent Stable Micro Systems texture analysis software (Godalming, UK) by following the method settings suggested by the manufacturer (Texture Analysis Application Areas, n.d.). A flat blade attachment was attached to the texture analyzer, and a probe was set to 25% strain. The cookies were placed on a support with a separation of 20 cm. Two compressions of the cookies were performed consecutively. Hardness was measured as maximum force at the first compression, and snapping force was measured as the amount of force required to fracture the cookie.

2.4.5. Nutrition Profile Calculations

For each formulation, Genesis R&D Food Development and Labeling Software from ESHA Research (Salem, OR, USA) was used to estimate the nutrition profile of the cookie treatments. Moisture loss during baking for each treatment was inputted into Genesis. Nutrition facts panels were generated for each formulation, as well as calories (kCal), protein, fat, carbohydrates, and added sugar per 100 g of cookie. A 100 g portion was used to allow comparison among treatments, as treatments varied in average cookie weights after baking.

2.4.6. Student Cookie Attribute Scoring

To understand the trends of the attributes, a group of 60 undergraduate students scored cookies on appearance, texture, flavor, sweetness, and aftertaste. Whole cookies from each treatment were placed on white plates for students to observe the appearance of each treatment. Students then consumed a small piece of one cookie from each treatment and rated each characteristic from 1 to 9 (1 = dislike extremely, 9 = like extremely). Students were provided water between sampling treatments. The samples were not blinded, meaning students knew which treatments they were rating. This may have contributed to some bias in the results but provides trends on cookie treatment attributes.

2.4.7. Statistical Analysis

Data were compiled, and the effects of the treatments were analyzed for their significance using one-way ANOVA followed by Tukey's multiple comparison test. An alpha level of 0.05 was used for all statistical analyses. All the statistical analyses were performed using Minitab 18[©] Software (State College, PA, USA).

3. Results and Discussion

3.1. Appearance

The type of sweetener affected the visual appearance of the cookies (Figure 1). The control cookies had a cracked and flat top typical for sugar-snap cookies. Sucrose crystallization causes the cookie's surface to dry and break as the cookie continues to expand during baking, producing the typical, cracked appearance of sugar cookies after baking [18]. The cookie formulations with alternative sweeteners yielded cookies with uneven, uncracked top surfaces after baking, and the cookies appeared thicker, except for treatment TR + S, which produced cookies with a more even surface. The uneven surface appearance was probably related to the higher initial moisture content of the alternative sweetener doughs compared to the control. The surface moisture lost during baking is replaced by the water that diffuses from the center of the cookie [18]. However, sucrose is able to recrystallize at the cookie's surface and no longer holds moisture at the cookie's surface. Alternative sweeteners are not able to recrystallize and continue to hold water on the cookie's surface, producing cookies without cracked surfaces. Indeed, at a similar moisture content, the presence of sucrose helped to obtain the desired surface appearance (samples e–f in Figure 1).

3.2. Spread

Cookie spread was measured using the W/T ratio. The treatment with the largest W/T ratio, and thus the largest spread, was C (7.7), followed by TR + S (6.6) and TR (5.7) (Table 3). C was bigger ($p < 0.05$) than all other treatments, while TR and TR + S were statistically similar and the closest to C, but they were significantly different from all the other treatments. This was probably related to the presence of sugar alcohol, erythritol, as a bulking agent in TR.

Table 3. Means and standard deviations for width-to-thickness ratios of each treatment.

Factor [1]	Controlled Mean
C	7.7 ± 0.5 [a]
SB	4.5 ± 0.3 [c]
EB	4.3 ± 0.4 [c]
TR	5.7 ± 0.4 [b]
SNL	4.2 ± 0.2 [c]
SB + S	4.5 ± 0.4 [c]
EB + S	4.5 ± 0.2 [c]
TR + S	6.6 ± 0.2 [b]
SNL + S	4.4 ± 0.2 [c]

[a–c] Tukey pairwise comparisons appear as superscripts by each mean. Means within a column that share a letter are not significantly different ($p > 0.05$). [1] C = control; SB = Splenda for baking; EB = Equal for baking; TR = Truvia; SNL = Sweet'N Low; SB + S = Splenda for baking + sucrose; EB + S = Equal for baking + sucrose; TR + S = Truvia + sucrose; SNL + S = Sweet'N Low + sucrose.

Figure 1. Top images of cookies from each treatment: control (**a**), Splenda for baking (**b**), Equal for baking (**c**), Truvia (**d**), Sweet'N Low (**e**), Splenda + sucrose (**f**), Equal + sucrose (**g**), Truvia + sucrose (**h**), Sweet'N Low + sucrose (**i**).

As the sucrose content in cookies increases, so does the diameter [8,9,19]. Sucrose slowly dissolves throughout the baking time, leading to an improved spread compared to sweeteners that are fully dissolved prior to baking [9]. The crystals of some alternative sweeteners do not share sucrose's ability to spread cookies during the baking process [20]. As the amount of dietetic sweeteners (mannitol and sorbitol) in cookies increases, the width of the cookies decreases [21].

3.3. Moisture Loss

Moisture loss among all treatments was not significant ($p > 0.05$). However, because alternative sweetener doughs required more water, the final moisture content was higher for all artificial sweetener treatments (Table 4). The moisture loss ranged from 5.34 to 6.58%.

Table 4. Means and standard deviations for percent moisture loss of each treatment.

Factor [1]	Controlled Mean
C	6.55 ± 0.27 [a]
SB	6.05 ± 0.32 [a]
EB	5.92 ± 0.99 [a]
TR	5.34 ± 0.32 [a]
SNL	5.51 ± 0.39 [a]
SB + S	6.58 ± 0.63 [a]
EB + S	6.28 ± 0.41 [a]
TR + S	6.58 ± 0.65 [a]
SNL + S	6.81 ± 0.77 [a]

[a] Tukey pairwise comparisons appear as superscripts by each mean. Means within a column that share a letter are not significantly different ($p > 0.05$). [1] C = control; SB = Splenda for baking; EB = Equal for baking; TR = Truvia; SNL = Sweet'N Low; SB + S = Splenda for baking + sucrose; EB + S = Equal for baking + sucrose; TR + S = Truvia + sucrose; SNL + S = Sweet'N Low + sucrose.

3.4. Color

Treatments C (80.842) and SNL + S (83.208) provided the only samples that had a difference in ΔE that was statistically significant ($p < 0.05$) (Table 5). Sucrose participates in the Maillard browning reaction by breaking down into reducing sugars fructose and glucose and reacting with amino acids present during baking [11,22]. Treatment SNL + S contains dextrose, which participates in browning along with the additional sucrose. The other treatments contain either maltodextrin or sugar alcohols as bulking agents that do not participate in browning.

Table 5. L*a*b*, ΔE, and standard deviations for each treatment.

Factor [1]	L*	a*	b*	ΔE
C	76.160 ± 1.690	4.383 ± 1.865	26.607 ± 1.188	80.842 ± 0.854 [b]
SB	77.927 ± 0.567	0.080 ± 0.525	24.623 ± 1.919	81.738 ± 1.022 [a,b]
EB	78.283 ± 0.045	0.257 ± 0.297	24.817 ± 0.641	82.125 ± 0.165 [a,b]
TR	77.757 ± 2.134	1.757 ± 0.272	25.737 ± 1.831	81.948 ± 1.449 [a,b]
SNL	77.773 ± 0.665	1.353 ± 0.140	25.340 ± 1.140	81.815 ± 0.531 [a,b]
SB + S	78.440 ± 0.857	0.150 ± 0.464	26.380 ± 0.427	82.759 ± 0.905 [a,b]
EB + S	78.930 ± 0.860	0.113 ± 0.356	25.360 ± 0.646	82.907 ± 0.766 [a,b]
TR + S	76.470 ± 1.469	4.080 ± 0.870	28.770 ± 0.944	81.816 ± 1.001 [a,b]
SNL + S	78.627 ± 0.788	0.837 ± 0.605	27.170 ± 1.943	83.208 ± 0.939 [a]

[a, b] Tukey pairwise comparisons appear as superscripts by each mean. Means within a column that share a letter are not significantly different ($p > 0.05$). [1] C = control; SB = Splenda for baking; EB = Equal for baking; TR = Truvia; SNL = Sweet'N Low; SB + S = Splenda for baking + sucrose; EB + S = Equal for baking + sucrose; TR + S = Truvia + sucrose; SNL + S = Sweet'N Low + sucrose

3.5. Texture Profile Analysis

The hardness of C (5268 N) was significantly higher ($p < 0.05$) than that of the other treatments (1055–3140 N), likely due to the final moisture content (Table 6). This is consistent with the results from previous studies, where the replacement of sucrose with erythritol in cookies resulted in no significant difference in hardness [23]. In a similar study, it was observed that a 50% replacement of sucrose with sugar alcohols (maltitol, sorbitol, or isomalt) produced cookies with similar textural properties to a control cookie [13]. Harder cookies require less force to fracture because they lack the flexibility of soft cookies. SNL had the greatest fracturability value (8667 N), and TR + S had the lowest fracturability value (2558 N) (Table 6). Regarding three-point fracturability, all partial sucrose replacement treatments (SB + S, EB + S, TR + S, and SNL + S) were statistically similar to C, SB, and TR. Of the total sucrose replacement treatments, only the three-point fracturability values for EB and SNL were significantly different from those of C. It could be hypothesized that these treatments are softer because of the level of maltodextrin or dextrose in the commercial alternative sweeteners used. In 2003, Gallagher et al. found that a relatively low replacement of sugar with raftilose®, an oligosaccharide, resulted in significantly softer cookies [24].

Table 6. Means and standard deviations for hardness and three-point fracturability of each treatment.

Factor [1]	Hardness (N)	Three-Point Fracturability (N)
C	5268 ± 1892 [a]	3561 ± 1351 [c]
SB	1649 ± 226 [b]	5342 ± 1099 [b,c]
EB	1665 ± 208 [b]	7802 ± 441 [a,b]
TR	3140 ± 525 [a,b]	2699 ± 524 [c]
SNL	2058 ± 186 [b]	8667 ± 2345 [a]
SB + S	1093.6 ± 89.6 [b]	3022 ± 819 [c]
EB + S	1055 ± 175 [b]	3193 ± 980 [c]
TR + S	2800 ± 621 [b]	2558 ± 272 [c]
SNL + S	1296.6 ± 79.8 [b]	4429 ± 639 [c]

[a–c] Tukey pairwise comparisons appear as superscripts by each mean. Means within a column that share a letter are not significantly different ($p > 0.05$). [1] C = control; SB = Splenda for baking; EB = Equal for baking; TR = Truvia; SNL = Sweet'N Low; SB + S = Splenda for baking + sucrose; EB + S = Equal for baking + sucrose; TR + S = Truvia + sucrose; SNL + S = Sweet'N Low + sucrose.

Traditional sugar-snap cookies made with sucrose have a characteristic snap when broken rather than bending. This snap occurs because of the recrystallization of sucrose during and after baking [18]. Additionally, the inclusion of erythritol in Truvia aided in giving cookies from treatment TR a harder texture most similar to that of C. Erythritol tends to crystallize, creating a harder texture for the final product [23]. Alternative sweeteners disrupt the recrystallization of sucrose, making softer cookies that are more difficult to fracture [18].

3.6. Nutritional Profile

Cookies from treatment C had the highest calorie amount of 410 kCal per 100 g, whereas cookies from treatments TR, SNL, and TR + S had the least calories at 350 kCal per 100 g (Table 7). Calories mostly remained consistent between the alternative sweetener types, irrespective of whether sucrose was partially or totally replaced. Regardless of the classification as a starch or a sugar, digestible carbohydrates have an energy value of 4 kCal per gram.

Table 7. Summary of nutrition profile information calculated using Genesis.

Factor [1]	Calories (kCal/100 g)	Fat (g/100 g)	Protein (g/100 g)	Carbohydrates (g/100 g)	Added Sugar (g/100 g)
C	410	14	5	65	28
SB	370	17	6	47	0
EB	370	17	6	47	3
TR	350	17	6	56	0
SNL	350	17	6	47	4
SB + S	370	15	5	54	15
EB + S	370	15	5	54	16
TR + S	350	14	5	55	14
SNL + S	370	15	5	54	16

[1] C = control; SB = Splenda for baking; EB = Equal for baking; TR = Truvia; SNL = Sweet'N Low; SB + S = Splenda for baking + sucrose; EB + S = Equal for baking + sucrose; TR + S = Truvia + sucrose; SNL + S = Sweet'N Low + sucrose.

Added sugar was the lowest for the cookies with total sucrose replacement, ranging from 0 to 4 g per 100 g. Partial sucrose replacement resulted in cookies with 14–16 g per 100 g of sugar. Total sucrose replacement decreased the average added sugar by 94%, and partial sucrose replacement decreased the average added sugar by 46%. Added sugar was more affected by sucrose level than carbohydrates. Total sucrose replacement resulted in an average carbohydrate decrease of 24%. Partial sucrose replacement resulted in an average carbohydrate decrease of 17%. Carbohydrate amount was less affected by sucrose replacement than added sugar because many alternative sweeteners contain bulking agents that contain carbohydrates. The alternative sweeteners used in SB and EB contain maltodextrin. SB, EB, and SNL contain dextrose. All treatment cookies contain some dextrose from the cookie formulation. While maltodextrin and dextrose contain no sugar, these ingredients do contribute to carbohydrate content. Protein and fat content remained very similar between treatments.

The Dietary Guidelines for Americans, 2020–2025, recommend no more than 50 g of added sugar per day for an individual on a 2000 kCal diet [25]. The Reference Amount Customarily Consumed (RACC) for cookies is equal to one 30 g cookie [26]. Our control cookie has approximately 8.4 g of added sugar. Replacing all sucrose with Splenda, Equal, Truvia, or Sweet-N Low yields cookies with 0 g, 0.84 g, 0 g, and 1.12 g of added sugar, respectively. Partial sucrose replacement yields a range of 3.9 g–4.4 g of added sugar. Therefore, consuming one cookie with total or partial sucrose replacement a day would more easily allow individuals to remain within the given dietary guidelines with respect to added sugar.

3.7. Student Cookie Attribute Scoring

Cookies from treatment C had the overall highest rating for each category: appearance, texture, flavor, sweetness, and aftertaste (6.89, 6.32, 6.57, 6.43, and 6.07, respectively) (Table 8). Treatment TR was rated the lowest of all treatments for appearance (2.84) and texture (3.30). Treatment EB was rated the lowest for all treatments for flavor (3.79) and sweetness (2.62). C and TR + S had similar aftertastes ($p > 0.05$), and SNL was rated the lowest for aftertaste (3.02). No treatment was rated above a seven in any category on the hedonic scale. However, the AACC Method 10-50.05 is optimized for the physical and chemical analyses of cookies, not for flavor or sensory acceptability. A different formulation should be considered to yield greater ratings. However, trends in the attributes of cookies made with alternative sugars can still be analyzed using this formulation.

Table 8. Means and standard deviations for hedonic ratings (out of nine) of sensory attributes for each treatment.

Factor [1]	Mean Appearance Rating	Mean Texture Rating	Mean Flavor Rating	Mean Sweetness Rating	Mean Aftertaste Rating
C	6.89 ± 1.42 [a]	6.32 ± 1.60 [a]	6.57 ± 1.54 [a]	6.43 ± 1.68 [a]	6.07 ± 1.59 [a]
SB	5.08 ± 1.60 [c,d]	5.54 ± 1.85 [a,b,c]	4.32 ± 1.81 [b,c,d]	4.57 ± 2.23 [c,d,e]	4.19 ± 2.02 [b,c]
EB	4.95 ± 1.66 [c,d]	4.61 ± 1.78 [c]	3.79 ± 1.78 [d]	2.62 ± 1.59 [f]	3.52 ± 1.50 [c,d]
TR	2.84 ± 1.51 [e]	3.30 ± 1.38 [d]	4.03 ± 1.95 [d]	4.33 ± 2.04 [d,e]	3.84 ± 1.86 [b,c,d]
SNL	4.35 ± 1.56 [d]	4.87 ± 2.04 [c]	3.86 ± 1.75 [d]	3.83 ± 1.92 [e]	3.02 ± 1.56 [d]
SB + S	5.46 ± 1.47 [b,c]	5.38 ± 2.05 [a,b,c]	5.12 ± 1.81 [b,c]	4.98 ± 1.67 [c,d]	4.75 ± 1.99 [b]
EB + S	5.77 ± 1.58 [b,c]	4.93 ± 1.64 [b,c]	4.13 ± 1.92 [c,d]	3.95 ± 1.95 [e]	4.36 ± 1.94 [b,c]
TR + S	5.41 ± 2.12 [c]	5.92 ± 1.71 [a,b]	6.43 ± 1.53 [a]	6.31 ± 1.64 [a,b]	6.05 ± 1.78 [a]
SNL + S	6.33 ± 1.45 [a,b]	6.12 ± 1.57 [a]	5.16 ± 1.73 [b]	5.39 ± 1.57 [b,c]	4.43 ± 2.16 [b,c]

[a–f] Tukey pairwise comparisons appear as superscripts by each mean. Means within a column that share a letter are not significantly different ($p > 0.05$). [1] C = control; SB = Splenda for baking; EB = Equal for baking; TR = Truvia; SNL = Sweet'N Low; SB + S = Splenda for baking + sucrose; EB + S = Equal for baking + sucrose; TR + S = Truvia + sucrose; SNL + S = Sweet'N Low + sucrose.

The appearance of TR + S was not significantly different from that of SB + S and EB + S ($p > 0.05$) (Table 8). Treatment C was not significantly different from SB, SB + S, TR + S, or SNL + S in terms of texture. Only C and TR + S had significantly similar flavor, sweetness, and aftertaste ratings. Treatment EB had a significantly different sweetness rating from all other treatments. Equal for baking is made of maltodextrin, aspartame, and acesulfame potassium (Ace-K) ("Equal Original Granular", n.d.). Aspartame is not a heat-stable sweetener [27]. While maltodextrin (bulking agent) and Ace-K are heat-stable, they are not present in great enough amounts to impart sweetness to cookies baked with them. Maltodextrin has very low sweetness, so a greater amount is required to impart noticeable sweetness. Ace-K is only included in Equal for baking at a 1:4 ratio with aspartame ("Equal Original Granular", n.d.). This ratio indicates that Equal for baking does not impart an acceptable sweetness level for cookies.

In this experiment, treatments that included sugar in the formulation were rated higher on the hedonic scale than cookies prepared using only alternative sweeteners. Sugar has better functional properties in baked goods compared to alternative sweeteners, including rheology, dough viscosity, and Maillard browning [22]. These improved functions contribute to the better texture, mouthfeel, flavor, and color of the final product compared to cookies made with alternative sweeteners. Specifically, alternative sweeteners alone lack the bulking ability of sugar and require additional ingredients to form dough with favorable rheological properties and texture [22]. To compensate, alternative sweeteners that are used for baking are often combined with maltodextrin. Maltodextrin is a bulking agent that does not impart a great amount of sweetness. Additionally, alternative sweeteners have different flavor and sweetness profiles from sucrose. Splenda for baking is made of maltodextrin and sucralose; Equal for baking is made of maltodextrin, aspartame, and Ace-K; Truvia is derived from the plant Stevia rebaudiana; and Sweet'N Low is made of saccharin [27]. These sweeteners are significantly sweeter than sucrose alone but have potentially unpleasant or bitter aftertastes [27]. Overall, treatment TR + S was the best reduced-sugar alternative, yielding cookies most similar to C in all attributes scored except for appearance.

4. Conclusions

The results of this study show that sucrose plays a critical functional role in sugar-snap cookies. Depending on the level and type of the alternative sweeteners used, the replacement of sucrose produced cookies with a decreased spread, color change, hardness, and an increased fracturability. Reduced (50%)-sugar cookies had improved characteristics when alternative sweeteners were substituted for only part of the sucrose in the formulation. However, more research is needed to determine the optimal level and type/s of alternative sweeteners in cookie formulations. Additionally, further research could explore functional ingredients, such as bulking agents, fiber, and fat sources, to improve cookie spread, color, and texture attributes. Overall, our study found that TR + sucrose (50% replacement) and SB (100% replacement) were the closest to the control in terms of physiochemical characteristics.

Author Contributions: Conceptualization, M.L.H., K.J.K.G. and U.Y.; data curation, M.L.H.; formal analysis, U.Y.; investigation, M.L.H.; methodology, M.L.H., K.J.K.G. and U.Y.; writing—original draft, M.L.H.; writing—review and editing, M.L.H., J.B., K.J.K.G. and U.Y. All authors have read and agreed to the published version of the manuscript.

Funding: This work was funded by the USDA National Institute of Food and Agriculture, Hatch project 1014344. This is contribution 22-269-J from the Kansas Agricultural Experiment Station, Manhattan, Kansas, USA.

Institutional Review Board Statement: The study was conducted in accordance with the Declaration of Helsinki, and approved by the Institutional Review Board of Kansas State University (protocol code 9631 and date of approval: 2/14/2019).

Informed Consent Statement: Informed consent was obtained from all subjects involved in the study.

Data Availability Statement: All data generated or analyzed during this research are included in this manuscript.

Acknowledgments: The authors would also like to thank Sierra (Savage) Hooten, Chris Miller, Hannah Otto, and Austin Weber for their assistance in research activities and the students of the Fundamentals of Food Processing class at Kansas State University for participating in cookie scoring.

Conflicts of Interest: The authors declare that they have no conflict of interest. The use of tradenames in this publication does not imply endorsement or criticism by Kansas State University.

References

1. Erickson, S.; Carr, J. The Technological Challenges of Reducing the Sugar Content of Foods. *Nutr. Bull.* **2020**, *45*, 309–314. [CrossRef]
2. World Health Organization. *Guideline: Sugars Intake for Adults and Children*; World Health Organization: Geneva, Switzerland, 2015; ISBN 978-92-4-154902-8.
3. Newens, K.J.; Walton, J. A Review of Sugar Consumption from Nationally Representative Dietary Surveys across the World. *J. Hum. Nutr. Diet.* **2016**, *29*, 225–240. [CrossRef] [PubMed]
4. Kelly, T.; Yang, W.; Chen, C.-S.; Reynolds, K.; He, J. Global Burden of Obesity in 2005 and Projections to 2030. *Int. J. Obes.* **2008**, *32*, 1431–1437. [CrossRef] [PubMed]
5. Department of Health and Human Services; Food and Drug Administration. *CFR—Code of Federal Regulations*; Title 21 Part 101 Section 101.9; Department of Health and Human Services: Washington, DC, USA, 2022.
6. Stanner, S.A.; Spiro, A. Public Health Rationale for Reducing Sugar: Strategies and Challenges. *Nutr. Bull.* **2020**, *45*, 253–270. [CrossRef]
7. International Food Information Council. *2021 Food and Health Survey: International Food Information Council*; International Food Information Council: Washington, DC, USA, 2021.
8. Kweon, M.; Slade, L.; Levine, H.; Martin, R.; Souza, E. Exploration of Sugar Functionality in Sugar-Snap and Wire-Cut Cookie Baking: Implications for Potential Sucrose Replacement or Reduction. *Cereal Chem.* **2009**, *86*, 425–433. [CrossRef]
9. Pareyt, B.; Talhaoui, F.; Kerckhofs, G.; Brijs, K.; Goesaert, H.; Wevers, M.; Delcour, J.A. The Role of Sugar and Fat in Sugar-Snap Cookies: Structural and Textural Properties. *J. Food Eng.* **2009**, *90*, 400–408. [CrossRef]
10. Slade, L.; Levine, H. Structure-Function Relationships of Cookie and Cracker Ingredients. In *The Science of Cookie and Cracker Production*; Faridi, H., Ed.; Chapman & Hall/AVI: New York, NY, USA, 1994; pp. 23–141.
11. Ameur, L.; Mathieu, O.; Lalanne, V.; Trystram, G.; Birlouezaragon, I. Comparison of the Effects of Sucrose and Hexose on Furfural Formation and Browning in Cookies Baked at Different Temperatures. *Food Chem.* **2007**, *101*, 1407–1416. [CrossRef]
12. Bingley, C. The Technological Challenges of Reformulating with Different Dietary Fibres. *Nutr. Bull.* **2020**, *45*, 328–331. [CrossRef]
13. Ho, L.-H.; Pulsawat, M.M. Effects of Partial Sugar Replacement on the Physicochemical and Sensory Properties of Low Sugar Cookies. *Int. Food Res. J.* **2020**, *27*, 557–567.
14. Luo, X.; Arcot, J.; Gill, T.; Louie, J.C.Y.; Rangan, A. Review of Food Reformulation of Baked Products to Reduce Added Sugar Intake. *Trends Food Sci. Technol.* **2019**, *86*, 416–425. [CrossRef]
15. Najjar, Z.; Kizhakkayil, J.; Shakoor, H.; Platat, C.; Stathopoulos, C.; Ranasinghe, M. Antioxidant Potential of Cookies Formulated with Date Seed Powder. *Foods* **2022**, *11*, 448. [CrossRef] [PubMed]
16. Zlatanović, S.; Kalušević, A.; Micić, D.; Laličić-Petronijević, J.; Tomić, N.; Ostojić, S.; Gorjanović, S. Functionality and Storability of Cookies Fortified at the Industrial Scale with up to 75% of Apple Pomace Flour Produced by Dehydration. *Foods* **2019**, *8*, 561. [CrossRef] [PubMed]
17. Vukušić Pavičić, T.; Grgić, T.; Ivanov, M.; Novotni, D.; Herceg, Z. Influence of Flour and Fat Type on Dough Rheology and Technological Characteristics of 3D-Printed Cookies. *Foods* **2021**, *10*, 193. [CrossRef] [PubMed]
18. Delcour, J.A.; Hoseney, R.C.; Hoseney, R.C. *Principles of Cereal Science and Technology*, 3rd ed.; AACC International: St. Paul, MN, USA, 2010; ISBN 978-1-891127-63-2.
19. Doescher, L.C.; Hoseney, R.C.; Milliken, G.A.; Rubenthaler, G.L. Effect of Sugars and Flours on Cookie Spread Evaluated by Time-Lapse Photography. *Cereal Chem.* **1987**, *64*, 163–167.
20. Laguna, L.; Vallons, K.J.R.; Jurgens, A.; Sanz, T. Understanding the Effect of Sugar and Sugar Replacement in Short Dough Biscuits. *Food Bioprocess Technol.* **2013**, *6*, 3143–3154. [CrossRef]
21. Pasha, I.; Butt, M.; Anjum, F.; Shehzadi, N. Effect of Dietetic Sweeteners on the Quality of Cookies. *Int. J. Agric. Biol.* **2002**, *4*, 245–248.
22. Davis, E.A. Functionality of Sugars: Physicochemical Interactions in Foods. *Am. J. Clin. Nutr.* **1995**, *62*, 170S–177S. [CrossRef] [PubMed]
23. Lin, S.-D.; Lee, C.-C.; Mau, J.-L.; Lin, L.-Y.; Chiou, S.-Y. Effect of Erythritol on Quality Characteristics of Reduced-Calorie Danish Cookies. *J. Food Qual.* **2010**, *33*, 14–26. [CrossRef]
24. Gallagher, E.; O'Brien, C.M.; Scannell, A.G.M.; Arendt, E.K. Evaluation of Sugar Replacers in Short Dough Biscuit Production. *J. Food Eng.* **2003**, *56*, 261–263. [CrossRef]
25. U.S. Department of Agriculture; U.S. Department of Health and Human Services. *Dietary Guidelines for Americans, 2020–2025*, 9th ed.; U.S. Department of Agriculture, U.S. Department of Health and Human Services: Washington, DC, USA, 2020.
26. Department of Health and Human Services; Food and Drug Administration. *CFR—Code of Federal Regulations*; Title 21 Part 101 Section 101.12; Department of Health and Human Services: Washington, DC, USA, 2022.
27. Chattopadhyay, S.; Raychaudhuri, U.; Chakraborty, R. Artificial Sweeteners—A Review. *J. Food Sci. Technol.* **2014**, *51*, 611–621. [CrossRef] [PubMed]

Article

Effects of Different Amylose Contents of Foxtail Millet Flour Varieties on Textural Properties of Chinese Steamed Bread

Shaohui Li [1,2,3], Wei Zhao [1,2,3], Guang Min [4], Pengliang Li [1,2,3], Aixia Zhang [1,2,3], Jiali Zhang [1,2,3], Yunting Wang [1,2,3], Yingying Liu [1,2,3] and Jingke Liu [1,2,3,*]

[1] Institute of Millet Crops, Hebei Academy of Agriculture and Forestry Sciences, No. 162, Hengshan St., Shijiazhuang 050035, China; lishaohui007@163.com (S.L.); zhaoweipg@163.com (W.Z.); lpl1937@126.com (P.L.); zhangaixia1977@126.com (A.Z.); zhangjiali416@126.com (J.Z.); wyt15731113407@163.com (Y.W.); lyy_133152333@163.com (Y.L.)

[2] National Foxtail Millet Improvement Center, No. 162, Hengshan St., Shijiazhuang 050035, China

[3] Minor Cereal Crops Laboratory of Hebei Province, No. 162, Hengshan St., Shijiazhuang 050035, China

[4] Grain and Oil Food Inspection Center of Wuhan, Wuhan 430021, China; lshh010@163.com

[*] Correspondence: liujingke79@163.com; Tel.: +86-311-87670703

Abstract: In order to improve the nutritional value and quality of steamed bread, and promote the industrial development of the whole-grain food industry, a texture analyzer was used to study the effects of cultivars of whole foxtail millet flour (WFMF) on the texture of Chinese steamed bread (CSB). Orthogonal partial least squares discriminant analysis (OPLS-DA) was also conducted. The addition of different cultivars of WFMF significantly altered the height–diameter ratio, specific volume, hardness, cohesiveness, gumminess, and chewiness of CSB ($p < 0.05$). Large amounts of foxtail millet flour significantly increased the hardness, gumminess and chewiness of the bread ($p < 0.05$), and the bread height–diameter ratio, specific volume, cohesiveness and springiness significantly decreased ($p < 0.05$). We screened sensory evaluation, chewiness, specific volume, and hardness as the signature differences in the quality components according to the variable influence on the projection (VIP) values. OPLS-DA could distinguish the addition levels of different samples.

Keywords: whole foxtail millet flour; amylose content; japonica and glutinous; Chinese steamed bread; texture properties

Citation: Li, S.; Zhao, W.; Min, G.; Li, P.; Zhang, A.; Zhang, J.; Wang, Y.; Liu, Y.; Liu, J. Effects of Different Amylose Contents of Foxtail Millet Flour Varieties on Textural Properties of Chinese Steamed Bread. *Processes* **2021**, *9*, 1131. https://doi.org/10.3390/pr9071131

Academic Editors: Yonghui Li and Xiaorong Wu (Shawn)

Received: 19 May 2021
Accepted: 26 June 2021
Published: 29 June 2021

Publisher's Note: MDPI stays neutral with regard to jurisdictional claims in published maps and institutional affiliations.

1. Introduction

Foxtail millet (*Setaria italica L. Beauv*) is rich in proteins, fats, carbohydrates, dietary fiber, vitamins and minerals. It is good for the spleen and stomach, and has been applied to food therapy for thousands of years [1,2]. After shelling and milling, millet is suitable for human consumption. However, the consumption of millet porridge alone cannot support the millet production and processing industry. The demand for whole grain food is growing worldwide. The sale of whole grain bread in the USA market has surpassed that of ordinary white bread [3,4]. Foxtail millet is an important whole grain that suffers relatively little loss of nutrients from processing and provides high nutritional value [5].

The raw materials used for whole grain food processing and their nutritional values are receiving great attention [6,7]. Chinese steamed bread (CSB) is a traditional staple food in China. It is convenient and nutritious [8]. The development of an improved CSB has been studied during the development of the whole-grain food industry. Quinoa flour [9,10], buckwheat flour [11,12], finger millet, and red kidney bean flour [13] have been added to CSB and the results have been documented. As whole grains have been proven to reduce the risk of diabetes, obesity, colorectal cancer, and cardiovascular disease [14], the use of composite flour to make bread is also a recent global development. Furthermore, due to the development of some social, economic, and health-related concepts, the use of whole-grain flour is a strategy for developing healthy food. Whole-grain cereals are the focus of much research and an important component of a healthy diet. Foxtail millet has also been studied, but the results are limited to a single cultivar of milled millet flour [15,16]. Since the millet bran contains a variety of nutrients, the nutritional value of millet is greatly reduced. Compared with commercial refined flour, there is a large amount of dietary fiber in whole grains [3], which affects product quality; thus, it is necessary to develop a whole-grain millet that combines the characteristics of being good for health and possessing good texture quality. There is no information on the production of CSB using a variety of cultivars of WFMF. The numerous foxtail millet varieties have different physical and chemical properties. Their characteristics must be evaluated to select the most appropriate varieties and additive amounts for their use in CSB.

Thus, we determined the effects of different varieties and amounts of WFMF via the height–diameter ratio and the specific volume of CSB. Texture characteristic changes in hardness, cohesiveness, springiness, gumminess and chewiness were analyzed. We also used OPLS-DA to identify differences between the varieties and determine the most suitable varieties for the production of CSB. This information could increase the commercial use of foxtail millet in CSB manufacturing.

2. Materials and Methods

2.1. Materials

Special first-grade multi-purpose wheat flour (WF) was obtained from Jinsha River Noodle Group Co., Ltd., Xingtai, China; Angel high active dry yeast was obtained from Angel Yeast Co., Ltd., Yichang, China. The materials (japonica: Jigu-19, Taixuan-17, Yugu-18, An11-5365, and Ji0626-4; glutinous: Chifeng-1, Fente-5, and N101) were supplied by the Institute of Millet Crops, Hebei Academy of Agriculture and Forestry Sciences, China, and planted in Mazhuang Experimental Station, Shijiazhuang, China (Table 1). Each foxtail millet sample was harvested, dried, and stored at −20 °C.

Table 1. Foxtail millet varieties.

Sample	Name Code	Amylose (%)	Type of Cultivar
Jigu-19	JG19	32.25 ± 0.68	Japonica
Taixuan-17	TX17	24.64 ± 0.37	Japonica
Yugu-18	YG18	15.19 ± 0.81	Japonica
An11-5365	A11-5365	25.50 ± 0.62	Japonica
Ji0626-4	J0626-4	12.36 ± 0.53	Japonica
Chifeng-1	CF1	7.40 ± 0.34	Glutinous
Fente-5	FT5	7.46 ± 0.65	Glutinous
N101	N101	5.25 ± 0.47	Glutinous

2.2. Preparation of WFMF

All of the samples were dehulled using a SY88-TH cereal huller (Korea Ssangyong Machinery Industry Co., Ltd., Incheon, Korea) and then milled into powder using a M3100 automatic cyclone mill (Perten Instruments, Hägersten, Sweden). They were then sieved through a 100 mesh screen sieve (Henan Xinxian Fasite Instrument Co., Ltd., Xinxiang, China) to prepare WFMF.

2.3. Amylose Content of WFMF

An Amylose/Amylopectin Assay Kit (Megazyme International Ltd., Bray, Ireland) was used to determine the amylose content of the samples.

2.4. CSB Preparation

CSB was prepared according to the method of Li et al. [17] with some modifications: 300 g amounts of mixed flour were prepared (each containing 0%, 10%, 20%, 30%, 40%, and 50% of WFMF) and mixed using a KVC3100 mixer (Kenwood, UK). Then, 150 mL of 35 °C water and 2 g yeast were added, and the mixing was conducted at a low speed for 1 min, and continued for another 9 min at a high speed. The smooth dough was fermented at a temperature of 35 °C and 75% RH in a constant temperature incubator (Percival Technology, Perry, IA, USA) for 60 min. The dough was divided to six equal portions and molded into semi-circular forms, which were placed in a plastic sealing box at room temperature for 10 min and then into a steamer and steamed for 30 min. The CSB was cooled at room temperature for 1 h before analysis. Each CSB treatment was prepared in triplicate.

2.5. CSB Evaluation

The specific volume (cm^3/g) by volume displacement method is discussed in [18], and the height–diameter ratio (height/diameter, cm/cm) of the CSB was measured. Sensory evaluation was determined according to a previously published study [19] with some modifications.

The texture analysis of CSB samples was performed using a TMS-Pro texture analyzer (Food Technology Corporation, Sterling, VA, USA) using texture profile analysis (TPA) test mode. A whole CSB (hemispherical, height × diameter = 4.5 cm × 7.5 cm) sample was placed in the center of the platform and compressed by a pressure plate probe $p/75$ mm at the speed of 30 mm/min to 50% of the original thickness, with a pre-test speed of 30 mm/min and a post-test speed of 30 mm/min. The TPA test starting point's trigger force was 1 N at a data acquisition frequency of 100 Hz. All of the experiments were performed three times at room temperature. The parameters obtained from the TPA experiment were cohesiveness, hardness, gumminess, springiness, and chewiness.

2.6. Statistical Analysis

All measurements were performed at least three times. The results were analyzed using SPSS 17.0 (Chicago, IL, USA) and are expressed as mean ± SD (standard deviation). We used one-way analysis of variance (ANOVA) to compare the treatment data. Duncan's test was used to compare the differences among means, and a $p < 0.05$ was considered to indicate a significant difference. Simca13.0 (Umetrics, Umea, Sweden) was used to perform orthogonal partial least squares discriminant analysis (OPLS-DA) and to calculate the variable influence on projection (VIP) values.

3. Results and Discussion

3.1. Height–diameter Ratio and Specific Volume of the CSB

The height–diameter ratio measurement results of CSB are shown in TPA 2. The different amounts of the different foxtail millet flour cultivars added to CSB affected the final product. The height–diameter ratio of CSB significantly differed ($p < 0.05$) in accordance with the different additions. The differences appeared in the height–diameter ratio of the same variety of millet and in different additive amounts, except for in the cases of TX 17 and A 11-5365. The height–diameter ratio represents the CSB's shape. The bread forms a uniform sponge-like hole during the steaming process, which props up its fullness without producing wrinkles [12]. The height–diameter ratio of the glutinous varieties was consistent, and that of the glutinous cultivar N101 was 0.82 when the addition was 20%, which was significantly lower than that of the other cultivars. This may have occurred due to the different contents of amylose and amylopectin in the powder. After gelatinization and expansion, the starch filled the gluten protein's skeletal structure in a non-uniform manner and, thus, could not support it, resulting in partial collapse and agglomeration. This made the shape of the CSB unacceptable and altered the height–diameter ratio.

The specific volume measurement results of CSB are shown in Table 2. The specific volumes of the different cultivars of WFMF added to CSB are significantly different ($p < 0.05$). Within the same cultivar, the CSB specific volume decreased with an increased amount of WFMF ($p < 0.05$). Specific volume indicates the degree of CSB fluffiness and reflects the gas production within the dough [20]. The specific volume of the commercially available CSB standard is generally >1.7 cm^3/g. The theoretical addition of WFMF is within 30%. The specific volume of CSB is related to the gas-holding property of the dough [21]. As the addition of WFMF increased, the gluten content in the dough decreased, ductility and springiness decreased, the internal voids of CSB were reduced and the specific volume decreased [22]. Usually, adding WFMF reduces the specific volume of CSB, and different cultivars of WFMF can change the specific volume of CSB. The results showed that the reduction in the glutinous cultivar was minimal, indicating that the gas-holding capacity of the japonica cultivars was lower than that of the glutinous cultivars.

The sensory evaluation results of CSB are shown in Table 2. The sensory evaluations of the different cultivars of WFMF added to CSB are not significantly different ($p < 0.05$). Within the same cultivar, the CSB sensory evaluation increased at first and then decreased with an increased amount of WFMF ($p < 0.05$). Almost every cultivar with the 30% addition showed better sensory evaluation, except when the additive amount was 0%. Sensory evaluation is a form of subjective, human evaluation, including appearance, smell, and taste. However, it reflects the preferences of consumers to a certain extent. In the sensory evaluation index, smell and taste are negatively correlated with hardness, and total acceptance is positively correlated with springiness and negatively correlated with hardness [18]. Moreover, because the color of steamed bread is bright, the varieties of whole grains with a minimal addition showed similarly high sensory evaluation scores, and it could be observed that the brightness of appearance is the main indicator.

3.2. CSB Texture Analysis

The hardness measurement results of CSB are shown in Table 3. The hardness values of the various CSB treatments were significantly different ($p < 0.05$). Hardness refers to the maximum force peak of the first compression cycle. Under the same added amount of different cultivars, the difference in the hardness of CSB of the same cultivar WFMF was significant ($p < 0.05$). With increasing amounts of WFMF, the hardness of most CSB samples increased by more than 50%. Among these, the hardness of YG 18, J 0626-4, and CF 1 reached (83.66 ± 9.09) N, (82.41 ± 10.10) N and (88.41 ± 8.26) N, respectively. In contrast, under the same addition, glutinous cultivars FT 5 and N 101 only showed a moderate increase in hardness. This may be related to the different contents of amylose, as a high content of amylose causes the CSB to become hard. CSB with high toughness and high elasticity has a better taste. Hardness is an important sensory indicator that reflects the quality of CSB, and it affects consumer evaluations. CSB hardness is generally negatively correlated with CSB quality [23]. Increased hardness lowers the texture quality of CSB and affects springiness and chewiness.

The results of the CSB cohesiveness measurement are shown in Table 3. Cohesiveness was significantly different among treatments when the same amounts of different cultivars of WFMF were added ($p < 0.05$). Cohesiveness is the contraction force inside the sample. It represents how well the product withstands a second deformation relative to its resistance under the first deformation (the larger the value, the stronger the cohesiveness and the less likely the sample will be destroyed by extrusion). Within the same cultivar of WFMF when the added amount was increased, the cohesiveness significantly decreased ($p < 0.05$). WFMF contains relatively less gluten protein than that of wheat flour [24,25], so the greater the amount added, the lower the stability of the gluten network structure in the dough. Therefore, as the addition of WFMF increased, the cohesiveness significantly decreased. Cohesiveness may be related to the synergistic effect of the gluten protein network structure and the gelatinized starch in the steaming process [26]. The glutenin and gliadin crosslink through the disulfide bond to generate the gluten network structure. The gelatinized starch can be better wrapped, and the gas produced is maintained. The addition of WFMF caused an imbalance of gluten and starch. The gluten protein was insufficient to wrap the gelatinized starch. The starch compromised the protein network structure during the second compression test and made it difficult for the CSB cross-linking structure to rebound to its original shape. As the addition of WFMF increased, the springiness and cohesiveness decreased.

Table 2. Effect of different cultivars and addition of whole foxtail millet flour on height–diameter ratio, specific volume, and sensory evaluation of CSB ($n = 3$).

Level (%)	Japonica					Glutinous		
	JG19	TX17	YG18	A11-5365	J0626-4	CF1	FT5	N101
Height–Diameter Ratio								
0	0.87 ± 0.05 [Aa]	0.87 ± 0.05 [Aa]	0.87 ± 0.05 [Aab]	0.87 ± 0.05 [Aa]	0.87 ± 0.05 [Aab]	0.87 ± 0.05 [Ab]	0.87 ± 0.05 [Aab]	0.87 ± 0.05 [Aa]
10	0.87 ± 0.01 [Aa]	0.90 ± 0.07 [Aa]	0.86 ± 0.02 [Aab]	0.85 ± 0.01 [Aa]	0.86 ± 0.03 [Aab]	0.86 ± 0.01 [Ab]	0.92 ± 0.03 [Aa]	0.82 ± 0.02 [Aab]
20	0.81 ± 0.03 [ABab]	0.82 ± 0.03 [ABa]	0.83 ± 0.01 [ABab]	0.84 ± 0.03 [Aa]	0.88 ± 0.05 [Aa]	0.89 ± 0.04 [Ab]	0.84 ± 0.03 [Ab]	0.82 ± 0.10 [ABab]
30	0.79 ± 0.03 [Cab]	0.85 ± 0.02 [BCa]	0.87 ± 0.02 [ABa]	0.81 ± 0.03 [BCa]	0.81 ± 0.01 [BCb]	0.93 ± 0.07 [Aab]	0.88 ± 0.01 [ABab]	0.84 ± 0.03 [BCab]
40	0.79 ± 0.02 [Bab]	0.86 ± 0.01 [Ba]	0.81 ± 0.03 [Bb]	0.82 ± 0.01 [Ba]	0.84 ± 0.03 [Bab]	0.95 ± 0.08 [Aab]	0.85 ± 0.03 [Bb]	0.86 ± 0.00 [Ba]
50	0.83 ± 0.02 [Cb]	0.83 ± 0.01 [Ca]	0.84 ± 0.02 [BCab]	0.82 ± 0.03 [Ca]	0.84 ± 0.02 [BCab]	1.00 ± 0.04 [Aa]	0.86 ± 0.01 [BCab]	0.89 ± 0.00 [Ba]
Specific Volume (cm^3/g)								
0	2.25 ± 0.15 [Aa]	2.25 ± 0.15 [Aa]	2.25 ± 0.15 [Aa]	2.25 ± 0.15 [Aa]	2.25 ± 0.15 [Aa]	2.25 ± 0.15 [Aa]	2.25 ± 0.15 [Aa]	2.25 ± 0.15 [Aa]
10	2.10 ± 0.10 [Aab]	2.09 ± 0.04 [Aab]	1.52 ± 0.12 [Bb]	1.96 ± 0.14 [Ab]	1.91 ± 0.17 [Ab]	2.02 ± 0.08 [Aab]	1.92 ± 0.11 [Ab]	2.03 ± 0.20 [Aab]
20	2.00 ± 0.15 [Aab]	1.93 ± 0.14 [ABb]	1.54 ± 0.11 [Cb]	1.97 ± 0.03 [ABb]	1.84 ± 0.12 [ABbc]	1.74 ± 0.08 [BCbc]	1.87 ± 0.03 [ABbc]	1.98 ± 0.04 [ABb]
30	1.90 ± 0.08 [Ab]	1.58 ± 0.08 [Bc]	1.58 ± 0.10 [Bb]	1.75 ± 0.06 [ABc]	1.88 ± 0.05 [Abc]	1.74 ± 0.18 [ABbc]	1.78 ± 0.03 [ABbcd]	1.87 ± 0.02 [Abc]
40	1.56 ± 0.07 [BCc]	1.50 ± 0.07 [Cc]	1.67 ± 0.08 [ABCb]	1.60 ± 0.06 [ABCc]	1.64 ± 0.07 [ABCcd]	1.75 ± 0.14 [Abc]	1.69 ± 0.01 [ABcd]	1.77 ± 0.04 [Abc]
50	1.46 ± 0.09 [BCc]	1.38 ± 0.08 [Cc]	1.52 ± 0.07 [ABCb]	1.60 ± 0.05 [ABc]	1.56 ± 0.02 [ABd]	1.65 ± 0.09 [Ac]	1.66 ± 0.05 [Ad]	1.68 ± 0.04 [Ac]
Sensory Evaluation								
0	88.10 ± 2.46 [Aa]	88.10 ± 2.46 [Aa]	88.10 ± 2.46 [Aa]	88.10 ± 2.46 [Aa]	88.10 ± 2.46 [Aa]	88.10 ± 2.46 [Aa]	88.10 ± 2.46 [Aa]	88.10 ± 2.46 [Aa]
10	76.67 ± 3.46 [Ac]	70.33 ± 4.63 [Acd]	76.70 ± 3.14 [Ab]	74.33 ± 5.61 [Ab]	72.00 ± 3.84 [Acd]	74.63 ± 2.80 [Ab]	74.20 ± 4.27 [Ab]	75.13 ± 5.46 [Acd]
20	75.90 ± 3.03 [ABc]	80.33 ± 3.88 [ABab]	78.27 ± 3.67 [ABb]	74.13 ± 2.25 [Bb]	79.77 ± 4.98 [ABabc]	80.77 ± 4.15 [ABab]	78.30 ± 3.73 [ABb]	81.40 ± 2.10 [Aabc]
30	84.13 ± 3.69 [Aab]	85.40 ± 0.95 [Aab]	85.77 ± 4.39 [Aa]	86.43 ± 4.61 [Aa]	81.93 ± 2.18 [Aab]	81.07 ± 3.88 [Aab]	85.4 ± 3.73 [Aa]	85.20 ± 5.66 [Aab]
40	80.47 ± 5.00 [Abc]	77.83 ± 5.51 [Abc]	74.90 ± 2.19 [Abc]	75.03 ± 3.56 [Ab]	73.97 ± 4.93 [Abcd]	75.23 ± 4.71 [Ab]	78.83 ± 2.74 [Ab]	78.67 ± 0.65 [Abcd]
50	66.10 ± 1.82 [Ad]	67.00 ± 6.08 [Ad]	69.93 ± 1.48 [Ac]	68.07 ± 6.20 [Ab]	66.27 ± 6.98 [Ad]	64.90 ± 6.07 [Ac]	74.03 ± 2.12 [Ab]	71.87 ± 3.37 [Ad]

Note: The data in the table are the mean ± SD. Different uppercase letters in the same row indicate that there is a significant difference in mean ($p < 0.05$). Different lowercase letters in the same column indicate that there is a significant difference in mean ($p < 0.05$).

Table 3. Effect of different cultivars and addition of whole foxtail millet flour on texture properties of CSB ($n = 3$).

Level (%)	Japonica					Glutinous		
	JG19	TX17	YG18	A11-5365	J0626-4	CF1	FT5	N101
	Hardness (N)							
0	36.17 ± 3.26 [Ad]	36.17 ± 3.26 [Ac]	36.17 ± 3.26 [Ae]	36.17 ± 3.26 [Ad]	36.17 ± 3.26 [Ae]	36.17 ± 3.26 [Ad]	36.17 ± 3.26 [Ac]	36.17 ± 3.26 [Ab]
10	38.16 ± 0.49 [Dd]	45.87 ± 3.47 [BCb]	40.97 ± 1.45 [CDde]	37.12 ± 2.41 [Dd]	40.70 ± 2.87 [CDde]	47.24 ± 3.00 [Bcd]	57.76 ± 5.38 [Aa]	36.69 ± 1.48 [Db]
20	48.23 ± 6.56 [Bc]	45.75 ± 3.76 [BCb]	48.61 ± 2.78 [Bcd]	47.29 ± 2.71 [BCc]	51.12 ± 2.96 [ABcd]	58.98 ± 7.87 [Abc]	49.00 ± 0.21 [Bab]	39.23 ± 2.78 [Cb]
30	75.30 ± 4.27 [Aa]	49.48 ± 3.27 [Cb]	55.87 ± 6.48 [BCbc]	52.39 ± 2.99 [Bbc]	54.28 ± 4.93 [Cc]	65.07 ± 2.42 [Bb]	48.75 ± 9.41 [Cab]	46.51 ± 6.38 [Cab]
40	72.44 ± 7.55 [ABab]	75.05 ± 5.97 [Aa]	60.80 ± 5.30 [BCDb]	59.23 ± 4.11 [BCDab]	69.05 ± 5.88 [ABCb]	84.01 ± 18.81 [Aa]	43.58 ± 7.01 [Dbc]	53.76 ± 8.58 [CDa]
50	63.56 ± 3.22 [BCb]	76.42 ± 5.49 [Aa]	83.66 ± 9.09 [Aa]	64.23 ± 9.87 [BCa]	82.41 ± 10.10 [Aa]	88.41 ± 8.26 [Aa]	39.43 ± 1.05 [Dbc]	52.86 ± 6.08 [CDa]
	Cohesiveness							
0	0.77 ± 0.01 [Aa]	0.77 ± 0.01 [Aa]	0.77 ± 0.01 [Aa]	0.77 ± 0.01 [Aa]	0.77 ± 0.01 [Aa]	0.77 ± 0.01 [Aa]	0.77 ± 0.01 [Aa]	0.77 ± 0.01 [Aa]
10	0.74 ± 0.01 [BCb]	0.74 ± 0.02 [BCb]	0.76 ± 0.01 [Aba]	0.77 ± 0.01 [Aba]	0.77 ± 0.02 [Aa]	0.75 ± 0.01 [ABCb]	0.74 ± 0.02 [Cb]	0.77 ± 0.01 [Aa]
20	0.71 ± 0.01 [Dc]	0.73 ± 0.01 [ABb]	0.74 ± 0.01 [Ab]	0.73 ± 0.00 [ABCb]	0.73 ± 0.01 [ABCDb]	0.72 ± 0.01 [CDc]	0.74 ± 0.01 [Ab]	0.72 ± 0.00 [BCDb]
30	0.67 ± 0.02 [Cd]	0.70 ± 0.01 [ABc]	0.72 ± 0.01 [Ac]	0.71 ± 0.00 [Ab]	0.68 ± 0.02 [BCc]	0.68 ± 0.01 [BCd]	0.69 ± 0.01 [Bc]	0.69 ± 0.00 [Bc]
40	0.61 ± 0.03 [Df]	0.65 ± 0.00 [ABCd]	0.67 ± 0.01 [ABd]	0.68 ± 0.01 [Ac]	0.64 ± 0.02 [CDd]	0.66 ± 0.03 [ABCd]	0.64 ± 0.02 [BCd]	0.65 ± 0.02 [ABCd]
50	0.64 ± 0.01 [ABe]	0.59 ± 0.02 [Ce]	0.61 ± 0.03 [BCe]	0.67 ± 0.04 [Ac]	0.59 ± 0.03 [Ce]	0.62 ± 0.02 [BCe]	0.61 ± 0.02 [BCe]	0.58 ± 0.01 [Ce]
	Springiness (mm)							
0	5.65 ± 0.08 [Aa]	5.65 ± 0.08 [Aa]	5.65 ± 0.08 [Aa]	5.65 ± 0.08 [Aa]	5.65 ± 0.08 [Aa]	5.65 ± 0.08 [Aa]	5.65 ± 0.08 [Aa]	5.65 ± 0.08 [Aa]
10	5.51 ± 0.14 [Aab]	5.52 ± 0.68 [Aa]	5.73 ± 0.07 [Aa]	5.30 ± 0.29 [Aa]	5.47 ± 0.28 [Aa]	5.75 ± 0.31 [Aa]	5.51 ± 0.12 [Aab]	5.51 ± 0.31 [Aab]
20	5.70 ± 0.21 [Aa]	5.50 ± 0.26 [Aa]	5.51 ± 0.66 [Aab]	5.78 ± 0.23 [Aa]	5.50 ± 0.24 [Aa]	5.62 ± 0.11 [Aa]	5.44 ± 0.19 [Aab]	5.44 ± 0.27 [Aab]
30	5.27 ± 0.16 [Abc]	5.48 ± 0.08 [Aa]	5.48 ± 0.30 [Aab]	5.63 ± 0.93 [Aa]	5.55 ± 0.27 [Aa]	5.47 ± 0.09 [Aa]	5.20 ± 0.17 [Ac]	5.35 ± 0.28 [Aab]
40	5.15 ± 0.11 [Ac]	5.04 ± 0.45 [Aa]	5.49 ± 0.27 [Aab]	5.43 ± 0.05 [Aa]	5.42 ± 0.30 [Aa]	5.37 ± 0.45 [Aa]	5.41 ± 0.03 [Abc]	5.18 ± 0.40 [Aab]
50	5.12 ± 0.16 [ABc]	5.03 ± 0.37 [ABa]	5.08 ± 0.17 [ABb]	5.23 ± 0.18 [Aa]	5.15 ± 0.16 [ABa]	5.31 ± 0.11 [Aa]	4.78 ± 0.12 [Bd]	4.96 ± 0.11 [ABb]
	Gumminess (N)							
0	27.91 ± 2.37 [Ad]	27.91 ± 2.37 [Ac]	27.91 ± 2.37 [Ad]	27.91 ± 2.37 [Ad]	27.91 ± 2.37 [Ac]	27.91 ± 2.37 [Ac]	27.91 ± 2.37 [Ab]	27.91 ± 2.37 [Ab]
10	28.43 ± 0.35 [Dd]	34.05 ± 1.94 [BCb]	31.29 ± 0.64 [CDcd]	28.32 ± 1.62 [Dd]	31.36 ± 1.61 [CDbc]	35.30 ± 2.32 [Bbc]	42.66 ± 3.03 [Aa]	28.21 ± 1.48 [Db]
20	34.28 ± 4.08 [BCc]	33.49 ± 2.87 [BCDb]	36.02 ± 1.88 [Bbc]	34.52 ± 1.87 [BCc]	37.12 ± 1.75 [ABb]	42.19 ± 5.14 [Ab]	28.95 ± 0.62 [CDb]	28.22 ± 2.01 [Db]
30	50.13 ± 1.88 [Aa]	34.68 ± 1.84 [DEb]	39.95 ± 4.15 [BCb]	37.28 ± 2.09 [CDbc]	37.13 ± 2.46 [CDb]	44.29 ± 2.14 [Bb]	33.82 ± 0.45 [DEb]	32.00 ± 4.45 [Eab]
40	43.80 ± 3.12 [BCb]	48.86 ± 3.63 [ABa]	40.97 ± 3.45 [BCb]	40.11 ± 2.49 [BCab]	43.82 ± 2.84 [BCa]	54.98 ± 10.37 [Aa]	27.89 ± 3.85 [Db]	34.73 ± 4.92 [CDa]
50	40.31 ± 2.17 [Cb]	45.35 ± 1.79 [BCa]	51.19 ± 7.28 [ABa]	42.72 ± 4.40 [BCa]	48.62 ± 4.66 [ABCa]	55.18 ± 4.03 [Aa]	29.54 ± 4.93 [Db]	30.76 ± 2.90 [Dab]
	Chewiness (mJ)							
0	157.70 ± 12.26 [Ad]	157.70 ± 12.26 [Ac]	157.70 ± 12.26 [Ad]	157.70 ± 12.26 [Abc]	157.70 ± 12.26 [Ac]	157.70 ± 12.26 [Ac]	157.70 ± 12.26 [Ab]	157.70 ± 12.26 [Aa]
10	156.75 ± 5.79 [Dd]	188.24 ± 29.09 [BCbc]	179.47 ± 4.49 [BCDbcd]	150.4 ± 16.86 [Dc]	171.34 ± 2.91 [CDc]	203.27 ± 23.87 [Bbc]	235.14 ± 19.58 [Aa]	155.54 ± 15.98 [Da]
20	188.58 ± 25.97 [BCDcd]	184.66 ± 24.01 [BCDc]	199.34 ± 34.01 [ABCcd]	199.88 ± 17.40 [ABCab]	203.85 ± 3.38 [ABb]	236.91 ± 26.68 [Aab]	157.52 ± 4.84 [CDb]	154.07 ± 18.12 [Da]

Table 3. *Cont.*

Level (%)	Japonica					Glutinous		
	JG19	TX17	YG18	A11-5365	J0626-4	CF1	FT5	N101
40	225.49 ± 18.13 ABb	246.84 ± 34.85 ABa	225.55 ± 28.9 ABab	217.96 ± 14.97 ABa	237.73 ± 23.95 ABa	298.12 ± 76.19 Aa	150.69 ± 20.29 Cb	179.92 ± 27.75 BCa
50	206.68 ± 17.35 Cbc	227.64 ± 7.49 BCab	260.61 ± 45.96 ABa	223.74 ± 28.44 BCa	250.22 ± 16.61 ABCa	293.33 ± 23.95 Aa	141.41 ± 26.97 Db	152.41 ± 14.98 Da

Note: The data in the table are the mean ± SD. Different uppercase letters in the same row indicate that there is a significant difference in mean ($p < 0.05$). Different lowercase letters in the same column indicate that there is a significant difference in mean ($p < 0.05$).

The springiness measurement results of the CSB are shown in Table 3. The springiness of CSB among the different cultivars was not significantly different ($p > 0.05$), except for when the additive amount was 50%. The difference in the japonica and glutinous cultivars was significant from 10% to 50% ($p < 0.05$), and the value was between 4.78 and 5.75. Springiness measures the elasticity extent of recovery between the first and second compressions and indicates the ability of a substance to return to its original shape after an external force has been removed. The springiness of CSB is a side effect of its softness. Depending on the gluten content and the gas-holding capacity of CSB [12], springiness affects the choice of consumers. The predominant gluten proteins in wheat flour are glutenin and gliadin. Glutenin determines the springiness and extensibility of gluten [25]. The protein content in millet is low, and adding WFMF reduced the total gluten content and decreased the CSB's springiness. This may be due to the increase in the additive amount, which increased the fiber content, and the decrease in gliadins and glutenins in dough, which led to a decrease in fermentation and ductility [27].

The results of the measurement of the gumminess of CSB are shown in Table 3. There was a significant difference in the gumminess of CSB when the same amounts of different cultivars of WFMF were added ($p < 0.05$). Gumminess describes the multiplication of hardness and cohesiveness and refers to the energy required to break down a semi-solid food before it is swallowed. As the addition of japonica WFMF increased, the gumminess significantly increased ($p < 0.05$). However, in the glutinous cultivar CSB, the gumminess value and the change trend in CF 1 were similar to those in the japonica CSB. The gumminess values of FT 5 and N101 were lower than those of japonica CSB. The gumminess of CSB may be related to the nature of the starch that it contains. The gelatinization properties of starch are mainly determined by its size, proportion, and content of amylose [28]. The starch granules gelatinized and expanded, and the gluten protein supported the structure of the CSB. A previous study found that the higher the amylopectin content, the lower the adhesiveness of brown rice bread [29]. Brown rice and oat substitution significantly increased the gumminess of CSB [30]. In a previous study, the final viscosity of buckwheat mixed powder was positively correlated with the hardness and gumminess of buckwheat steamed bread [31], which is consistent with the results of this study.

The chewiness results of CSB are shown in Table 3. The different cultivars of WFMF added to CSB produced significantly different chewiness values ($p < 0.05$). The japonica cultivars produced better results than those of the glutinous cultivars. There was a significant difference related to the amount of added millet flour ($p < 0.05$). Chewiness is the energy required to chew a solid sample. With the increase in the additive amount of WFMF, the chewiness of the CSB increased in different cultivars. As the amount of WFMF from cultivar FT 5 increased, the CSB's chewiness first increased and then decreased. The chewiness decreased to 141.41 mJ when the amount added was 50%. Chewiness reflects the amount of energy required to chew a food into a swallowable state and it is correlated with cohesiveness and springiness [30]. The addition of whole quinoa flour is positively correlated with the chewiness of steamed bread [9,32], findings which are consistent with the results of the current study. Previous findings regarding the chewiness of buckwheat bread are also consistent with the results of the current study. A 30% brown rice and oat substitution significantly increased the gumminess of CSB [30]. The addition of a certain proportion of WFMF maintains the water absorption and water-holding capacity of dough [33]. The gluten network structure has good cross-linking properties, suitable hardness and acceptable chewiness. However, when the addition of WFMF is excessive, too many millet starch granules will compete with the protein and starch in the wheat flour, thereby weakening the gluten network structure. This results in an increase in hardness and chewiness.

3.3. OPLS-DA of CSB

First, we performed OPLS-DA, as shown in Figure 1a, which is an objective method by which classification can be conducted and the trend among samples can be observed. We constructed this model by adding the texture properties and evaluation of the steamed bread to each group. The model fit the independent variable R^2X (cum) = 0.996, R^2Y (cum) = 0.365, and Q^2 (cum) = 0.196, which indicates that it has a reliable predictive ability. The results show that the model was stable. The obvious aggregation tendency is reflected in the sample population, achieving better separation.

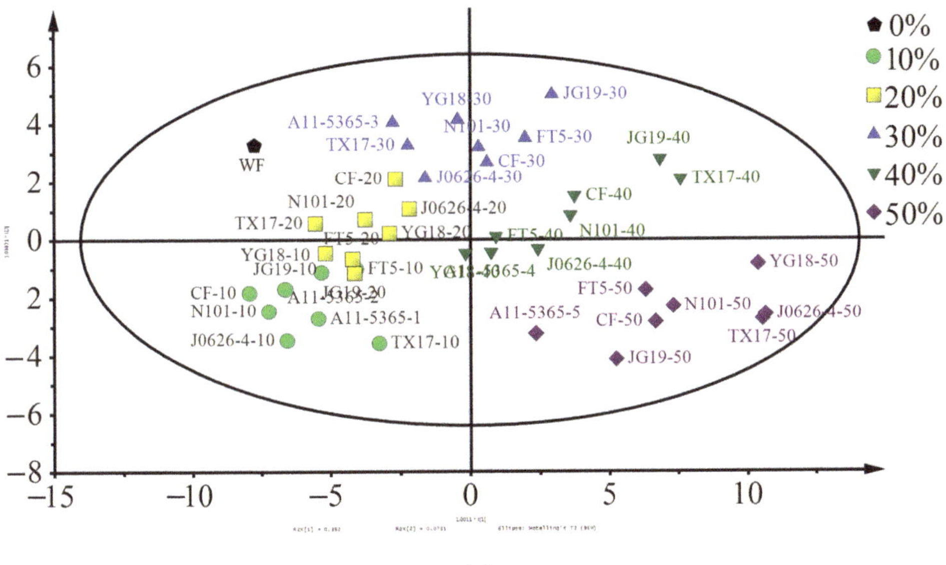

(a)

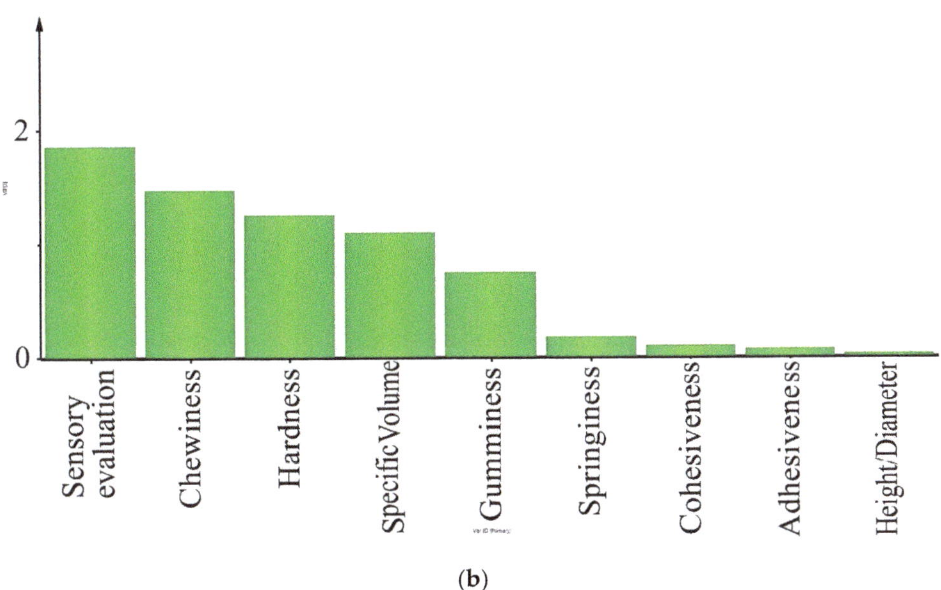

(b)

Figure 1. *Cont.*

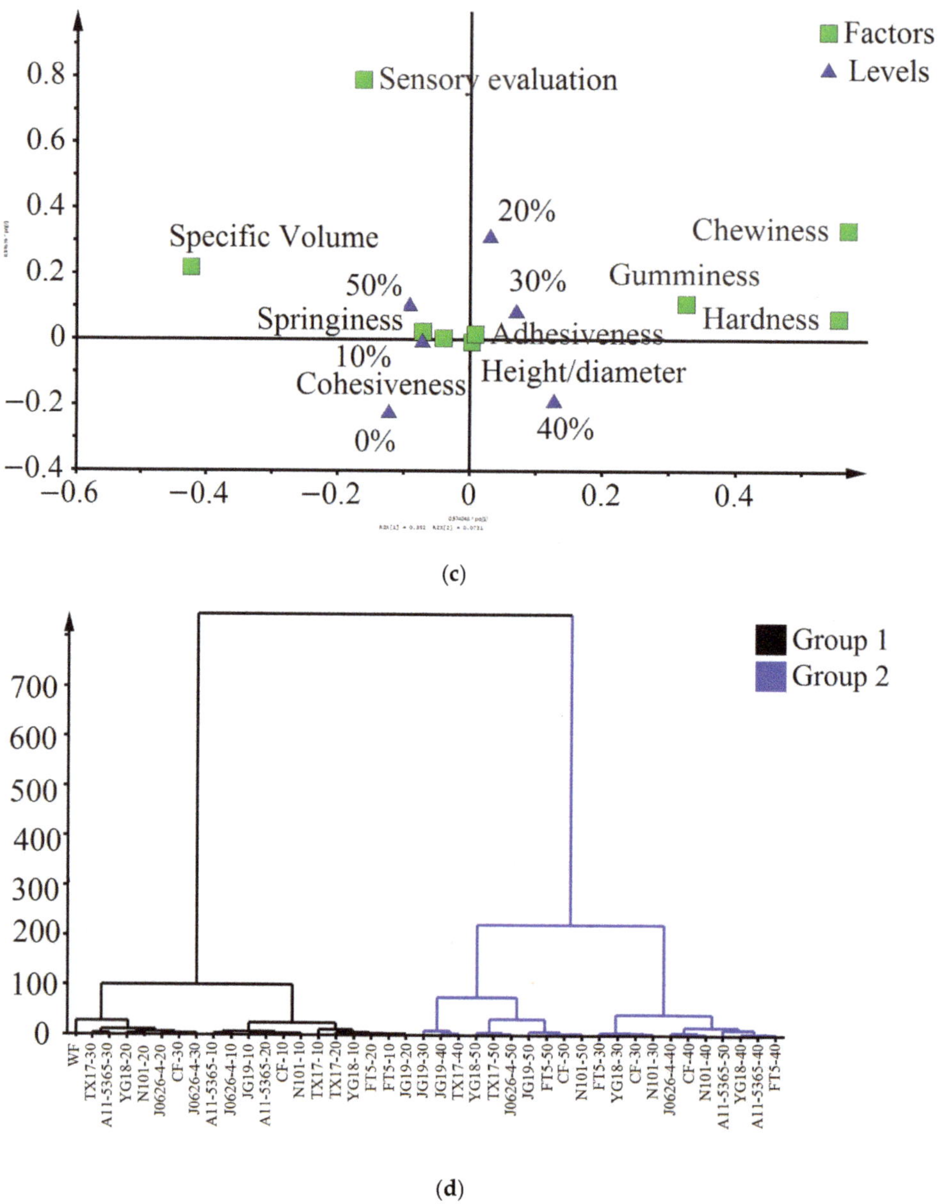

Figure 1. (**a**) OPLS-DA score plot; (**b**) VIP; (**c**) loading plot; (**d**) clustering results.

The VIP value was detected to further determine which variables significantly contribute to the OPLS-DA prediction model. Sensory evaluation, chewiness, specific volume, and hardness (VIP > 1) were considered to be contribution indicators of CSB (Figure 1b). The loading-plot diagram in Figure 1c demonstrates that the addition level indicators gathered near the origin, whereas some evaluation indicators that had a considerable contribution to the prediction of the model classification were scattered at the two ends of the plot. The loading-plot diagram supports the determination of the VIP value.

The clustering results of different cultivars and different additive amounts, as shown in Figure 1d, indicate that they could be divided into two groups: CSB with mostly 10–30% cultivars and that with mostly addition levels of 40–50%. This indicates that the addition level is the key to the production of CSB. With an increase in the total amount of WFMF, all cultivars can cause the volume to decrease. The molecular structure of amylopectin destroyed the network structure during the manufacturing process for CSB [34] due to the higher content of amylopectin in the glutinous cultivars. Compared with the japonica cultivar, glutinous CSB exhibited increasing hardness and chewiness and a large reduction in volume [35]. When the amount of WFMF added was 20–30%, CSB had improved evaluation and texture properties. When the addition of all millet cultivars exceeded 40%, the CSB's specific volume and texture were considerably degraded.

4. Conclusions

The effects of cultivars and the addition of various amounts of WFMF on the quality characteristics of CSB were studied. The specific volume and score of sensory evaluation significantly decreased as the amount of WFMF increased and the height–diameter ratio significantly changed. As the amount of WFMF increased, the hardness and chewiness of CSB increased, and the cohesiveness and the springiness gradually decreased. The texture properties of CSB were different to those of the millet cultivars and addition levels. This might be because the large addition causes the dough network structure to lack gluten, and because of the amylose content of the different cultivars. OPLS-DA was performed to carry out the classification and examine the trend among the samples. Four indicators (VIP > 1), sensory evaluation, chewiness, specific volume, and hardness, were considered to be the contribution indicators of CSB. All samples were divided into two groups by cluster analysis. The cultivars at an addition level of 20–30% featured improved quality and resulted in good sensory evaluation, which could provide a theoretical basis for the industrialization of steamed bread made of whole foxtail millet flour.

Author Contributions: Conceptualization, S.L. and J.L.; methodology, P.L.; software, J.Z.; validation, W.Z. and Y.L.; formal analysis, J.Z. and A.Z.; investigation, S.L.; resources, Y.W.; data curation, G.M.; writing—original draft preparation, S.L.; writing—review and editing, S.L. and J.L.; visualization, W.Z.; supervision, J.L.; project administration, J.L.; funding acquisition, J.L. All authors have read and agreed to the published version of the manuscript.

Funding: This research was funded by the China Agriculture Research System (CARS-06-13.5-A29), Innovation Project of Hebei Academy of Agriculture and Forestry Sciences (2019-2-2) and Projects of Science and Technology in Hebei Province (17227108D).

Institutional Review Board Statement: Not applicable.

Informed Consent Statement: Not applicable.

Data Availability Statement: Data generated or analyzed during this study are included in this published article.

Conflicts of Interest: The authors declare no conflict of interest.

References

1. Fujita, S.; Sugimoto, Y.; Yamashita, Y.; Fuwa, H. Physicochemical studies of starch from foxtail millet (*Setaria italica Beauv.*). *Food Chem.* **1996**, *55*, 209–213. [CrossRef]
2. Zhang, L.Z.; Liu, R.H. Phenolic and carotenoid profiles and antiproliferative activity of foxtail millet. *Food Chem.* **2015**, *174*, 495–501. [CrossRef] [PubMed]
3. Mozaffarian, R.S.; Lee, R.M.; Kennedy, M.A.; Ludwig, D.S.; Mozaffarian, D.; Gortmaker, S.L. Identifying whole grain foods: A comparison of different approaches for selecting more healthful whole grain products. *Public Health Nutr.* **2013**, *16*, 2255–2264. [CrossRef] [PubMed]
4. Rezaei, S.; Najafi, M.A.; Haddadi, T. Effect of fermentation process, wheat bran size and replacement level on some characteristics of wheat bran, dough, and high-fiber Tafton bread. *J. Cereal Sci.* **2019**, *85*, 56–61. [CrossRef]
5. Chen, J.; Duan, W.; Ren, X.; Wang, C.; Pan, Z.; Diao, X.; Shen, Q. Effect of foxtail millet protein hydrolysates on lowering blood pressure in spontaneously hypertensive rats. *Eur. J. Nutr.* **2017**, *56*, 2129–2138. [CrossRef]

6. Okarter, N.; Liu, R.H. Health benefits of whole grain phytochemicals. *Crit. Rev. Food Sci. Nutr.* **2010**, *50*, 193–208. [CrossRef]
7. Schmiele, M.; Jaekel, L.Z.; Patricio, S.M.C.; Steel, C.J.; Chang, Y.K. Rheological properties of wheat flour and quality characteristics of pan bread as modified by partial additions of wheat bran or whole grain wheat flour. *Inter. J. Food Sci. Technol.* **2012**, *47*, 2141–2150. [CrossRef]
8. Shan, S.; Li, Z.; Newton, I.P.; Zhao, C.; Li, Z.; Guo, M. A novel protein extracted from foxtail millet bran displays anti-carcinogenic effects in human colon cancer cells. *Toxicol. Lett.* **2014**, *227*, 129–138. [CrossRef]
9. Iglesias-Puig, E.; Monedero, V.; Haros, M. Bread with whole quinoa flour and bifidobacterial phytases increases dietary mineral intake and bioavailability. *LWT Food Sci. Technol.* **2015**, *60*, 71–77. [CrossRef]
10. Wang, S.; Opassathavorn, A.; Zhu, F. Influence of Quinoa Flour on Quality Characteristics of Cookie, Bread and Chinese Steamed Bread. *J. Texture Stud.* **2015**, *46*, 281–292. [CrossRef]
11. Fang, Y.; Qing, H.; Yu, J.; Gu, X.; Wang, M. Comparison of Tartary Buckwheat Flour and Sprouts Steamed Bread in Quality and Antioxidant Property. *J. Food Qual.* **2014**, *37*, 318–328. [CrossRef]
12. Guo, X.N.; Yang, S.; Zhu, K.X. Influences of alkali on the quality and protein polymerization of buckwheat Chinese steamed bread. *Food Chem.* **2019**, *283*, 52–58. [CrossRef]
13. Bhol, S.; Bosco, S.J.D. Influence of malted finger millet and red kidney bean flour on quality characteristics of developed bread. *LWT Food Sci. Technol.* **2014**, *55*, 294–300. [CrossRef]
14. Topping, D. Cereal complex carbohydrates and their contribution to human health. *J. Cereal Sci.* **2007**, *46*, 220–229. [CrossRef]
15. Burešová, I.; Tokár, M.; Mareček, J.; Hřivna, L.; Faměra, O.; Šottníková, V. The comparison of the effect of added amaranth, buckwheat, chickpea, corn, millet and quinoa flour on rice dough rheological characteristics, textural and sensory quality of bread. *J. Cereal Sci.* **2017**, *75*, 158–164. [CrossRef]
16. Cordelino, I.G.; Tyl, C.; Inamdar, L.; Vickers, Z.; Marti, A.; Ismail, B.P. Cooking quality, digestibility, and sensory properties of proso millet pasta as impacted by amylose content and prolamin profile. *LWT Food Sci. Technol.* **2019**, *99*, 1–7. [CrossRef]
17. Li, S.H.; Zhao, W.; Li, P.L.; Min, G.; Zhang, A.X.; Zhang, J.L.; Liu, Y.Y.; Liu, J.K. Effects of different cultivars and particle sizes of non-degermed millet flour fractions on the physical and texture properties of Chinese steamed bread. *Cereal Chem.* **2020**, *97*, 661–669. [CrossRef]
18. Wu, C.; Liu, R.; Huang, W.; Rayas-Duarte, P.; Wang, F.; Yao, Y. Effect of sourdough fermentation on the quality of Chinese Northern-style steamed breads. *J. Cereal Sci.* **2012**, *56*, 127–133. [CrossRef]
19. Sun, R.; Zhang, Z.; Hu, X.; Xing, Q.; Zhuo, W. Effect of wheat germ flour addition on wheat flour, dough and Chinese steamed bread properties. *J. Cereal Sci.* **2015**, *64*, 153–158. [CrossRef]
20. Zhu, F.; Sakulnak, R.; Wang, S. Effect of black tea on antioxidant, textural, and sensory properties of Chinese steamed bread. *Food Chem.* **2016**, *194*, 1217–1223. [CrossRef]
21. Almeida, E.L.; Chang, Y.K. Effect of the addition of enzymes on the quality of frozen pre-baked French bread substituted with whole wheat flour. *LWT Food Sci. Technol.* **2012**, *49*, 64–72. [CrossRef]
22. Marchini, M.; Carini, E.; Cataldi, N.; Boukid, F.; Blandino, M.; Ganino, T.; Vittadini, E.; Pellegrini, N. The use of red lentil flour in bakery products: How do particle size and substitution level affect rheological properties of wheat bread dough? *LWT Food Sci. Technol.* **2021**, *136*. [CrossRef]
23. Liu, C.; Liu, L.; Li, L.; Hao, C.; Zheng, X.; Bian, K.; Zhang, J.; Wang, X. Effects of different milling processes on whole wheat flour quality and performance in steamed bread making. *LWT Food Sci. Technol.* **2015**, *62*, 310–318. [CrossRef]
24. Sharma, N.; Niranjan, K. Foxtail millet: Properties, processing, health benefits, and uses. *Food Rev. Int.* **2017**, *34*, 329–363. [CrossRef]
25. Wang, X.Y.; Guo, X.N.; Zhu, K.X. Polymerization of wheat gluten and the changes of glutenin macropolymer (GMP) during the production of Chinese steamed bread. *Food Chem.* **2016**, *201*, 275–283. [CrossRef]
26. Annor, G.A.; Tyl, C.; Marcone, M.; Ragaee, S.; Marti, A. Why do millets have slower starch and protein digestibility than other cereals? *Trends Food Sci. Technol.* **2017**, *66*, 73–83. [CrossRef]
27. Fu, J.-T.; Chang, Y.-H.; Shiau, S.-Y. Rheological, antioxidative and sensory properties of dough and Mantou (steamed bread) enriched with lemon fiber. *LWT Food Sci. Technol.* **2015**, *61*, 56–62. [CrossRef]
28. Chen, J.S.; Fei, M.J.; Shi, C.L.; Tian, J.C.; Sun, C.L.; Zhang, H.; Ma, Z.; Dong, H.X. Effect of particle size and addition level of wheat bran on quality of dry white Chinese noodles. *J. Cereal Sci.* **2011**, *53*, 217–224. [CrossRef]
29. Kokawa, M.; Suzuki, Y.; Suzuki, Y.; Yoshimura, M.; Trivittayasil, V.; Tsuta, M.; Sugiyama, J. Viscoelastic properties and bubble structure of rice-gel made from high-amylose rice and its effects on bread. *J. Cereal Sci.* **2017**, *73*, 33–39. [CrossRef]
30. Hsieh, P.-H.; Weng, Y.-M.; Yu, Z.-R.; Wang, B.-J. Substitution of wheat flour with wholegrain flours affects physical properties, sensory acceptance, and starch digestion of Chinese steam bread (Mantou). *LWT Food Sci. Technol.* **2017**, *86*, 571–576. [CrossRef]
31. Lin, L.-Y.; Liu, H.-M.; Yu, Y.-W.; Lin, S.-D.; Mau, J.-L. Quality and antioxidant property of buckwheat enhanced wheat bread. *Food Chem.* **2009**, *112*, 987–991. [CrossRef]
32. Machado Alencar, N.M.; Steel, C.J.; Alvim, I.D.; de Morais, E.C.; Andre Bolini, H.M. Addition of quinoa and amaranth flour in gluten-free breads: Temporal profile and instrumental analysis. *LWT Food Sci. Technol.* **2015**, *62*, 1011–1018. [CrossRef]
33. Ning, J.; Hou, G.G.; Sun, J.; Wan, X.; Dubat, A. Effect of green tea powder on the quality attributes and antioxidant activity of whole-wheat flour pan bread. *LWT Food Sci. Technol.* **2017**, *79*, 342–348. [CrossRef]

34. Liu, X.; Mu, T.; Sun, H.; Zhang, M.; Chen, J.; Fauconnier, M.L. Influence of different hydrocolloids on dough thermo-mechanical properties and in vitro starch digestibility of gluten-free steamed bread based on potato flour. *Food Chem.* **2018**, *239*, 1064–1074. [CrossRef]
35. Zhang, D.; Mu, T.; Sun, H. Effects of starch from five different botanical sources on the rheological and structural properties of starch-gluten model doughs. *Food Res. Int.* **2018**, *103*, 156–162. [CrossRef] [PubMed]

MDPI
St. Alban-Anlage 66
4052 Basel
Switzerland
Tel. +41 61 683 77 34
Fax +41 61 302 89 18
www.mdpi.com

Processes Editorial Office
E-mail: processes@mdpi.com
www.mdpi.com/journal/processes

9 783036 570662